Equidosimetry

NATO Security through Science Series

A Series presenting the results of scientific meetings supported under the NATO Programme for Security through Science (STS).

The Series is published by IOS Press, Amsterdam, and Springer Science and Business Media, Dordrecht, in conjunction with the NATO Public Diplomacy Division.

Sub-Series

A. **Chemistry and Biology**   (Springer Science and Business Media)
B. **Physics and Biophysics**   (Springer Science and Business Media)
C. **Environmental Security**   (Springer Science and Business Media)
D. **Information and Communication Security**   (IOS Press)
E. **Human and Societal Dynamics**   (IOS Press)

Meetings supported by the NATO STS Programme are in security-related priority areas of Defence Against Terrorism or Countering Other Threats to Security. The types of meeting supported are generally "Advanced Study Institutes" and "Advanced Research Workshops". The NATO STS Series collects together the results of these meetings. The meetings are co-organized by scientists from NATO countries and scientists from NATO's "Partner" or "Mediterranean Dialogue" countries. The observations and recommendations made at the meetings, as well as the contents of the volumes in the Series, reflect those of the participants in the workshop. They should not necessarily be regarded as reflecting NATO views or policy.

Advanced Study Institutes (ASI) are high-level tutorial courses to convey the latest developments in a subject to an advanced-level audience

Advanced Research Workshops (ARW) are expert meetings where an intense but informal exchange of views at the frontiers of a subject aims at identifying directions for future action

Following a transformation of the programme in 2004 the Series has been re-named and re-organised. Recent volumes on topics not related to security, which result from meetings supported under the programme earlier, may be found in the NATO Science Series

www.nato.int/science
www.springeronline.com
www.iospress.nl

Equidosimetry – Ecological Standardization and Equidosimetry for Radioecology and Environmental Ecology

edited by

F. Bréchignac

International Union of Radioecology, Secretariat General,
Centre d'Etudes de Cadarache,
Saint-Paul-lez-Durance, France

and

G. Desmet

Former President of the International Union of Radioecology,
Zemst, Belgium

Published in cooperation with NATO Public Diplomacy Division

Proceedings of the NATO Advanced Research Workshop on
Ecological Standardization and Equidosimetry for Radioecology and Environmental
Ecology
Kiev, Ukraine
14–20 April 2002

A C.I.P. Catalogue record for this book is available from the Library of Congress.

ISBN-10 1-4020-3649-3 (PB)
ISBN-13 978-1-4020-3649-1 (PB)
ISBN-10 1-4020-3648-5 (HB)
ISBN-13 978-1-4020-3648-4 (HB)
ISBN-10 1-4020-3650-7 (e-book)
ISBN-13 978-1-4020-3650-7 (e-book)

Published by Springer,
P.O. Box 17, 3300 AA Dordrecht, The Netherlands.

www.springeronline.com

Printed on acid-free paper

TABLE OF CONTENTS

Part 3. Radioecology and Ecotoxicology in General Ecology

Part 4. Problems of Estimation of Risks from Different Factors

Part 5. Problems of Synergism of Different Pollutants

Part 6. Genetic Factors, Environment and Toxicants

Part 7. Applied Radioecology and Ecotoxicology

Part 8. Possibility of Standardization of Radionuclides and Chemotoxicants

Preface

Obviously, to understand the dynamics of ecosystems and their biodiversity patterns, the assessment of the influence of the effects of physical factors (light, temperature, radiation etc.), of chemical factors (various chemical pollutants - heavy metals and other), and of biological factors (viruses, phages, parasites and etc.) to biota is paramount.

In the last decades, researches on an equidosimetric evaluation of the influence of these various factors on the viability of the biological nature were carried and their ecological standardisation initiated.

A clear theoretical and a practical method of equidosimetry and consequently, an ecological normalization of the different factors in unified uniform units are necessary. Such an equidosimetric method will enable to evaluate and to describe condition and behaviour of the ecosystems under influence of the different factors and intensity of their effect, and to conduct an adequate ecological normalization of the factors and their combinations.

In radioecology and radiation protection, methods of **radiation dosimetry** are key for dose assessment.

Today researches on selected biological parameters are being conducted with the purpose to compare effects from radiation and from chemical factors on ecosystems. For a number of heavy metals it was possible to put forward effect equivalences.

Comparative evaluations and researches are conducted at the most different levels from molecular to cell level to the level of organisms and also to ecosystems level. Predictions can be made about regressions and loss of separate populations and their successions, through model calculations and full-scale researches.

The discussions concentrate around the following main problems and questions:

1. What criteria are to be proposed for the equidosimetrical evaluation and prognosis of the condition of biota in the ecosystems under a complex of harmful effects and how can these effects then be translated for a reasonable ecological normalization to be reached?

2. Which features of the model objects are most adequate for equidosimetric evaluations?

3. How can the outcome of equidosimetric evaluations of the action of the different factors be used for predicting the survival and overall dynamics of biodiversity, and ecological quality of the ecosystems?

4. How can "equidosimetry" be used in the context of Environmental Decision Support Systems (EDSS) for the environmental management of affected ecosystems?

5. How can "equidosimetry" be used for the assessment of the quality of ecosystems for the use of the human population?

To solve these and other problems a Seminar has been organised devoted on "Ecological standardization and equidosimetry for radioecology and environmental ecology".

Considerable experience with radioecological and related ecological researches on terrestrial and aquatic ecosystems, especially after the accident on ChNPP and other environmental accidents, has been reached. The complicated combined effects of the radiation, chemical and biological factors, after the accident and during the post-accident countermeasures have highlighted the need for equidosimetrical evaluations of influence of the various factors and the need for their ecological normalization.

Gilbert Desmet
Former IUR President

There is today a revitalised attention in Radioecology devoted to the so-called "multipollution context" which this Advanced Research Workshop, held in Kiev in April 2002, has certainly contributed to stimulate. A rather long publication process of these proceedings, however, has been necessary due to important changes in the managing Board of IUR, and especially the retirement of its former President, G. Desmet, who particularly promoted this development.

After some time on duty, as new General Secretary of IUR, it became obvious to me that publication would not reach reality without a strong personal commitment in pursuing the editorial task up to completion and in compliance with Kluwer Press standards. On top of many other duties, I therefore engaged in carrying over, mostly during spare time, because I had reached several convictions: the scientific pertinence of the topic, the merit of the contributing authors, the interest to the wider scientific community, and the moral responsibility of IUR as a professional Knowing Society. Above all, I am convinced that progress towards sustainable development, especially when focusing to technogenic substances that humankind gradually introduces within its environment, will not be possible without an integrated view and understanding of the combined effects of the various and concomitant sources of stress. This was precisely the focus of this pioneer Workshop.

It is with much pleasure that I see today this commitment outcome reaching reality, yielding an exciting collection of focused scientific findings and results as well as a concrete reward to the contributing authors.

F. Bréchignac
IUR General Secretary
Director of scientific assessment, IRSN, France

Part 1.

Approaches to Ecological Standardization

PROBLEMS OF ECOLOGICAL STANDARDIZATION AND DOSIMETRY OF ACTION OF DIFFERENT FACTORS ON BIOTA OF ECOSYSTEMS

Yu. KUTLAKHMEDOV, P. BALAN
Taras Schevchenko National Kiev State University, Departament Radiobiology and Radioecology, 01004, Kiev, Ukraine, Volodimirskaya, 64, Kiev, 01033, UKRAINE

V. KUTLAKHMEDOVA-VISHNYAKOVA
International scientific-informational centre of Diagnostics and rehabilitation of biological and ecological systems MES and NASU, Kiev, UKRAINE

1. Introduction

In the case of radionuclides releases and disposal into the environment it is important to assess the maximum admissible values of input of radionuclides into an ecosystem, where there are not yet noticeable biological changes as the result of ionising radiation. The natural boundary for the estimation of maximum permissible disposal of radionuclides into ecosystems is the dose commitment or the annual absorbed dose rate. G. Polikarpov and V. Tsytsugina have proposed a scale of dose commitments on ecosystems consisting of four basic dose limits. From the given scale it follows that the real dose limit for release and accidental "disposal" of radionuclides in ecosystems and their components is the dose rate that exceeds 0.4 Gy y^{-1} for terrestrial animals and 4 Gy y^{-1} for hydrobionts and terrestrial plants. At such values of dose rates it is possible to expect the beginning of the development of evident ecological effects. Dose commitments from α^-, β-, γ-radiation are not difficult to assess for the radionuclides composition of the Kyshtym and the Chernobyl releases. According to our assessments the calculated total dose (estimation by B. Amiro) equalling from 0.4 Gy y^{-1} to 4 Gy y^{-1} corresponds to a concentration of ^{137}Cs of about 100-1000 kBq l^{-1} (kg^{-1}) in the ecosystem or in its elements (terrestrial plants and hydrobionts). The total dose 4 Gy y^{-1} corresponds to a ^{137}Cs concentration of about 1000 kBq l^{-1} (kg^{-1}) for the fresh water ecosystem. The maximum permissible releases of radionuclides into the ecosystems could be assessed on the basis of the above-mentioned models and equations, using an assessment of maximum permissible concentration of radionuclides in the components of the ecosystem.

1. For benthos in bottom sediments of a freshwater water body, the maximum permissible release of radionuclides in a water –L body should not exceed – Nk (6,7). Where L is the limit of concentration of radionuclides in the aquatic population 1000 kBq/l, S is the surface area of a water body, F- the radiocapacity factor, h – depth of active bottom sediments, k – transfer factor – water – bottom sediment.

2. For the water column population (pleuston, neuston, plankton, nekton) maximum permissible releases of radionuclides should not exceed –Nb. For a specific freshwater water body, where S=2 km^2, H=4 m (depth of lake), K_b=1000 (transfer factor

1

F. Brechignac and G. Desmet (eds.), Equidosimetry. 1–18.

water-biota), F=0.7, the maximum permissible release of radionuclides (2) is N_b<17.1 TBq in the water of the whole water body. At the same time the maximum permissible release of radionuclides for its benthos into a water body was estimated using equation (1) N_k< 0.18 TBq. This magnitude is 70 times less than the permissible release of radionuclides, which was assessed for the population of the water column of a water body.

2.Theory of Radiocapacity of Ecosystems

2.1.DEFINITION AND MODELS OF RADIOCAPACITY

Radiocapacity is the maximum amount of radionuclides, which can be contained, in a given ecosystem, without damaging the main trophic properties, i.e. productivity, conditioning and reliability. Special attention will be given to the concept of the radiocapacity factor. Methods of calculation of the radiocapacity of water and over-land ecosystems, and also agrocenoses and factors stipulating these magnitudes will be considered.

The measure of radiocapacity and also of the radiocapacity factor is convenient and universal, and reflects the main properties of ecosystems. Using mathematical means of stationary and dynamic models is quite simple and is suitable for ecosystems of any complexity. This approach allows acceptance of the important prognostic evaluations of the quality and condition of ecosystems .

The fundamental property of ecosystems is their ability to accumulate and to keep radionuclides inside themselves. The radiocapacity factor could serve as the measure of this property. This measure is most conveniently characterized by the ratio of the amount of radioactivity strongly sorbed by components of the ecosystem, to the whole radioactivity, which is contained in the given ecosystem.

2.2.MODEL OF RADIOCAPACITY OF NONFLOW FRESHWATER RESERVOIR

A conclusion that could be drawn from equation (1) is very important for the calculation of the radiocapacity factor of a reservoir (F). Ratio [1].

$$F = \frac{kh}{H + kh} \tag{1}$$

The ratio (1) shows which amount of radionuclides contained in the reservoir is the part of bottom sedimentations (F), and which is the part of water (1-F). F can be defined as the radiocapacity factor of a reservoir.

2.2.1.MODELLING OF THE ROLE OF BIOTA AS REPOSITORY OF RADIONUCLIDE ACCUMULATION

Radionuclides concentration reaches a maximum value within a few minutes in zooplankton, a few days in multicellular algae, and a few months in fishes. An average total accumulation factor of radionuclides by these organisms is a constant magnitude and equals to about 10^3 [2].

Which is the role of flora and fauna of reservoirs in radionuclides distribution?

We should consider a role of biota as repository of radionuclides in water-cooling pond at a significant concentration of biota. It is possible to do this as follows.

Let the concentration of biota in a unit of reservoir water volume be P (g m^{-3}), and let the average accumulation factor of radionuclides by biota be K. Then the combined content of radionuclides in the biota of a reservoir will be:

$$A_b = P \cdot C \cdot K \cdot S \cdot H \qquad (2)$$

(A_b = Accumulation in biota).

The radiocapacity factor for the biotic component of a reservoir can be evaluated by using the following equation [3].

$$F_b = \frac{PKH}{(H + kh + PKH)} \qquad (3)$$

Let us for an example calculate F_b for a real situation, when P makes 10 g m^{-3} of water with an average accumulation factor $K = 10^4$, an average depth of reservoir $H = 6$ m, $h = 0.1$ m., $k = 800$. We shall obtain a radiocapacity value F_b close to 0.9, when 90 % of radionuclides arriving into the reservoir reach the biomass of the biota. It is necessary to take this into consideration, in a range of real situations in reservoirs or in separate rising zones of reservoirs where high concentration of biota are recorded.

2.2.2. MODEL OF RADIOCAPACITY IN A CASCADE OF FRESHWATER RESERVOIRS

Thence the equation of the whole radiocapacity cascade is as follows:

$$F_k = 1 - \prod_{i=1}^{n} (1 - F_i) \qquad (4)$$

Analysis of the equation shows, that the greater the number of reservoirs built into the cascade, the higher is the factor of its radiocapacity. The common radiocapacity of the cascade is always higher than the radiocapacity of the best reservoirs included in it.

Since the cascade of reservoirs of the Dnieper represents a system of slowly flowing reservoirs, we could apply a simple equation (4) to calculate common radiocapacity. From this equation it follows that the radiocapacity factor of the cascade is equal to $F_c = 0.9994$ for Cs-137. This magnitude expresses an extremely high degree of radiocapacity of the cascade, which is much higher than even the maximum radiocapacity of Kremenchug reservoir

2.2.3. MODEL OF RADIOCAPACITY OF SLOPE ECOSYSTEMS

The developed prime model of an evaluation of slope ecosystem radiocapacity, allows assessing dynamics, time and place of expected concentration of radionuclides in some elements of slope ecosystem. The factor of radiocapacity of slope ecosystem may be assessed by the equation (5).

$$F_s = 1 - \prod_{i=1}^{k} P_i = 1 - P_s \tag{5}$$

Where, P_i is probability of a flow of radionuclides from an appropriate element of slope (slanting) ecosystem in a year. P_1 = flow from forest ecosystem. P_2 = flow from stony plot. P_3 = flow from meadow ecosystem, P_4 =from terrace in a lake.

This models permit to make a short term and long-term estimation and a prognosis of state and distribution and redistribution of radionuclides in different types of ecosystems.

3. Ecological Standardization of Permissible Radionuclide Pollutions of Ecosystems on the Basis of Theory of Radiocapacity

3.1. PROBLEMS OF ECOSYSTEM ECOLOGICAL STANDARDIZATION [4].

In the case of radionuclide releases and disposal into the environment it is important to assess the maximum admissible values of income of radionuclides into an ecosystem, where there are not yet noticeable biological changes as the result of ionizing radiation.

The natural boundary for the estimation of maximum permissible disposal of radionuclides into ecosystems is the dose commitment or the annual absorbed dose rate. G. Polikarpov and V. Tsytsugina have proposed a scale of dose commitments on ecosystems consisting of four basic dose limits [5], which is presented in Table 1.

Table 1. The scale of dose commitments onto ecosystems [5]

ZONE	ABSORBED DOSE RATE (Gy y^{-1})
Zone of radiation well-being*	< 0.001-0.005
Zone of physiological masking**	0.005 - 0.05
Zone of ecological masking:	
for terrestrial animals	0.05 - 0.4
for hydrobionts and terrestrial land plants	0.05 - 4
Zone of evident ecological changes:	
Dramatic:	
for terrestrial animals	>> 0.4
for hydrobionts and terrestrial plants	>> 4
Catastrophic:	
animals and plants	>> 100

* - 0.001 Gy y^{-1} is the dose limit for human population;
** - 0.020 Gy y^{-1} is the dose limit of professional irradiation during normal practical work

From the given scale follows that the real dose limit for release and accidental "disposal" of radionuclides in ecosystems and their components is the dose rate that exceeds 0.4 Gy y^{-1} for terrestrial animals and 4 Gy y^{-1} for hydrobionts and terrestrial plants. At those dose rates values it is possible to anticipate the beginning of development of visible ecological effects in ecosystems. Dose commitments from α, β-, γ-radiation are not difficult to assess for the radionuclides composition of the Kyshtym and the Chernobyl releases. According to our assessments the calculated total dose (estimation by B. Amiro [6]) being between 0.4 Gy y^{-1} to 4 Gy y^{-1} corresponds to a concentration of ^{137}Cs of about 64-640 kBq l^{-1} (kg^{-1}) in the ecosystem or in its elements (terrestrial plants and hydrobionts). The total dose 0.4 Gy y^{-1} corresponds to a ^{137}Cs concentration of about 64 kBq l^{-1} (kg^{-1}) for an ecosystem with terrestrial animals.

3.2.ASSESSMENT OF PERMISSIBLE DUMPING AND SELF-DISPOSAL OF RADIONUCLIDES IN A FRESHWATER RESERVOIR

For a freshwater reservoir the equation for the evaluation of the biocenosis radiocapacity factor in the water column is represented by formula (3).
Where p is the hydrobiont a biomass in water (a value of biomass from 1 to 10 g m^{-3} is accepted as reasonable),then K$_b$ is the concentration factor of water body biocenosis and its components, which may reach 1000 to 100000 units.
In this case, F$_b$ can range from the small value 0.05 up to a very large radiocapacity value 0.97, where practically all radionuclides are concentrated in the biotic component of a water body.
The maximum permissible releases of radionuclides into the ecosystems could be assessed on the basis of the above-mentioned models and equations on the basis of an appraisal of the maximum permissible concentration of radionuclides in the components of the ecosystem.
1. For benthos of bottom sediments of a freshwater water body, the maximum permissible release of radionuclides in a water body (N$_k$) should not exceed:

$$N_k < \frac{LhS}{kF} \tag{6}$$

Where L is the limit of concentration of radionuclides in the aquatic population 640 kBq kg^{-1}, S is the surface area of a water body (the remaining labels were mentioned above.).
2. For the water column population (pleuston, neuston, plankton, nekton) the maximum permissible releases of radionuclides (N$_b$) should not exceed:

$$N_b < \frac{LHS}{K_b(1-F)} \tag{7}$$

where the labels of equations (1, 2, 3) are used. For a specific freshwater water body, where $S=2$ km^2, $H=4$ m, $K_b=1000$, $F=0.7$, the maximum permissible release of radionuclides is $N_b<17.1$ TBq in the water of the whole water body. At the same time the maximum permissible release of radionuclides into a water body for its benthos was estimated using equation (6) $N_k< 0.18$ TBq. This magnitude is 70 times less than the permissible release of radionuclides, which was assessed for the population of the water column of a water body.

In general , the ratio of assessments of maximum permissible releases of radionuclides into a water body is determined by the following equation through two critical links (population of water column and benthos):

$$\frac{N_k}{N_b} = \frac{hK_b(1-F)}{HkF} \tag{8}$$

3. Similar evaluations of maximum permissible releases of radionuclides can be carried out also for other types of ecosystems. Particularly, in a system of cascades of reservoirs (such as the Dnieper cascade), the first reservoir (the Kiev reservoir) is critical for the dose commitment. In bottom sediments of the upper part of Kiev reservoir, the levels of radionuclide content in bottom sediments reach about 370 kBq kg^{-1} and more. In fact, this means that at the upper part of the reservoir the level of effective release of radionuclides reaches the maximum permissible value. In the population of benthos, it is therefore possible to expect noticeable ecological consequences. The theoretical calculated maximum permissible release of radionuclides in the Kiev reservoir is estimated to be a total of 83 TBq, while the real content of ^{137}Cs in bottom sediments is estimated on location measurements at a 200 TBq. This significantly exceeds the maximum permissible concentration [7].

4. In marine ecosystems, the bioproductivity develops mainly in shallow coastal water. On an average concentration 10 g m^{-3} of biota in water forms, the radiocapacity reaches 0.9-0.99. In that case, the release of large amounts of water with radionuclide content of 1.2-12 kBq l^{-1} can result in radionuclide contamination of the community up to 115 kBq kg^{-1}. This is higher than an ecological permissible level.

3.3. ASSESSMENT OF MAXIMUM PERMISSIBLE RELEASES AND DISPOSAL OF RADIONUCLIDES IN SLOPE ECOSYSTEMS

The developed prime model of an evaluation of slope ecosystem radiocapacity permits assessment of dynamics, time and place of expected concentration of radionuclides in some elements of the slope ecosystem. The factor of radiocapacity of the slope ecosystem may be assessed by the equation (5).

Where, P_i is the probability of a flow of radionuclides from an appropriate element of the slope ecosystem in a year. P_1 = flow from forest ecosystem. P_2 = flow from stony area. P_3 = flow from meadow ecosystem, P_4 = from terrace in a lake.

As an example of a slope ecosystem, the simplest version: Forest $\Rightarrow$ Stony area $\Rightarrow$ Meadow $\Rightarrow$ Terrace $\Rightarrow$Lake was selected. The computing curves in dynamics of long-lived radionuclides redistribution in such a model of ecosystem are represented in

Figure1. The initial contamination of the forest was 3.7 TBq. (For the sake of simplicity radioactive decay was not taken into consideration).

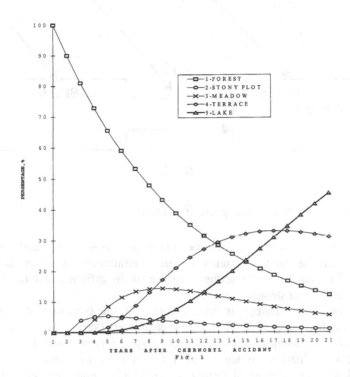

Fig. 1. Radiocapacity of Slope Ecosystems.

It is visible precisely, that the permissible levels of radionuclide contamination decrease noticeably in accordance with the approach of the place of contamination to a lake. Strictly speaking, the ecological norm depends on the ecosystem character. The closer to the critical unit (lake) a subsystem is situated and the higher the value of probability of drain, the lower the ecological specification of permissible contamination.

4. Existing available approaches to evaluating of the effect of the various toxic factors on the condition of ecosystems

Let us consider the basic concept of our approach for an evaluation of the effect of the toxic factors (radiation, chemical, biological and other pollutants) on of ecosystem biota. This goes about a comparison of effects of the different factors, through a modification of the parameters of radiocapacity.

Let the system consist at all of two blocks: an environment (water, for example) and biota. Then the chamber model looks as follows:

8

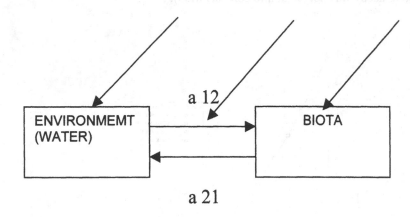

Fig. 2. Box-model of simplest of ecosystem.

Biota in ecosystem can be one-way, two-specific, and multispecific. Ecosystems can be model systems (maize, Lemnaceae etc.) or actual natural ecosystems. In order to evaluate the influence of the different factors of an external medium on biota some strategies are used:

1. The main strategy is an evaluation of the condition of biota. For this purpose, in the literature, up to 30 different parameters (variety of type of ecosystems, number at use, the biomass of the different ecosystem types, velocity of a duplication etc.)(8). This evaluation strategy of the condition of biota as response on contamination by a multitude of factors is used widely among ecologists. This strategy differs by this property, that the measurable results of toxic effects (depressing of kinds, loss of kinds, succession etc.) are observed in ecosystems after significant times after the start of the impacting factors. Herewith is dealt with the remote visible results of an action of contaminants on biota. Thus in case such an approach is needed to evaluate effects on biota, there is a possibility to determine already existing effects, which as a rule, are irreversible and result in depressing and - or losses biota or part of it. This approach allows only to stating an eventual result of influence of pollutants on biota.

2. The second method of evaluating the influence of the toxic factors on biota in ecosystems is based on the right choice of a highly sensitive biological test - object, which is capable to react noticeably and sharply in a medium of toxic substances. Such type by indicators of a toxic effect can be among the different types of plants, animal (mouse) and fish (gold fishes) etc. It is a highly sensitive method for testing toxicity of different pollutants. It is also known that the type of indicator can be highly sensitive to the of one pollutant and not to react on an other pollutant. On the one hand, this method allows to show a probable toxic effect at a low level for one factor, and on the other hand be not suitable for testing of other factors. A whole system of different types of indicators is required to assess the problem.

3. A third approach is also possible; it is to evaluate the interaction of biota with their environment. This comes to finding a method for evaluating of rates of exchanges between a medium and biota, this means the rate of absorption (nutritious substances for example) and rate of outflow of substances from biota into the environment.

The approach is based on the known fact that any biologically significant effect on biota, finds its reflection in a modification of the parameters of absorption and outflow of substances between an environment and biota..

The worse the condition , it is the more possible to suppose, proceeding from the block diagram, that the toxic effect is necessarily exhibited through a modification of parameters **a12** and **a21**, or both simultaneously. The worse the condition of biota after an external exposure, the more this effect should reduce parameter **a12** and - or increase the parameter of outflow - **a21**.. Especially noticeable modifications can come about on the magnitude of the relation of parameters -**a12/ a21**.

Let for an example a simple model system be considered: a hydroponics plant culture, where some radioactive tracer, for example Cs-137 is added. With this tracer, the redistribution (absorption and - or outflow) should reflect a condition of biota. The parameters of the relation of biota with their environment (in this case with water) show the balance between the input of nutrients to biota (**a12**) and their outflow to their environment - **a21**. In this ecosystem model - the system of differential equations, in a box- model, is represented by: (MAPLE 6):

$$difl := \frac{\partial}{\partial x} y(x) = a21\, z(x) - a12\, y(x)$$

$$dif2 := \frac{\partial}{\partial x} z(x) = a12\, y(x) - a21\, z(x)$$

$$(9)$$

$$inicond := y(0) = 1,\ z(0) = 0$$

Where y (x)- is the contents of the tracer Cs-137 in water, and z(x)) the content of tracer in biota of the model ecosystem. Here at the entry conditions–at the initial time (χ = 0) the total content of the tracer is in the water, and the content in the biota is equal = 0. The solution of this set of equations is represented below (10):

$$(10)$$

$$y(x) = \frac{a21}{a12 + a21} + \frac{a12\, e^{(-(a12 + a21)x)}}{a12 + a21},\quad z(x) = \frac{-\dfrac{a12\, e^{(-(a12 + a21)x)}\, a21}{a12 + a21} + \dfrac{a12\, a21}{a12 + a21}}{a21}$$

Using the above, the size of parameters **a12** and **a21,** deducted from the experiment, it is not difficult to constructing the graph of the dynamics of distribution of tracer in the boxes:

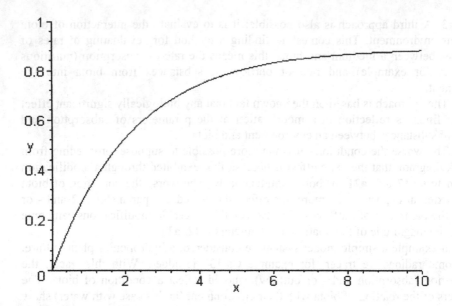

Fig. 3. Dynamics of redistribution of tracer Cs-137 in biota in 2-box-model ecosystem.

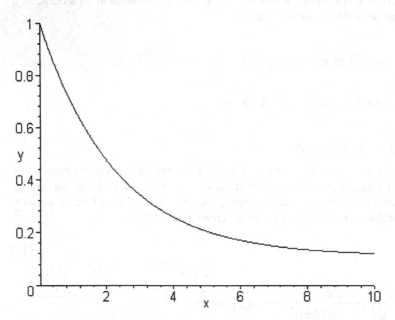

Fig. 4. Dynamics of redistribution of tracer Cs-137 in water in 2-box-model ecosystem.

It is interesting also to analyse the graph of the relation - **Z/Y**:

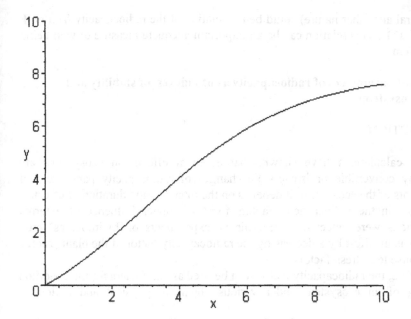

*Fig. 5. Dynamics of redistribution of relation (**Z(x)** / **Y(x)**) radiocapacity of tracer Cs-137 in 2-box-model ecosystem.*

The outcome of the modelling allows to evaluating a number of the important parameters of the model ecosystem, for long time observation time ranges. In particular, the magnitude of the radiocapacity of biota and water, is determined as the magnitude of the contribution of the tracer radionuclides, which remains in water or biota.

From the equations (9,10) it is possible to see that for a range of $x \to \infty$, the magnitudes of radiocapacity can be determined by the following formulas:

F w = a21 / a21 + a12 ; F b = a12 / a21 + a12;

And their relation by:

F b / F w = a21 / a21   (11)

It is not difficult to show, that in the given simple case, the factor of tracer (Cs-137accumulation by biota) **Tf** $\cong$ **a12 / a21 + a12**. Thus **Tf** - **Fb** - factor of radiocapacity of the biota model ecosystem is proportional.

Thus, if the external toxic effect on biota of ecosystems is important, it will be reflected in a diminution of exchange rate parameters between boxes (biota - environment) Then a modification of magnitude of the factors of a radiocapacity **F b** and **F w**, will entail, and as a corollary the sizes Tf to biota which are directly connected with parameters of a radiocapacity will be changed y. The most sensitive measure of a modification of the condition of a biota ecosystem, in response to external factors (chemical, natural and other nature) could be the relation of the radiocapacity factors **F b / F w = a21 / a21** . This relation can be an important adequate measure of well being of biota ecosystem.

5. Assessment of parameters of radiocapacity as of indexes of stability and reliability of ecosystems

5.1. INTRODUCTION

Model calculations have shown, that essential effects on ecosystems are accompanied by convertible or irreversible changes of radiocapacity parameters of biotic components of the ecosystem. It depends on the intensity and duration of effect.

Changes in the radiocapacity factors for Cs^{137} under influence of external gamma-irradiations were detected in preliminary experiments on hydroponics' crop cultivation. It was manifest by a decreasing the radiocapacity factor of the plants rooted system in response to a stress-factors.

Therefore, the radiocapacity factors can be used as parameters for the condition and well-being of an ecosystem under various chemical, physical and biological influences.

The theory and the models of ecosystems radiocapacity have allowed to formulate and to define the approaches to the substantiation of ecological standards on permissible levels of ecosystems and the contamination of their components, and also on permissible releases of radionuclides [4].

The present hypothesis consists of the following elements: we presume that some rather few long-lived radionuclides have contaminated an ecosystem in its steady;. the radionuclides are arranged in compartments of the ecosystem and are described by appropriate values of the radiocapacity factors; the external harmful factors (toxic, radiation and other nature) act on the ecosystem. If the effect is essential, it may affect the ecosystem and its parameters. It will bring about sharp changes of the values of the radiocapacity factors of the ecosystem components [9, 10].

We assume that the noticeable radionuclide redistribution has taken place in an ecosystem with known initial distribution of radionuclide-tracer (or tracers). It signifies that values of radiocapacity factors have changed (thence values of ecosystem radiocapacity). A predicting evaluation of an ecosystem condition, its stability and reliability is supposedly to be made at measurable changes of radiocapacity parameters. The predicting evaluation can be conducted long before a response on effect of integral parameters (variety of species, biomass, number and rate of reproduction of species in the ecosystem) is noticed.

5.2. MODELLING OF CHANGE OF RADIOCAPACITY PARAMETERS OF HARMFUL EFFECTS IN THE ELEMENTARY ONE-SPECIES SYSTEM

As an initial model we shall consider a simple fresh-water ecosystem such as a reservoir (lake) with one species of algae, representing the phytoplankton, and determining bioproductivity of the lake. Parameters of the reservoir are: V (volume) $=10^{10}$; S(surface) = 2 km^2; H (depth) = 5 m; concentration of algae $\rho=10$ $g\,m^{-3}$, transfer factor (T_f) of algae $K_b=10^4$. In the first moment after a peak release in the reservoir of $3.7 \cdot 10^{10}$ Bq ^{137}Cs (in summer period) the factor of radiocapacity of bottom sediments will become 0.7 ($k=10^3$, $h = 0.1$ m, and factor of radiocapacity of a water after release = 0.3 (1)). It is known, that the redistribution of radionuclides between water and bottom sediments (1) in such reservoir will take place very quickly, practically in the course of number of days [7]. Thus pursuing to our approaches the radiocapacity factors of the components are estimated by the formulas (12):

$$
\begin{cases}
F_{bs} = \dfrac{kh}{kh + H + H\rho K_b} \\[3mm]
F_w = \dfrac{H}{kh + H + H\rho K_b} \\[3mm]
F_b = \dfrac{H\rho K_b}{kh + H + H\rho K_b}
\end{cases}
\tag{12}
$$

An estimation can now be made of the parameters of the considered reservoir coming to : $F_{bs}=100/605= 0.165$; $F_w= 5/605 = 0.008$; $F_b=500/605= 0.826$.
By definition the total radiocapacity factor of a reservoir is equal to 1.

$$
F = F_{bs} + F_w + F_b = 1
\tag{13}
$$

Situation changes resulting from stress-effect:

$$
\frac{dF}{dt} = \frac{dF_{bs}}{dt} + \frac{dF_w}{dt} + \frac{dFb}{dt} = 0
\tag{14}
$$

Obviously the stress-factor impacts on biota. However, as a rule, there is no direct influence to water and bottom sediments of such stress-factors . Then it is possible to observe a noticeable decrease of F_b, and at the expense of it an increase of remaining values.
Then:

$$
\text{If } \ \frac{dF_b}{dt} \langle 0 \ , \text{and} \ \frac{dF_w}{dt} + \frac{dF_{bs}}{dt} \rangle 0
$$

$$
\frac{dF_b}{dt} = -\left(\frac{dF_w}{dt} + \frac{dF_{bs}}{dt} \right)
\tag{15}
$$

Now we revolve to the formula (12) and take a derivative from F_b:

$$\frac{dF_b}{dt} = \left(\frac{(kh+H)H\left(\dfrac{d(\rho K_b)}{dt}\right)}{2(kh+h+\rho K_b h)} \right) \langle 0 \tag{16}$$

It is right to assume, that in initial moment of stress-effect the integral index of a species biomass density will not yet has time to react. Change of F_b will happen at the expense of change (decreasing) Kb (factor of accumulation of biota).Because of stress-effect in ecosystem F_b shows a 10 times decrease and comes to 10^3. Consequently the structure of the radiocapacity factors of ecosystem elements will vary and will come to : (In parentheses the values before change)

$F_b = 0.32$ (0.85), $F_{bs} = 0.67$ (0.165), $F_w = 0.03$ (0.008).

It should be recalled that at a rather small change of F_b (coefficient of biota accumulation) the radiocapacity factors of ecosystem elements changes essentially (3-5 times). The ecosystem, through change of the factors of radiocapacity (factors of accumulation), witnesses already to significant ecological events when is not yet marked by essential changes of integral indexes (for example, density of a population).

It is possible to generalize the considered above versions of responses of the ecosystems radiocapacity factors to the case of multispecies ecosystems. Such process is essentially capable to change the initial tracer distribution on the ecosystem species, and so to change the values F_i for various biota species. In multispecies ecosystem a predictable change of the radiocapacity factors may take place much more sharply and more appreciably, alongside the fact that changes in values of the radiocapacity factors at separate species can be small.

5.3. EXPERIMENTAL INVESTIGATION.

There are not many experimental data on the influence of the external factors on parameters of ecosystem radiocapacity. Some data will be given.

The researches were carried out on experimental models, such as hydroponics' cultures of peas and maize.

Half litre glass vessels containing growing plants (4-5 twenty-four hours) were irradiated by external gamma-irradiation at doses of 1-2 Gy. It was shown, that this dose provokes depressing of growth of the main root and additional roots. Thus it is possible to observe decrease of volume of a rooted system and absorption surface. A small amount of ^{137}Cs (1000 Bq l^{-1}) is introduced into these culture vessels . It has been found for this experimental model that exactly the volume and surface of a rooted system determines the accumulation of the radionuclide in plants, and determines the factor of biota radiocapacity.

It was shown in the experiments *(Fig 6,7)*, that the external gamma-irradiation provokes a noticeable decrease of the accumulation of the tracer (^{137}Cs) in rooted and over ground biomass of plants during growth. It turns out that the larger the dose, the smaller is the biota radiocapacity factor in this experimental model. A precise relation between the degree of decline of biota and its radiocapacity factor was established.

These researches have shown, that similar effects are reached at adding heavy metals (for example Cd) in salt solutions*(Figs 6,7)*.

Prolongation of researches in this direction will supposedly allow extending these ideas and approaches to the most various experimental models. Study of the action of the physical, chemical and biological factors on parameters of biota radiocapacity and comparison of a level of changes of radiocapacity with an action of external gamma-irradiation doses will allow to enter an independent method of ecological dosimetry of the indicated factors.

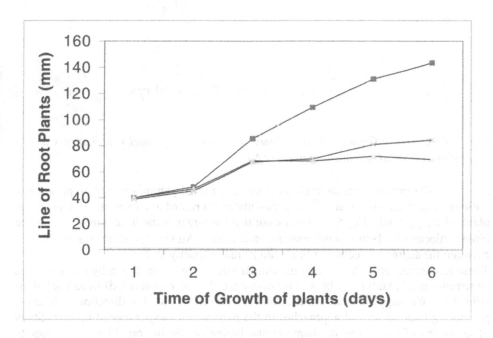

Fig 6. Dynamics of Growth of Root Plants in Control (□), under action of Cd (X) and after action of gamma-irradiation (Δ)

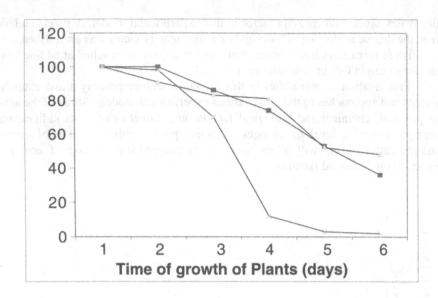

Fig 7. Dynamics of Growth of Root Plants in Control (X), under action of Cd (□) and after action of gamma-irradiation (Δ)

The experiments have shown, that the external gamma-irradiation provokes a noticeable decrease of tracer (^{137}Cs) accumulation in rooted and overground biomass of plants during growth.*(Fig 6,7)*. It turns out that the larger is the dose, the smaller is the biota radiocapacity factor in this experimental model. An exact relation was established between the degree of decline of biota and its radiocapacity factor

These researches have shown, that the similar effects are reached both by thermal stress (warming-up of plants) and by adding heavy metals (for example Cd) in salt solutions *(Fig 6,7)*. We suppose, that an extension of researches in this direction will allow promoting these ideas and approaches to the most various experimental models. Study of an action of the physical, chemical and biological factors on biota radiocapacity parameters and comparison of the level of changes of radiocapacity through external gamma-irradiation will allow to introduce an independent method of ecological dosimetry of the indicated factors., Application of similar simple model objects will presumably allow to substantiate and to concretise the given assumption.

The quantitative equivalents of dose loads on ecosystems for different factors can be found by constructing by means of model objects the relation between the change of the radiocapacity factors at a gamma-irradiation and the one of other harmful factors.

Such researches will allow shifting from a dead point a problem of the quantitative description of ecosystems and "dosimetry" of various effects on them.

6. Conclusions

1. This theory of radiocapacity of ecosystems permits description of regularities of radionuclides distribution for different types of aquatic and terrestrial ecosystems (9,10).

2. The scale of dose commitment on ecosystems permits evaluation of the maximum permissible concentrations of radionuclides. The higher these concentrations rise, the more noticeable influence is expected on structure, on biological characteristics and on radiocapacity parameters.

3. Regularities of radionuclide redistributions in different types of ecosystems, which are described by radiocapacity models, permit the definition of maximum permissible releases of radionuclides in particular "compartments" of ecosystems on the basis of ecological standardization.

4 In selected ecosystems (e.g. pond, water-cooling pond, forest) general ecological maximum permissible releases of radionuclides into the ecosystem or its components are determined:

a) by initial radionuclide contamination of the ecosystem and its elements;

b) by dynamics of redistribution of radionuclides;

c) by radiocapacity parameters of the ecosystems.

5. The proposed method of assessment of ecological maximum permissible radionuclide contamination of ecosystems and their components may be used as a theoretical basis for the System of Ecological Standardization of radionuclide releases from NPPs in normal and accidental conditions.

6. It was shown, that in conditions of noticeable effects at a physiological level on ecosystem by factors such as stress, suppression and/or depressing of one species, a predictable decline of ecosystem and changes in the values of the radiocapacity factors and radiocapacity of ecosystem as a whole are necessarily to be expected.

7. The following remarks are to made though: . There are the essential concentrations of radionuclides in biota and in bottom sediments (roughly $3.7*10^5$ Bq l^{-1}) in the ecosystems of 10-km zone of the Chernobyl Nuclear Power Plant (NPP). Presented here are all sorts of stress-effects (chemical, biological and other nature), capable to provoking releases of radionuclides in environment (in a water, for example). Then the biota radiocapacity of the ecosystem (one or several species) will sharply decrease. It will cause in turn an increase of the content of radionuclides in water, bottom sediments and in biomass of other species of biota. Such a process is capable to continue, down to full ecosystem destruction (the lake Karachay in the Ural is an example) .

8. Thus, any effect on ecosystems affecting Tf_i and P_i and other indexes of an ecosystem condition can be reflected as a noticeable change (decreasing or possibly increasing) of F_i (radiocapacity factor) and of the value of radiocapacity. Such a situation will be reflected in a reduction of permissible releases in ecosystem, for example, in the reservoir-cooler of NPP. In this case the initial well-being in ecosystem will be quickly lost. In other words the strategy directed on preservation and/or increase the radiocapacity of ecosystem and the radiocapacity factors of composed biota species of ecosystem is an optimum method of ecological-ethical management of the ecosystems.

9.Thus well-being and viability of ecosystem bear witness of its high radiocapacity. To the contrary, high radiocapacity, and stable high values of the radiocapacity factors of ecosystem biota species bear witness of the well-being and reliability of ecosystem.

10.The control of radiocapacity and the radiocapacity factors and especially of their time changes and after-effects (gamma-radiation , hard metals –Cd) can serve as an objective predicting criterion and method of an assessment of the well-being any ecosystems (water, continental, wood and marine ecosystems).

11.The presented method of application of tracers can be successfully applied for research and characterisation of the status and well-being of ecosystems. There are some Chernobyl's tracers to be assumed in an ecosystem. The artificial introduction in natural ecosystem of the radionuclides ^{134}Cs, ^{137}Cs for example can also take place. The condition of the radiocapacity factors in the ecosystems can be assessed by estimating their dynamics over years and seasons. If after various antropogenic actions and countermeasures a noticeable reduction of the radiocapacity factors of some components of biota are observed, this condition can serve as a predicting index for a possible decrease of ecosystem viability, and of the decrease of its radiocapacity.

12.If the observation with tracers shows stability and especially increase of the radiocapacity factors of biota species, we the general well-being of ecosystem may be considered not to be uncertain, although some changes can certainly take place, especially in case of the use of countermeasures. At the same time the steady decrease in the value of the radiocapacity factor of one species only, not been compensated by the increase of the radiocapacity factors of other species, can be an sign of danger for a species and ecosystem as a whole. It is essential that the similar decrease of F_i (radiocapacity factors) has a predicting character. It is important, that there is possibility to carry out similar experiments in nature.

7. References

1. Agre A.L.and Korogodin V.I. (1960). About Distribution of Radioactive Pollutions in Non-flowing Reservoir. Med. Radiology, N1, p.67-73 (in Russian).
2. Polikarpov G.G. (1964). Radioecology of Marine Organisms. Moscow, Atomisdat, 296 p. (in Russian).
3. Antropogenic Radionuclides Anomaly and Plants (1991). Kiev, Lebid, 160 p.(in Russian).
4. Kutlakhmedov Y., Polikarpov G, and Kutlakhmedova-Vyshnyakova (1997): Radiocapacity of Different types of Natural Ecosystems and Ecological Standartization Principles// J. Radioecology.- V.6(2),p/15-21
5. Polikarpov G.G. and Tsytsugina V.C. (1995): After Effects of Kyshtym and Chernobyl Accidents on Hydrobionts, Radiation Biology. Radioecology., Vol.35 N4: 536-548 (in Russian).
6. Amiro B.D. (1992): Radiological Dose Conversion Factors for Generic Non-human Biota. Used for Screening Potential Ecological Impacts, J. Environ. Radioactivity Vol.35, N1, : 37-51.
7. Radioactive and Chemical Pollution of Dnieper and it Reservoirs after Chernobyl Accident (1992): Kiev, Naukova Dumka, 195 p.
8. Kutlakhmedov Y.A., Polikarpov G.G. and Korogodin V.I. (1988): Principles and Methods of Ecosystems Radiocapacity Estimation. In: D.Grodsinsky (Ed.) Heuristic of radiobiology. Kiev, Naukova dumka: pp. 60-66. (in Russian).
9. Kutlakhmedov Y., Korogodin V. and Kutlakhmedova-Vyshnyakova (1997): Radiocapacity of Ecosystems, J. Radioecology, V5(1) 1997 pp.25-35.
10. Kutlakhmedov Y.A. and Korogodin V.I. (2002): Bases of Radioecology, Kiev, High school, 420 p, (in press).

THE PRINCIPAL APPROACHES TO STANDARDIZATION OF TECHNOGENIC CONTAMINATION OF ENVIRONMENT

G. PEREPELYATNIKOV
Ukrainian Radiological Training Centre,
Mashinostroitelej str., 7, Chabany, Kiev prov.
UKRAINE

Ecologists and hygienists elaborated a lot of approaches to the standardisation of technogenic pollutants founded on various principles. For standards elaboration several characteristics such as the state of soil kenosis and biochemical characteristics of soil [1], or phytotoxicity of pollutants taking into account the of metabolite ion stability constant [2], or buffer capacity of soil and its capacity of auto-cleaning. Sometimes, crops yield quality and its losses due to pollutant action or intensity of migration through food chains is taken into account as a base. No single approach exists to toxicant speciation in soil and their synergetic actions .

Integrated principles of technogenic pollutants standardisation are absent nowadays. Almost all methods of toxic substances (pollutants) content standardisation in the environment and foodstuffs can be divided into medical, hygienic, landscape, biotic and soil ones. A common methodology or approaches to ecological standardisation are not found.

This methodology elaboration is especially important for such technogenic pollutants as heavy metals, their geochemistry being bound to anthropogenic activity. The trouble with elaboration of common methodological approaches exist in the large differences between properties of various systems to be standardised. For example, water and air are considered as relatively homogeneous system, and soil is a heterogeneous one.

As soil is a basic natural factor to supply the population of the Earth with food practically to the full extent, the importance of elaboration of technogenic pollutant norms for soils becomes exactly clear. A range of principal approaches to elaboration of such norms was proposed [3, 4, 5, 6].

Firstly, the presence of in safe amounts Heavy metals in soil in addition to the natural ones should be taken as a fact. In other words the fact of impossibility of soil restoration into its original state has to accepted.

Secondly, the level of soil contamination can be used as a criterion when the natural soil properties are damaged or changed significantly. Both quantitative and qualitative changes of natural soil functions should be used for standardisation.

Thirdly, ecological standardisation being a complete undertaking needs participation specialists in different disciplines such as ecologists, soil scientists, physicians, hygienists, microbiologists and chemists.

The introduction of Limited Permissible Concentrations of heavy metals in soil, water, air, forages, and foodstuffs is assumed by some current concepts of standardisation. Soil-and-ecological approach to heavy metals standardisation in soil is based on the prevailing features of conserving soil properties, or at least to achieving of a high quality production (good for consumption) [7].

F. Brechignac and G. Desmet (eds.), Equidosimetry, 19–23.
© 2005 *Springer. Printed in the Netherlands.*

The principal object of ecological standardisation is the limitation of harmful actions of technogenic pollutants on biosphere and humanity by introducing of elaborated standards of their presence in the biogeosphere or trophic human chains.

To achieve the object the following tasks should be set on:
- groups of specialists in principal directions of standardisation should be formed;
   terms and definitions of standardisation should be harmonised;
- or integrated principles and directions of standardisation should be defined, range of tasks should be outlined;
- methodological approaches to standardisation should be harmonised as far as possible;
- a system of standard actions should be elaborated.

On the basis of the experience gained during the elimination of the consequences of Chernobyl catastrophe, several conclusions can be drawn and used for the creation of methodological approaches to the standardisation of technogenic pollutants in natural- and agrocenoses in the area of standardisation of radioactive pollutants.

Firstly, natural- and agrocenoses include critical landscapes or links of migration chains, which accumulate small quantities of technogenic pollutants in special conditions everywhere. For instance, herbage of wet meadows subjected to radioactive contamination is characterised by high specific content of ^{137}Cs everywhere.

Secondly, during elaboration of standards of technogenic contamination, a value of critical link or critical landscape contamination can be bound to the final standard of critical foodstuff or critical animal fodder contamination by a direct proportional dependency. Principle of soil contamination standardisation by meat as a critical product can be considered as an example (fig.1).

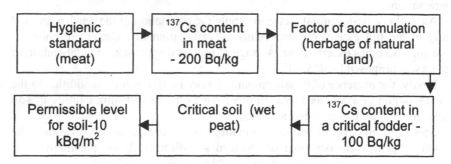

Fig. 1. Principle scheme of ^{137}Cs standardisation in soil.

Meat and milk with a radionuclides content within PL-97 (no more then 200 Bq/kg and 100 Bq/l, respectively) are allowed to be collected from peat wet soil (TF for grasses is equal 10) with density of ^{137}Cs contamination 10 kBq/m^2 when the above mentioned scheme of standardisation is used. This standard of a critical link contamination can be established when contamination of a final product is getting below the hygienic standard. A control of other links of the migration chain is no longer relevant.

Concrete critical links should be found for resolving of standardisation tasks and for defining their area of   application. For example, critical links defining radioactive contamination of agricultural production can be found in agricultural ecosystems (fig. 2).

The basic methodological approaches to ecological standardisation can be brought to consequent performing of some necessary works and fulfilling some conditions.

At the beginning of the elaboration of standards, boundaries of homogeneous equilibrium critical biogeocenoses (ecosystems) as geochemical provinces, regions, zones etc. should be defined because correct application of any standard is possible for homogeneous systems only. Ranging of landscapes and biocenosis by criticacy should be carried out within homogeneous equilibrium critical biogeocenoses. That allows to guaranteeing the standard holding in chains of biogeocenoses in case of standard elaboration for a critical chain.

Standardisation should be performed for various pollutants or their groups independently but taking into account their synergism. It seems useful to find a critical indicative pollutant for separate migration links of trophic chain. In any case pollutants have to be ranged by danger or criticacy.

During elaboration of standards, creation of block-schemes describing pollutants migration through trophic chains of man with accounting of migration links as complete as possible is an important task.

Estimation of migration fluxes of pollutants in links of man trophic chains is an important task also. Comparative analysis of pollutant fluxes in links of human trophic chains would allow distinguishing critical links and ranging them by importance and danger extent.

One of the standardisation tasks is the calculation and prediction of the relative value for maximum introduction of a pollutant into the human organism from critical link (links) of a trophic chain. Quantitative estimation of the) influence of a link (links on pollutant entering into the human organism would allow to reduce expenses to their control and make it more efficient.

Real and predicted values of a pollutant coming in into the human body from a critical link (links) of the trophic chain should be compared with sanitary-and-hygienic standards or limited permissible entering, when a harmful action of pollutants on human or biota is not defined. Such a comparison allows to solve the reverse task of pollutant standardisation in an initial link of biogeocenoses or trophic chain (detection of standardised pollutant content in a critical link of biogeocenoses or trophic chain) and establish a well justified standard.

The whole methodology for standard elaboration may be presented by the following scheme (fig.3).

22

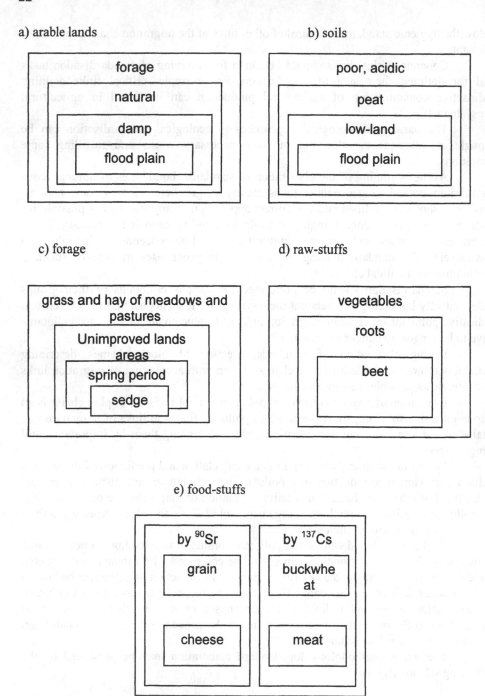

Fig. 2 Determination of critical links in agricultural ecosystems by ^{137}Cs.

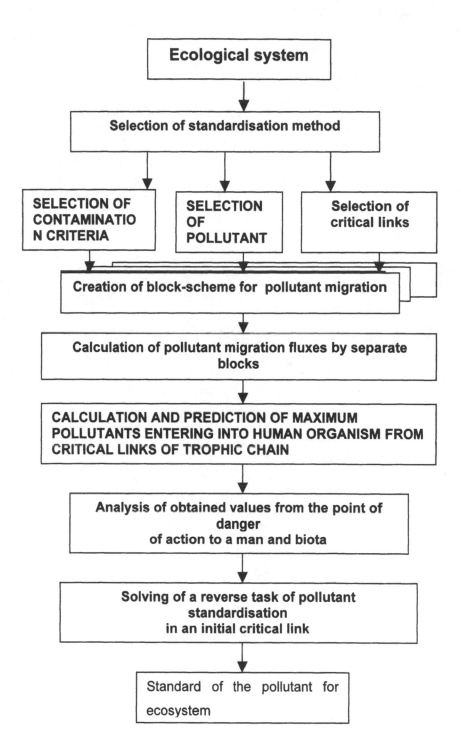

Fig. 3. Methodoology for standard elaboration.

ESTIMATION OF PARAMETERS OF RADIOCAPACITY OF BIOTA IN ECOSYSTEMS; CRITERIA OF THEIR WELL-BEING

Yu. KUTLAKHMEDOV
Institute of cellular biology and genetic engineering UAAS
UKRAINE

P. BALAN
Kiev Shevchenko National University, Vladymyrskaya str, 60, Kiev,
UKRAINE

1. Abstract

The history of the Chernobyl accidents knows many examples, where the behaviour radionuclides in ecosystems in an obvious way reflected the course and laws of ecological phenomena, this is for example this phenomenon of active washout of nuclides from the surface of the Dnieper reservoir, in particular, from territories of Chernobyl zone of 30 km, into the territory of Ukraine and Byelorussia. Rate and parameters of such drain (0,01-2 % in year for Cs-137 and 0,1% - 4 % in year for Sr-90), define themselves as fundamental drains characteristic for big ecosystems like the Dnieper reservoir areas. The dramatic example of those phenomena of sorbptioning and desorbption of radionuclides (especially for Cs-137) in the closure of a reservoir cooler, the Kiev water basin and other superficial reservoirs, reflects the fundamental hydrological properties of water ecosystems.

The following conclusion can be drawn. At any severe hydrological, biological and technological process in ecosystems polluted by radionuclides, an appreciable redistribution of radionuclides - tracers in biotic and a-biotic components ecosystems is observed. Also the inverse conclusion is possible therefrom. Therefore it is countable count that research of distribution and redistribution radionuclides on biotic and abiotic components of ecosystems may serve as an important tool to research of dynamics and to forecasting the condition of an ecosystems and its biota.

The theory of radiocapacity of ecosystems developed hereby has permitted to assess the integrated characteristics of radionuclides distribution in ecosystems, to establish laws of radioecological processes and to determine the location of increases of radionuclides in ecosystems. The theory and models of radiocapacity of ecosystems have allowed to define approaches to a substantiation of ecological specifications on permissible levels of pollution in an ecosystem and their elements, and to permissible dumps and radionuclides emissions in ecosystems [1-3].

Our researches on reliability of ecosystems have shown a line of parameters, such as a variety of type of ecosystems; of biota; of number of species in ecosystems and rate of duplication of species. They form eventually the necessary "integrated parameter", suitable for an estimation of the general stability and reliability of ecosystems [2, 3, 4,5].
The definition however of such parameter demands a huge amount of initial data on contributing parameters at different stages of the life of ecosystems. Already on noticeable changes of the status of the ecosystem an estimation of their well -being and reliability can be made. Clearly, when in ecosystem as a whole, the variety of

F. Brechignac and G. Desmet (eds.), Equidosimetry, 25–31.
© 2005 Springer. Printed in the Netherlands.

species decreases, the biota, number and speed of duplication of dominant kinds is reduced, then it is not difficult to observe a damage and/or an oppression of ecosystems. Such affirmative research is useful, but it is not enough, as it does not allow the making of an outstripping forecast of the condition of ecosystems nor to offer counter-measures. Radiocapacity - a sensory characteristic for the plant ecosystem's well-being is preferable. It is a non-dimensional quantity that characterises the part of radionuclides, which can be retained by the ecosystem without having deteriorated neither the whole system nor its parts. The only considerable effect on the ecosystem to be considered would the effect on the of radiocapacity rate factor. Separately, we based oneself on the theory, according to which the negative influence on the ecosystem would give rise to deterioration of its well-being; and, accordingly, - would be observe a reduction of the radiocapacity of the system. On the other hand, - the improvement of the ecosystems well-being would induce an increase of its of radiocapacity rate factor.

2. Radiocapacity as outstripping parameter of a condition ecosystems

The radiocapacity of an ecosystem is a measure of the quantity of radionuclides, which an ecosystem may contain without consequences for itself. It is obvious, that suppression and/or oppression of an ecosystem (reduction of a variety of kinds because of their destruction, reduction of a biota, number and speed of duplication) will be reflected at once in the size of the radiocapacity and of the ratio of radiocapacity factors. Radiocapacity factors determine the relative share of radionuclides in each component of a ecosystem (inert and biotic).

The hypothesis presented here is based on a large amount of empirical data, consisting of the following principles: we assume an ecosystem in steady condition contains relatively small amount long-living radionuclides (without serious impact on the condition of the ecosystem). Radionuclides are allocated to ecosystem compartments and are described by appropriate values of the factors of radiocapacity.

An external harmful factor (toxic, radiating and of any other nature) impacts on a given well-defined ecosystem. If its influence is essential it will affect the ecosystem and its parameters; ,if not essential it is not relevant to consider it,. In particular the first reaction of an ecosystem on the action of a harmful factor may be a condition of stress. Initially it will be expressed by a physiological suppression reaction and/or growth inhibition, and an outflow release of ballast substances in an environment. Then, if the action of the harmful factor proceeds and/or accrues, reductions of other parameters, down even to oppression and destruction of the most sensitive species are possibly to be expected. If this concept is true, and it is being based on a very large quantity of experimental data (on stresses in ecosystems) at stressful levels of impact on an ecosystem, when ballast substances are released in an environment, then there will be also a trace of this dump in a species in the biota that is reacting to stress to the population such that this component of ecosystem will express a sharp change of the magnitude s of radiocapacity factors.

If also o the inverse is true: if in ecosystem with known initial distribution there was an appreciable redistribution trace or traces, i.e. change of values of s of radiocapacity factors (so also the radiocapacity of the ecosystem as a whole) then an essential effect on the ecosystem has taken place. Similar events may mean, that the tested ecosystem is witnessing the influence of serious stress -, is in an unstable condition. The given concept provides an opportunity to assessing the condition of the ecosystem, its stability and reliability with regard to appreciable changes of radiocapacity parameters, long before any reaction of the influence of integrated parameters is noticeable .

As an initial model a simple fresh-water ecosystem (lake) - a reservoir with one kind of plankton determining the bio-efficiency of the given lake will be considered.

The parameters of a given reservoir are V (volume) 1E+10 l (S - the area 2 km2, Í - depth of 5 m, concentration seaweed ð = 10 g/m3, Tf - factor of accumulation seaweed 1E+4. Initially, after a peak release of 3,7 E+10 Bq Cs-137 (summer period) in a reservoir), the of radiocapacity factor of the sediment suspension will be - 0,7 (k = 1E+3, h = 0,1 m, the factor of radiocapacity of water initially right after the release =0,3 [1]. It is known, that in such reservoir, practically within several days, there will be a very fast redistribution of radionuclides between water and ground suspension [6]. Thus according to our analysis radiocapacity factors of components are estimated by the following formulas:

Fg = kh / (kh+H+pTf * H) - for ground suspension;
Fw = H / (kh+H+pTf * H) - for water;   (1)
Fb = pTf H / (kh+H+pTf * H) - for biota.

The parameters of the considered reservoir allow to estimate: Fg = 100/605 = 0,165; Fw = 5/605 = 0,008; Fb =500/605 = 0,826. In total the radiocapacity factor of a reservoir by definition is equal to 1.

F = Fg + Fw + Fb = 1.   (2)

At stress influence there is a change of the situation:

dF/dt = dFg/dt + dFw/dt + dFb/dt = 0.   (3)

Thence it is possible to observe appreciable reduction Fb, and due to it an increase of sizes of other contributing factors.
Then

If dFb/dt < 0, and dFw/dt + dFg/dt > 0;   (4)

That dFb/dt = dFw/dt + dFg/dt.

take a derivative from Fb will be drawn from the formula (1)

$$ dFb / dt = \frac{(kh + H) \cdot H (dpKb / dt)}{(kh + h + pKb\ h)2} \prec 0 . $$ (2.13)

It is possible to assume, that at the initial moment of stress the integrated parameter of density of a species of a biota will not yet have had in the time to react, and all change Fb will occur due to change (reduction) of Tf - the biota factor accumulation. Is realistic to assume, that because of stress on an ecosystem there is reduction of Tf of 10 times which makes Tf = 1E+3. Thus the structure of radiocapacity factors of elements of the ecosystem will change: Fb = 0,32 (was 0,85), Fg = 0,67 (was 0,165), Fw = 0,03 (was 0,008).

At a rather small change of Tf, there is an essential, up to 3 - 5 times, a change of radiocapacity factors of contibuting elements of the ecosystem. Thence, the ecosystem, through change of radiocapacity factors witnesses of significant ecological events, when essential changes of integrated parameters (density of a population, for example) are not yet noticed.

Visibly, the presented approach allows to make an performant estimation of the condition and reaction of ecosystems. In particular, in the given concrete reservoir deterioration of water quality is observed. That in itself demands control and assessment. Deterioration of parameters of radiocapacity ecosystem may mean changes and restrictions water use.

2. The methodology of experiment

In order to check the models, a special type of experiment has been developed. As the object for the researches – a hydropic culture corn of plants was chosen. For the experiment three-day sprouts of plants of corn were selected and put on hydropic containers where they grew for 15 days. The plants were divided into 3 groups. One group of plants before putting on the containers was irradiated with a strong dose of 15 Gy (the dose was established in preliminary experiments as being oppressing growth). The second group of plants was placed on a solution containing appreciable (slightly toxic concentration - as it was determined in preliminary experiments) a concentration of 50 $\mu M/l$ of $CdSO_4$. A third group of plants is used as a control. In each variant of the experiment three containers with 25 plants on each was used. Radioactive caesium(Cs-137 as $CsCl_2$) was added to all containers. The initial water radioactivity in banks was identical and has made up to 47900 Bq/l. All containers with plants were placed under an optimum level of illumination (5000 lux). Regularly, in a day, water samples were taken to measure the residual radioactivity, and plant root lengths were also measured. The experiment was set up in a way that simultaneously characteristics of radiocapacity of water culture of plants, the absorption of Cs-137 might be observed and a degree of oppression growth reactions of plants under action of harmful factors be defined.

The remaining of radioactivity (tracer) in the water samples was made liquid scintillation ("RackBeta"), and the length of the growth measured routinely (every other day). By doing so, it was possible to look after the variation of the characteristics of radiocapacity and of the growing processes depending on the influence of the pollutants.

3. The basic results of researches

The researches have shown the factors chosen for the experiment with a combination of heavy gamma irradiation of sprouts at a dose of 15 Gy together with slightly toxic concentration of salt of cadmium (50 $\mu M/l$) demonstrated appreciable oppressing influence on dynamics growth reaction of root plants (Fig 1).

It is shown, that in conditions of appreciable influences at a physiological level on ecosystem, such as stress, suppression and/or oppression of even one species of organisms, it is necessary to expect an measurable deterioration of an ecosystem–and changes of values of factors of radiocapacity and radiocapacity of the ecosystem as a whole.

Any serious influence on ecosystem, touching Tfi and pi and other parameters of the condition of an ecosystem (water and ground biota), right away can be reflected as an appreciable change (reduction, or increase of Fi – the factors of radiocapacity. Such a situation will demand a reduction of values of allowable releases and emissions in ecosystem, for example in a cooling reservoir of an atomic power station. Calculations show, that the ecological specification with regard to allowable levels of releases of radionuclides, due to influence on biota ground suspension (benthos), should be reduced to a 100 times. In case of appreciable change of radiocapacity ecosystem through the influence of harmful factors, initial well-being may be quickly lost.

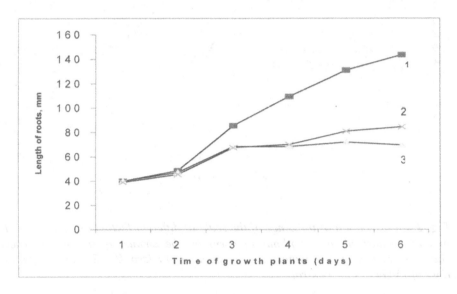

Fig. 1. Influence of gamma irradiation in a dose 15 Gy and entering of salt Cd (50 microM/l) on dynamics of growth of roots of corn in water culture: 1 - the control; 2 - gamma irradiation; 3 - entering salt cadmium.

In figure 2 the result of influence of a combined action of gamma radiation (15 Gy) and Cd (50 μM/l) on radiocapacity of biota (water culture plants of corn) is shown. It is clear, that the influence of harmful factors influences the radiocapacity of the ecosystem. Visibly the tracer content in water at toxic action is increased, and the radiocapacity of biota (water culture of plants) drops. An unequivocal suppression by the toxic elements on the radiocapacities of the biota in the ecosystem can be noticed, reflecting a suppression of their physiological growth processes being also reflected in a reduction of the radiocapacity factor of biota. It is visible, that influence on a given simple test - ecosystem of two so different factors as defined as gamma irradiation and a toxic heavy metal, crates an approximately equal biological effect both on oppression of growth of roots, and on reduction of the factor of radiocapacity of plants.

The basic opportunity to using the present model, the test - object, together with the approach for equidosimetric estimations of influence of different toxic factors on an ecosystem is shown. For the first time a precise method for equidosimetric estimations of action of various harmful factors at an ecosystem level, through calculation of their influence on parameters of radiocapacity ecosystems is offered.

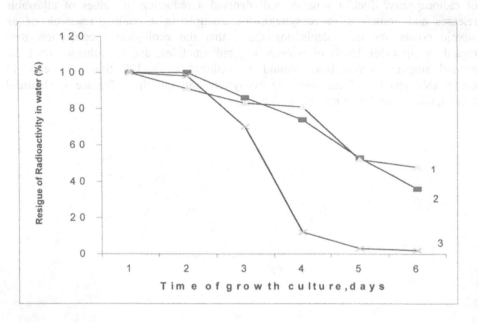

Fig. 2. An estimation of influence of the scale of different factors on radiocapacity of biota (water culture of plants of corn) on the amount of a residual tracer radioactivity (Cs-137) in water: 1 - t cadmium (50 microM/l); 2 – an irradiation dose of 15 Gy; 3 - the control.

The certificate for well-being and viability of an ecosystem, in the widest sense, speaks about its high radiocapacity and vice-versa. On the other hand, high radiocapacity and stable high values of radiocapacity factors of the species of the biota in an ecosystem informs about well-being and ecosystem reliability. The control of radiocapacity and of radiocapacity factors, especial of their changes in time after an impact, may serve as a serious objective outstripping criterion and a method of an estimation of well-being of any ecosystem (water, continental, wood and sea).

4. Conclusions

From the presented results follows that controlling both radiocapacity and its factors could be an important prediction criterion for the ecosystem condition assessment. That index is much more sensitive to ecotoxicants influence comparing to such well-known criteria as species biogeocenoses diversity and speed of biomass increasing.

5. References

1. *Grodzinsky D.M., Kutlahmedov J.A., Miheev A.N. etc.* Anthropogenous radionuclides anomaly and plants. - Kiev: Lybid, 1991. - 160 p.
2. Kutlahmedov *Yu, Polycarps G., Korogodin V.I., Kutlahmedova-Vishnyakova V. Yu.* Methodology and methods of research radionuclide and others pollutants in ground and water ecosystems (manual). - Kiev: Medecol, 1997.
3. *Kutlakhmedov Y., Polikarpov G., Kutlakhmedova-Vyshnyakova V.Yu.* Radiocapacity of Different Types of Natural Ecosystems (without man) and their Ecological Standardization Principles//J. Radioecol.- 1997. - V.6 (2.) - P. 1521.
4. *Kutlakhmsedov Yu., Davydchuk V., Kutlakhmsedova-Vyshnyakova V. Yu.* Radiocapacity of Forest Ecosystems//Contaminated Forests / àd. Igor Linkov, William R. Schell. (Nato Sci. Ser.). - Kluver Acad. Publ., 1998.
5. *Kutlakhmedov Yu., Davydchuk V., Kutlakhmsedova-Vyshnyakova V. Yu.* Radiocapacity of Forest Ecosystems//Contaminayed Forests / àd. Igor Linkov, William R. Schell. (Nato Sci. Ser.). - Kluver Acad. Publ., 1998.
6. *Agre À.. L., Korogodin V.I.* About distribution of radioactive pollution in slow exchanged a reservoir// Medical radiology. - 1960. - N 1. - P. 67- 73.
7. *Trapeznikov A.V., Pozolotina V.N., Jushkov P.I. etc.* Researches of a radioecological situation in the rivers Techa and Iset, polluted with dumps PO Mayak // Problems of radioecology and boundary disciplines. N 2. – Zarechniy. 1999. -P. 20 - 67.
8. *Kutlakhmedov Yu., Davydchuk V., Kutlakhmedova-Vyshnyakova V. Yu.* Radiocapacity of Forest Ecosystems//Contaminated Forests / àd. Igor Linkov, William R. Schell. (Nato Sci. Ser.). Kluwer Acad. Publ., 1998.
9. *Kutlahmedov Y.A., Polykarpov,G. G., Zotov V.P. etc.* Medical and biologic consequences of Chernobyl accident. Remote radioecological both radiobiological problems and the analysis of efficiency of counter-measures on protection bio-and ecosystem from consequences of Chernobyl accident (in 2 parts). Kiev, Medecol, ÌNIC BIO-ECOS. 2000, 293 p.
11. *Kutlahmsedov Y.A., Korogodin V.I., Kol'tover V.K.* Bases of radioecology. Kiev, Vishaya school, 2003 (in a seal).

Part 2.

Approaches to Equidosimetry in Ecology

EQUIDOSIMETRY: A REFLEXION ON RISK ASSESSMENT

H. MAUBERT
SPR, CEA Cadarache, 13108 St-Paul-lez-Durance cedex
FRANCE

1. Introduction

The subject of equidosimetry is somewhat puzzling. What could that be? It is natural to refer to radioactive doses, which have been derived mainly in the framework of man radioprotection, extended to the environment. Also, we could try to apply the same concepts to other pollutants as chemicals. But is that all? Many other hazards threaten mankind. In this paper we will try to overlook some of them in order to define a priority in the extension of the radioprotection concept of doses to other fields.

2. Assessment of risk linked to radionuclides

Radiations coming from radionuclides possess a certain amount of energy. Through interaction with matter this energy may be released. Risk assessment [1] for radionuclides is subsequently based on the evaluation of energy released by radiation in human living matter. It is the concept of absorbed dose which unit is the gray (Gy), 1 Gy being defined by the release of 1 J in one kg of matter.

From experience coming from Hiroshima, Nagasaki and accidental irradiations, it has been observed that for global external irradiation, and without treatment, a lethal dose at 50 to 60 days is between 3.5 and 4 Gy. Below 1 Gy functional signs cannot be externally observed and concern only blood white cells. After a while these symptoms disappear. On the contrary doses above 10 Gy is always very serious and often fatal even with treatment and hospitalization. Observed effects shortly after an irradiation are called deterministic effects.

The concept of released energy into living tissue has been further refined with the derivation of equivalent dose depending on the tissue and the type of radiation. The unit for equivalent dose is the Sievert (Sv). A considerable literature has been published on the subject.

At low doses radiation may cause cancers, or rather increase the probability of getting a cancer. An increase in cancer rate has been proved above 100 mSv for adults (survivors of Hiroshima and Nagasaki) and above 10 to 20 mSv for children. It is still unknown if there is a linear relationship between dose and cancer occurrence. To day, the philosophical base for radioprotection is based on a linear relationship between dose and effect, although there is some indication that a threshold does exist

the risk assessment is based on the assumption that whatever the radiation dose delivered to an organism a stochastic (probabilistic) effect will be possible, and that the apparition frequency will depend on the dose amount.

In other words, at some dose an effect may observed and an extrapolation is made towards low doses.

The subject of this paper is "equidosimetry". the basis for radioprotection is anthropocentric. Environment is taken into consideration only as far as it satisfies a human need. Historically it is considered that if man is protected from artificial

F. Brechignac and G. Desmet (eds.), Equidosimetry, 35–41.

irradiation, then the environment is also protected. starting in the years 90, following the conferences of Rio and Tokyo, the need for an environmental protection politic does emerge. there are many programs in this direction : ICRP working group on environmental protection, IAEA program on radiological protection of the environment, etc… UNSCEAR in his 1996 report gives dose rate criteria: for natural communities of plants and animals a dose rate of 0.1 mGy/h on a small number of individuals would be without effect. For those exposed individuals the radiosensitivity indicator would be the reproduction capacity, unlike for humans where the apparition of cancer is the indicator.

3. Assessment of risk linked to chemical substances

For chemical substances we do not have the convenient energy release evaluation to determine the adverse effects. Non-carcinogenic and carcinogenic effects are considered separately.

For non-carcinogenic effects and a given substance, a critical effect is determined, and then, derived from experiments or case studies, or epidemiological studies reference doses are determined, either for oral intake or inhalation. In case of oral intake, reference doses (RfD) are of a daily exposure to the human population (including sensitive subgroups) that is likely to be without an appreciable risk of deleterious effects during a lifetime. A value above the RfD does not imply that a deleterious effect will necessarily occur.

These RfDs [2] derive from a NOAEL (Non Observed Adverse Effect Level) or a LOAEL (Lowest Observed, Adverse Effect Level is determined). (IRIS database). This value is divided by an Uncertainty Factor (UF) intended to account for (1) the variation in sensitivity among the members of the human population, i.e., interhuman or intraspecies variability; (2) the uncertainty in extrapolating animal data to humans, i.e., interspecies variability; (3) the uncertainty in extrapolating from data obtained in a study with less-than-lifetime exposure to lifetime exposure, i.e., extrapolating from subchronic to chronic exposure; (4) the uncertainty in extrapolating from a LOAEL rather than from a NOAEL; and (5) the uncertainty associated with extrapolation from animal data when the data base is incomplete.

This method for risk evaluation differs from that used for radionuclides. The RfD is based on the assumption that thresholds exist for certain toxic effects such as cellular necrosis. For some substances this may be complicated by the fact that they are necessary to life in small amounts and that both too high and too low doses are deleterious. It is for example the case of iodine of manganese…

For carcinogenic effects a risk probability is the result of a low-dose extrapolation procedure. It may be expressed as a relationship between the probability of cancer and some daily intake, or concentration in water, food or air.

This approach is very similar to that used in the case of radionuclides. From an observed effect a low dose extrapolation, usually linear is assumed.

As far as ecotoxicity is concerned values for NOEC (No Observed Effect Concentration) may be found for some species, derived from in situ observations or, more often, derived from experimentations in simplified ecosystems and controlled environment.

4. Discussion on pollutants

Basically, for polluting substances, or toxicants, either radioactive or not, the approach is more or less always the same:
➢ a contamination level is determined at which an adverse effect is observed, usually leading to a maximum level at which no adverse effect is observed.
➢ from this value a "safe" level is determined using a kind of safety factor sometimes called uncertainty factor,
➢ for carcinogenic effects a probabilistic function is derived, often being a low dose extrapolation.

These evaluations are mainly anthropocentric, little being available to assess environmental risk of small quantities of pollutants. Many example can be found of the effect of heavy local pollution either chronic of accidental such as excess hydrocarbon releases or spills, heavy metals, the effect of sulfur atmospheric emissions on lichens, etc. But for light pollution extending on vast areas, it is almost impossible now to determine whether or not there is an effect, for example on the biocoenosis, even less on the morbidity of plants or animals. Sometimes illness, rarefaction or proliferation of such or such species is observed, but what these phenomena are the consequences of is not always clear.

The interaction of pollutants is another question. If it is possible to determine a risk evaluation for the exposition to radionuclides, or to a chemical pollutant such as cadmium or benzene, what if both are present? when a tool such as a probability function is available, it is possible to add the probabilities, but do they really add ? and in nature how does a radioactive pollution combine with some atmospheric pollution caused by transport for example?

The basis of toxicological evaluations are well fit to determine possible adverse effects in Man, much less for the environment, the concept of equidosimetry being still in the area of research.

5. What about pollutants that are not pollutants?

Major changes may occur from the release of substances that are not really pollutants such as carbon dioxide, which is well known to be a greenhouse effect gas. It is not yet fully proved whether the increase in CO_2 in the atmosphere will induce an increase in the world average temperature, but many indications show that atmospheric releases concern now the whole planet.

The first description of the greenhouse effect [3] has been made by JB Fourier in 1827, and the first analysis of the phenomenon involving CO_2 has been made by the Swedish chemist S. Arrhenius. But the phenomenon has been completely underestimated until the fifties, when systematic measurements of CO_2 started in Alaska and Hawaii. The first world conference on climatic change was held in Geneva in 1979.

It is now considered as a fact that global temperatures are rising. Observations [4] collected over the last century suggest that the average land surface temperature has risen 0.45-0.6°C in the last century, while sea level has risen worldwide approximately 15-20 cm. Approximately 2-5 cm of the rise has resulted from the melting of mountain glaciers. Another 2-7 cm has resulted from the expansion of ocean water that resulted from warmer ocean temperatures. In the same time, precipitation has increased by about 1 percent over the world's continents. High latitude areas are tending to see more significant increases in rainfall, while precipitation has actually declined in many tropical areas.

It is not yet know, exactly what will be the consequences of global warming. Some areas could become warmer, but not all, or dryer or wetter. Sea level could rise endangering lowlands. Effects could be observed on biocœnosis, as many species are closely dependant on environmental parameters such as temperature and humidity.

There is no quantified scale to judge of these changes, neither how to compare them with toxicological data, or even probabilistic evaluations.

In this category of pollutants that are not pollutants, we may add the consequences of human activity, when excessive. For example, pig breeding and subsequent spreading of manure in the fields, may be a good practice as long as the number of pigs is low enough, that is to say fed more or less from local feed production. Then the manure is not a pollution. But when it becomes intensive breeding supported by massive imports of feed, nature cannot absorb a too big input of manure and long-term pollution of water resources by nitrates occur. It is now the case in Brittany where in many rural communities tap water is no longer suitable for drinking.

6. Overuse of natural resources

We are now about 6 billion people. In 1830 the world population was estimated at 1 billion people and, a hundred years ago, it was about 1,6 billion people. It is obvious that this population increase is a tremendous strain on the planet.

Often ecological efforts are aimed at conservation. But a state of primary environment, such as it might be dreamed of, does not exist anymore anywhere. Not to speak of ancient ages, like the end of the secondary era, more and more species become extinct, a phenomenon which is co-incidental with the spreading of man, the two not being completely correlated. Before Man could master fire he lived in harmony with nature, its pressure on environment remaining very low. But hundred of thousand years ago the use of fire to scare game into traps already destroyed primitive forest in large areas in Africa. In the same manner, in order to give more space to the population of buffalo, the American prairie was extended by fire about ten thousand years ago.

It is well known that Hannibal (217 BC) had elephants when he invaded Italy in his war against Rome. These elephants came from North Africa. Sylla after his war against Jugurtha (around 100 BC), still in North Africa, spend some time relaxing and hunting local elephants and lions. Now these animals are absent from these countries. Anyway presently North Africa could hardly support the life of wild elephants as a reason of aridity and demographic pressure. More recently the over-hunting of whales is another illustration, but in this case conservation efforts may prevent the disappearance of some species at least for a while. Over fishing is another problem, probably modifying already the ecological balance in the Ocean at large.

Agriculture has been the secondary large technological revolution in the history of Man. In large areas, the primitive forest has been replace by pasture and the by cultivated land. Presently only 5% of the area of China is covered by forest, although 90 % of the country was forested at the beginning of the Neolithic ages [5]. The "discovery" of new lands such as new Zealand or Australia by Europeans lead to the disappearance of large terrestrials birds, and other animals, either by hunting or by the introduction of new species such as rabbits, sheep and dogs. Besides disappearance of species, the proliferation of one species at the detriment of biodiversity is another indication of the degradation of the ecosystem.

In the middle ages in Europe it could be talked of "horror sylvanum", the forest being seen as dangerous, unknown and generally bad to Man. Think of the fairy tales involving frightening adventures in dark, hostiles forests.

The third technological revolution, the industrial one, is characterized by the addition in the biosphere of substances that are not naturally present or only in minute amounts, like mercury, for example, and the accumulation of not degradable wastes. But the most prominent feature of the industrial revolution is the massive consumption of energy mainly from fossil not renewable sources.

If we go back to our concept of equidosimetry, what can we find for the modification of the ecosystem? In a project like the European ELISA [6] project a whole series of indexes have been developed, such as Habitat coherence, Structural complexity, Habitat extent, Habitat connectivity, and for Species presence, trends, diversity, populations, Flagship species... In these indicators, we may find the number of species in a definite surface area, the number of endangered species, the number of protected species, or the area of protected natural parks, and so on.

However, all changes in the environment, even if they are losses, are not necessarily directly or immediately bad for man. Only a massive deforestation could ensure enough food production for an expanding population. After all, one can very well live without Blue Whales in the Ocean, or without Elephants in North Africa. Agricultural landscapes are often very suitable and pleasant, and conservation may sometimes appear as a romantic idea of a lost paradise.

Possible hazards which are not easily quantified, may be referred as the loss of possible resources, and the harm to the interdependence of all nature.

The loss of possible resources may the disappearance of a particular endangered species only capable of synthesizing a molecule which could prove useful drugs used in medical applications. The rarefaction of the global genetic patrimony could also impair future life development on the planet.

But the studying ecology teaches us that everything is interdependent. An old examples of this is the pollinisation of plants by insects, known since antiquity. The dependence of the predators on their preys is obvious. What may be less obvious, is that the well-being and health of the preys depends also on the predator. When great predators are absent, proliferation, aging and decay of herbivorous population may be sometimes observed. Phenomena like El Nino in the Pacific Ocean show how a single change in water temperature can modify deeply the vegetal and animal populations, and finally affect resources available to Man.

The virtual disappearance of untouched areas, the ever growing energy consumption, the modifications of natural cycles such as the one of carbon or sulfur, or even nitrogen, may be triggers that will bring majors unexpected changes. All this is yet to evaluate in a concept of equidosimetry.

7. Other hazards

If we leave the field of environment and come a little more on that of specifically human hazards, among them poor education, endemic diseases such as malaria, lack of adequate food, and so on. This is a vast subject, and since our starting point was radioactivity, we will take the example of iodine.

There have been some nuclear events [7] leading to a spreading of radio-iodine:

The "bravo" test of a nuclear weapon in Bikini in 1954, where 250 atolls resident were exposed with an average global dose of 1,9 Gy. Thyroid deficiency and cancer of the thyroid occurred. The frequency of these diseases increased with the dose, 10 Gy leading to 50 % of thyroid deficiency. The occurrence decreased with age. On the Rongelap atoll 5 of the 6 children under 1 developed thyroid deficiency, 13 % for the children above 10.

The atmospheric tests in Nevada, Semi-Palatinsk, and the Windscale accident also led radio-iodine exposure but for these events, effect were either not observable, either there are no consolidated data.

The Tchernobyl accident led to the exposure of hundred thousand of people, with doses to the thyroid reaching or even exceeding 1 Gy for 1 y old children. More than 15000 children could have received more than 1 Gy, 5000 more than 2 Gy and 500 more than 10 Gy. Starting around the year 1990 a marked increased in the thyroid cancer among children below 15 at the time of the accident was observed. 1800 cases of cancer have been counted between 1986 and 1998, instead of a few tens expected.

All this is very serious, and of course, it deserves a close attention, not only to give adequate treatment to the sick, but also to have the means of deriving an adequate dosimetry, which we probably have, and most of all to avoid new episodes involving the release of radioactive pollutants.

But if we talk about equidosimetry, there is another plague involving iodine, it is the lack of this (stable) element. Iodine is an element present in small quantity (15 to 20 mg) in the human body, mainly in the thyroid, and it is necessary to the making of thyroid hormones which are essential to the metabolism and to the process of growth. The needs [8] for iodine are evaluated to be of 90 µg/day for the child between 0 and 6, 120 µg/day between 6 and 12, 150 µg/day for the teenagers, and 200 µg/day for the pregnant and nursing woman. At the level of a population, the lack of iodine leads to dysfunction of the thyroid, goiter, fertility trouble, increase in postnatal mortality, and mental retardation. It has been estimated that in 1990, about 1572 million people (28 % of the total world population) was exposed to a lack of iodine. 11,2 million people was affected by endemic cretinism, the extreme form of mental retardation due to the lack of iodine, and that 43 million people suffered from some kind of mental retardation. All this is potentially avoidable, provided that a source of iodine is made available to the concerned populations, generally under the form of iodine added to table salt, cattle feeding salt or alimentary industry salt. In 1990 only 5 % of the concerned populations had access to iodine complemented salt, and. about 68 % in 1990, thanks to the combined actions of concerned countries, and international organisms such as WHO, World Bank UNICEF…. Although this is a public health success a lot is left to do for some countries, and to avoid a new degradation of the situation resulting from international politics, conflicts…

There may be in the world over majors health problems, but the point here is to show that so far there is no kind of dosimetry to evaluate their consequences, and no tool like the calculation of radioactive doses, which lead to a satisfactory exposure – effect relationship.

8. Conclusion

Starting from the familiar background of radiotoxicity, this brief outlook showed us that there is not so far a common scale to evaluate and compare the potential adverse effects of pollutants, diseases or others to the environment and man.

We live in a paradox where, at least for a part of the world population, never in the past the conditions of life have been as favorable, with controlled environment, sufficient food of great quality, education, medical care and so on.

But at the same time it is easy to see that this so-called western way of life leads to an over–consumption of natural resources that may be problematic to sustain in the future, and that for this reason is hardly extendable to mankind at large. In all cases, it is difficult to imagine that the world population will continue its growth without very serious difficulties in the future.

Therefore even if the tool for equidosimetry is not ready, at least it is time to lay flat all the environmental problems and to rank them by order of maximum threat to the future and try to treat them. But a country alone cannot do this, because it may involve a major strategic change, in a world where no one is isolated anymore.

Finally, we have not talked about a powerful engine for the development of protection tools like radioprotection, estimation of the consequences of pollutants, and others. Although it is wise when a danger occurs to seek the ways to evaluate it and learn how to protect ourselves from it, we have seen that there is no equivalence of evaluation between all the potential hazards.

Radioprotection seems to be a relatively accurate science, with adequate dosimetry, well known dose – effect relationships, eventually medical treatment, etc. This is no doubt due to the fact that the adverse effects of newly discovered radioactive products were identified in a developed and scientific world. It is also due to the fact that the effects of the atom bomb were so astonishingly devastating that it induced in the population a fright that pushed the politic and scientific community into precise evaluation of the consequences of nuclear industry. It was a good thing, and the thought process behind radioprotection could be usefully spread.

But fear, and its consequence, stress, is not always a good counselor. The huge cattle slaughters deriving from the mad cow disease may appear as an silly over reaction.

When the energy consumption is ever growing, the fear of radioactivity that leads countries to abandon clean nuclear power, thus dilapidating precious fossil fuel resources, aggravating greenhouse gases emission and disruption in great natural cycles, should be tamed.

Before trying to go further and extending the concept of dose, we should answer these questions:

Is it possible to consider the problems globally because we know it now, since we have seen the picture of the earth from the moon, we live in a finite world?

And

Is a controlled change still possible?

It is therefore more important to concentrate on what is dangerous, than on what is frightening.

9. References

[1]   Henri Métivier; Les bases de la protection radiologique; cours de l'Institut National des Sciences et Techniques Nucléaires, 2002

[2]   IRIS database. Internet http//www.epa.gov/iris/ ; accessed on April 12 2002

[3]   Mission interministérielle de l'effet de serre. Site internet http//www.effet-de-serre.gouv.fr ; accessed on 19 March 2002.

[4]   Internet site http//www.epa.gov accessed on 20 March 2002

[5]   François Ramade; Eléments d'écologie appliquée ; McGraw Hill, 1978.

[6]   Site internet http://www.ecnc.nl/doc/projects/elisa.html, accessed on 12 March 2002

[7]   A. Flüry-Hérard et F. Ménétrier ; Conséquences sur la santé des iodes radioactifs libérés en cas d'accident nucléaire. Journée les iodes radioactifs ; l'iode stable, 15 février 2002, Institut Curie, Paris.

[8]   François Delange; Nutrition iodée et développement du jeune enfant ; Journée les iodes radioactifs ; l'iode stable, 15 février 2002, Institut Curie, Paris

AN EQUI–DOSIMETRIC APPROACH TO THE COMPARISON OF RADIATION AND CHEMICAL EFFECTS ON NATURAL POPULATIONS OF AQUATIC ORGANISMS

V. TSYTSUGINA

The A.O. Kovalevsky Institute of Biology of the Southern Seas, NAS, Nakhimov Prospekt, 2, Sevastopol, 99011, UKRAINE

1. Introduction

Cytogenetic studies of natural populations of aquatic organisms show that chromosome damage is induced by radiation and chemical factors [1–3]. Therefore it is important to provide equi–dosimetric assessment of effects caused by ionising radiation and chemical pollutants [4-7].

We propose an approach which makes it possible to determine deleterious factors and to compare their efficiency. The approach is based on comparing cytogenetic effects in *the mutagen equivalent doses.* In this method several cytogenetic criteria (a % of cells with chromosome aberrations, a distribution of aberrations in the cells, a number of aberrations per aberrant cell) are used.

2. Distribution of chromosome aberrations in the cells

2.2. DISTRIBUTION OF CHROMOSOME ABERRATIONS IN CELLS IN THE EXPERIMENTAL MUTAGENESIS

On the basis of the experimental study of separate and combined effects of ionising radiation and chemical mutagens on the aquatic organisms (Table 1) a general conclusion was made concerning distribution of chromosome aberrations in the cells [3]:

Mutagen	Distribution of chromosome aberrations in the cells
R (only) & (R' + C)	The best fit to the Poisson distribution
	$P(n) = (m^n \cdot e^{-m}) / n!$
C (only) & (C + R)	The best fit to the geometric distribution
(C'+R)	$P(n) = (1 - q) \, q^n$

where:

R is ionising radiation, C is chemical mutagen, R' & C' indicate the dominant impact in the combination, m is the mean number of chromosome aberrations per cell, n is the number of chromosome aberrations in a cell, q is the fraction of aberrant cells.

It is shown that the observed distribution of radiation – induced chromosome aberrations in the cells fits the Poisson distribution more closely than the geometric distribution. For the distribution of aberrations produced by chemical mutagens the reverse is true, i.e. the geometric distribution provides the best fit to the data. Under the impact of both factors, the observed distribution fits well the Poisson

F. Brechignac and G. Desmet (eds.), Equidosimetry, 43–49.

Table 1. Distribution of chromosome aberrations in cells of crustacean and fish embryos (experimental mutagenesis) [1, 3]

Species	Irradiation or chemical mutagen	Dose or concentration	Cell numbers with number of aberrations				χ^2-criterion of concordance with distribution	
			0	1	2	3	Poisson	Geometric
Crustaceans								
Idotea baltica	$^{90}Sr - ^{90}Y$	3 Gy	1950	173	5	0	0.83	6.44
Idotea baltica	$^{90}Sr - ^{90}Y$	0.5 Gy	1731	100	3	0	0.003	1.16
Gammarus olivii	$^{90}Sr - ^{90}Y$	3 Gy	8114	819	28	0	3.79	32.90
Gammarus olivii	$^{90}Sr - ^{90}Y$	1.5 Gy	1756	113	4	0	0.03	1.27
Gammarus olivii	^{137}Cs	0.5 Gy	611	31	1	0	0.05	0.20
Gammarus olivii	$^{90}Sr - ^{90}Y + ^{137}Cs$	2 Gy	1204	93	2	0	0.68	3.62
Gammarus olivii	External gamma	5 Gy	2640	371	27	3	2.26	10.87
Gammarus olivii	External gamma	3 Gy	740	90	4	0	0.34	3.92
Gammarus olivii	Pb acetate	1 mg.l^{-1}	926	38	5	0	15.14	5.18
Gammarus olivii	Chlorphene	0.2 mg.l^{-1}	1041	52	5	0	7.98	1.70
Gammarus olivii	^{137}Cs + chlorphene	0.5 Gy + 0.2 mg.l^{-1}	531	25	2	0	2.64	0.42
Gammarus olivii	$^{90}Sr - ^{90}Y + ^{137}Cs$ + Pb acetate + chlorphene	2 Gy + 1 mg.l^{-1} + 0.2 mg.l^{-1}	1190	86	2	0	0.39	2.64
Gammarus olivii	External gamma + Pb acetate	3 Gy + 1 mg.l^{-1}	720	94	5	1	1.83	3.60
1. FISHES **SCORPAENA PORCUS**	External gamma	6 Gy	1460	232	11	1	2.56	17.50
Scorpaena porcus	External gamma	1 Gy	1541	193	13	0	0.07	4.50
Scorpaena porcus	DDT	0.1 mg.l^{-1}	807	77	14	2	20.19	3.95

Table 2. Distribution of chromosome aberrations in cells of aquatic organisms from natural populations

Species	Location	Year	Cell number with aberrations				χ^2 – criterion of concordance with distribution	
			0	1	2	3	Poisson	Geometric
Engraulis encrasicholus (embryons)								
	The Aegean Sea	1968	989	52	5	0	7.12	136
Mysidacea sp. (embryos)	The Atlantic Ocean	1978	247	28	3	1	2.12	0.03
Anchylomera Brossevilei (embryos)	The Mediterranean Sea	1985	1293	60	4	0	3.94	0.40
Monodacna caspia (juveniles)	The Black Sea	1986	408	34	2	1	1.21	0.05
Pontogammarus robustoides (embryons)	The Dnieper River	1992	1127	69	5	1	4.91	0.38
Mullus barbatus (embryons)	The Black Sea	1999	411	42	6	1	5.96	0.62
Stylaria lacustris (somatic cells)	Water-body in Strakholes'e (30 km from the ChNPP)	1991	830	49	8	0	17.48	5.58
2. WATER-BODIES IN THE CHERNOBYL ZONE								
Gammarus sp. (embryos)	Kopachy	1991	1339	111	9	0	3.60	0.23
Stylaria lacustris (somatic cells)	Kopachy	1991	1946	160	11	0	1.74	0.67
Stylaria lacustris (somatic cells)	The Pripyat River	1991	550	78	9	0	1.53	0.47
Stylaria lacustris (somatic cells)	The Chernobyl NPP cooling pond	1993	392	34	1	0	0.14	1.18
Stylaria lacustris (somatic cells)	Krasnyansky Starik	1993	540	17	1	0	1.74	0.10
Stylaria lacustris (somatic cells)	Lake Glubokoe	1993	967	40	0	0	0.87	1.68

distribution if ionising radiation has greater influence. The observed distribution provides the best fit to the geometric distribution if both factors are comparable in effect or chemical mutagens have a greater influence.

2.3. DISTRIBUTION OF CHROMOSOME ABERRATIONS IN CELLS OF THE AQUATIC ORGANISMS FROM NATURAL POPULATIONS

A study was made of the distribution of chromosome aberrations in the cells of marine and freshwater organisms before and after the ChNPP accident [1-3] (Table 2). It can be seen that before the Chernobyl accident the distributions of chromosome aberrations in cells of the aquatic organisms (crustaceans and fishes) fit well the geometric distribution. It is probably, therefore, that chromosome damage in these aquatic organisms was induced by chemical pollutants.

After the Chernobyl accident the distributions of the chromosome aberrations in cells of the aquatic organisms fit the Poisson distribution only in water bodies with a higher level of radioactive pollution (the Chernobyl 10-km zone). In other water bodies in the Chernobyl zone as well as in the Dnieper River and the Black Sea the distributions of chromosome aberrations in the cells of worms, mollusks, crustaceans and fishes were closer to the geometric distribution. Therefore it may be concluded that in these cases chromosome aberrations were induced either by chemical pollution or by the combined (radioactive and chemical) pollution but the contribution of chemical pollution to the total damage of natural populations of the aquatic organisms was great.

3. Comparison of the effects in mutagen equivalent doses

In order to assess radiation influence in the cases when both factors (radiation and chemical) are comparable in the effect or chemical factor has a greater influence we propose to compare effects in *mutagen equivalent doses* [8, 9]. *The mutagen equivalent dose* is estimated as a % of cells with chromosome aberrations and an additional criterion is the number aberrations per aberrant cell.

Figure 1 presents data on the number of aberrations per aberrant cell under separate and combined influence of ionising radiation and chemical mutagens in *the mutagen equivalent doses* (% of cells with chromosome aberrations) on crustacean and fish embryos. Figure 1 shows that the number of aberrations per aberrant cell can be a good indicator of radiation influence in the cases of combined radiation and chemical effects.

3.1. EQUI-DOSIMETRIC DATA IN THE EXPERIMENTAL MUTAGENESIS

Comparison of the effects in *the mutagen equivalent doses* makes it possible to assess the equivalence of ionising radiation and chemical mutagens in the experimental mutagenesis. Equivalence means the ability of ionising radiation and chemical mutagens to induce an equal number of cells with chromosome aberrations.

Data on Figure 1 allow to assess the ecological Gy-equivalent of the effects of metals and chloroorganic compounds (Gy / μmol $.l^{-1}$) [6]

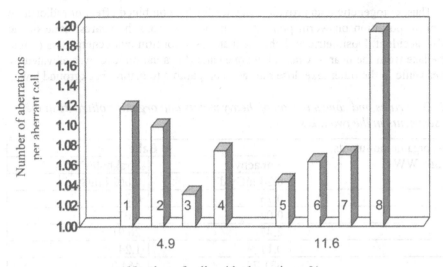

Fig.1. Number of aberrations per aberrant cell under separate and combined effects of ionizing radiation and chemical mutagens in mutagen equivalent doses on crustacean and fish embryos.

1 – Gammarus olivii, $Pb^{(2+)}$ (1.9 µmol.l^{-1}); 2 – G. olivii, chlorphene (0.4 µmol.l^{-1}); 3 – G. olivii, ^{137}Cs (0.5 Gy); 4 – G. olivii, ^{137}Cs (0.5 Gy) + chlorphene (0.4 µmol.l^{-1}); 5 – G. olivii, external gamma (3 Gy); 6 – Scorpaena porcus, external gamma (1 Gy); 7 – G. olivii, external gamma (3Gy) + Pb^{2+}(1.9 µmol.l^{-1}); 8 – S. porcus, DDT (0.2 µmol.l$^-$

$$^{137}Cs / Pb^{(2+)} = 0.3$$

$$^{137}Cs / chlorphene = 1.3$$

$$External\ gamma / DDT = 5.0$$

These results are similar to those obtained in the aquatic microcosm test [10].

3.2. IDENTIFICATION AND EQUI-DOSIMETRIC ASSESSMENT OF RADIOACTIVE AND CHEMICAL POLLUTION

The method of comparing effects in *mutagen equivalent doses* was also used for the cytogenetic study of natural populations of oligochaeta *Stylaria lacustris* in the lake located in the nearest Chernobyl zone (village Kopachy) and in the lake in 30 km to the South of the Chernobyl NPP (village Strakholes'e). In cells of worms from both lakes the distributions of chromosome aberrations were closer to the geometric distribution (Table 2). In worms from these lakes the mean numbers of cells with chromosome aberrations were almost equal (about 8%) but the numbers of aberrations per aberrant cell were greatly different (Fig.2).

Thus, cytogenetic data provide evidence for the combined effect of radioactive and chemical pollution on worm population in the lake from the nearest zone of the Chernobyl accident. Dosimetric and chemical analyses confirm cytogenetic data (Table 3). In the lake from the nearest Chernobyl zone natural irradiation level was elevated by 100 times while in the other lake dose rate was comparable to natural background level.

Table 3. Dose rates and concentration of heavy metals and organic pollutants in the bottom sediments in the two lakes

Metals, organic pollutants, (mg.kg⁻¹ WW)	Lake	
	Kopachy	Strakholes'e
	0.24 mGy.d⁻¹	0.0024 mGy.d⁻¹
Pb	0.73	0.83
Cd	0.04	0.07
Zn	2.47	26.91
Cu	0.11	1.24
Fe	631.73	464.05
Mn	8.53	52.21
Ni	1.65	0.70
Co	0.55	0.28
Heptachlorcyclohexane	0.0104	0.0046
PCBs	0.111	0.120

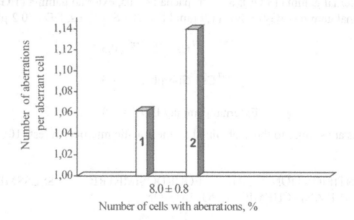

Fig. 2. Number of aberrations per aberrant cell in worms Stylaria lacustris from two lakes 1 – Kopachy, 2 – Strakholes'e.

Therefore from the data obtained it may be concluded that the combined effect of chronic ionising irradiation (dose rate 0,24 mGy.d⁻¹) and chemical pollution on oligochaeta Stylaria lacustris in the lake from the nearest Chernobyl zone (Kopachy) is equivalent to chemical effect on these worms in the other lake (Strakholes'e)

4. Conclusion

Approaches to the identification and the equi-dosimetric assessment of radiation, chemical and combined effects on natural populations of the aquatic organisms are proposed. These approaches are based on generalization of our previous experimental data concerning separate and combined effects of ionising radiation and chemical mutagens (heavy metals and chlororganic compounds) on crustacean and fish embryos, as well as on the results of field studies carried out before and after the ChNPP accident. For the identification and the equi-dosimetric assessment of the damaged factors it is proposed to use the following criteria: the comparison of cytogenetic effects in *mutagen equivalent doses*, the distribution of chromosome aberrations in cells and the number of aberrations per aberrant cell.

5. References

1. V.Tsytsugina . Chromosome mutagenesis in population of aquatic biota in the Black Sea, Aegean Sea and Danube and Dnieper Rivers, 1986-1989 Proc. of Seminar on comparative assessment of the environmental impact of radionuclides released during three major nuclear accidents: Kyshtym, Windscale, Chernobyl. Luxemburg, 1-5 October 1990, V.II, pp. 895-904.
2. G.Polikarpov,, V.Tsytsugina. Radiation effects in the Chernobyl and Kyshtym aquatic ecosystems. Radioecology and the restoration of radioactive–contaminated sites (F.F.Luykx, M.J. Frissel, eds.). Kluwer Academic Publishers. Dordrecht /Boston/ London. 1996. pp. 269-277.
3. V.Tsytsugina. An indicator of radiation effects in natural populations of aquatic organisms. Radiation Protection Dosimetry, 1998. V.75, N 1-4, pp. 171-173.
4. G.Polikarpov. Biological aspects of radioecology: objective and perspective. Proc. of the international workshop on comparative evaluation of health effects of environmental toxicants derived from advanced technologies. 1998. Chiba, January 8 – 30. Tokyo: Kodansha Scientific Ltd., pp. 3 – 15.
5. G.Polikarpov. Effects of nuclear and non-nuclear pollutants on marine ecosystems. Marine pollution. Proceeding of a Symposium held in Monaco, 5-9 October 1998. IAEA – TEC DOC –1094. IAEA – SM – 354/22. International Atomic Energy Agency. 1999, pp. 38-43.
6. G.Polikarpov, V.Tsytsugina. Comparison of cytogenetic and ecosystem efficiency of radioactive and chemical mutagens in the hydrobiosphere. Reports of National Academy of Science of Ukraine, 1999, '6, pp. 199-202 (in Russian).
7. G.Polikarpov. The future of radioecology: in partnership with chemo-ecology and eco-ethics. J. Environ. Radioactivity, V.53, ' 1, pp. 5-8.
8. Ch.Auerbach. Mutation research. Problems, results and perspectives. London-New York, 1976, P. 450.
9. L.J.Stadler, G.F.Sprague. Contrasts in the genetic effects of ultraviolet radiation and X-rays. Science, 1937, V.85, pp.57-58.
10. S. Fuma, K. Miyamoto, H. Takeda, K. Yanagisawa, Y. Inoue, N. Sato, M. Hirano & Z. Kawabata. Ecological effects of radiation and other environmental stress on aquatic microcosm. . Proc. of the international workshop on comparative evaluation of health effects of environmental toxicants derived from advanced technologies. 1998. Chiba, January 8 – 30. Tokyo: Kodansha Scientific Ltd, pp. 131 – 144.

DETERMINATION OF HETEROGENEITY BIOTA AS A PERSPECVE PARAMETER OF ECOLOGICAL STANDARDIZATION

G.V. TALALAEVA
Institute of Plants & Animals Ecology, RAS, Urals Division, 8 Marta str. 202, Ekaterinburg, 620144, RUSSIA

1. Introduction

Problems of ecological standards in technogenic areas and estimation of adaptive processes in residents of anthropically polluted regions have been receiving widespread attention for a long time.

However, the general theory of ecological standards and co-evolution of live and inanimate nature in conditions of technogenesis has not been developed so far. One of the reasons may be the unfinished process of technogenesis of industrial areas. It is precisely this process that quickly and irreversibly changes interrelations between live and inanimate systems and provokes transitional states in biosystems when the basis of ecological standards – the stability of indices of human and biota vital functions – loses sense.

The paper considers the following aspects of the problem.

1. Evaluation of the methodological approaches to the problem of ecological standardization during the twentieth century.
2. Anthropogenic environmental factors as catalysts of natural selection and accelerators of human technogenic evolution.
3. External (geoecological) and internal (psycho-social) factors modifying the human long-term adaptation processes.
4. Possible artificial acceleration of human evolutionary processes by technogenic factors (magnetic fields, low radiation doses).
5. Activation of quantification processes, bifurcation phenomena, increased heterogeneity in the adaptive state indices of residents of anthropically changed areas in the Urals.

The report is based on 20-year of observations on adaptation processes and on the adaptive ability of mid-latitude residents to technogenic transformation. Special attention was given to patients with clinical sings of decompensated adaptive processes (cardiological and post-traumatic stress disorders): residents from various geological zones and geochemical regions in the Urals, liquidators of the Chernobyl accident consequences, immigrants from the Trans Polar region, residents of the East-Ural radioactive zone and their descendants of the I-II generations, participants of the Afghan and Chechen wars living in the Urals, senior school children from zones of ecological disaster (Kamensk-Uralsky, Nizhny Tagil), etc. Data from the database of the Sverdlovsk Regional Department of the Russian State Medico-Dosimetrical Register and the Report on the Cardiological Aid in Sverdlovsk Region in 2000 have been analyzed.

F. Brechignac and G. Desmet (eds.), Equidosimetry, 51–56.

2. Evolution of the methodological approaches to the problem of ecological standardization during the twentieth century

The historical analysis shows that during the last century the character of technogenic impact on the biota and social consequences of the technogenic pressure on Man changed at least three times.

Simultaneously, the notion of ecological standards and security, of the adaptive success of live forms inhabiting technogenic areas was made more precise.

In the first half of the twentieth century ecological zoning was based on the morphocentrical approach, the notion of "standard" was considered as structural stability of a live system, homeostasis – as a way to maintain physiological constants; the interaction between the live system and the environment was considered in the model "stimulus - response", technogenic effects on the biota were estimated from linear correlations "dose - effect"; maximal permissible values of pollutants which did not significantly change the structure of human organs and tissues were the criteria for ecological zoning.

Medico-ecological safety of areas was examined by cases of professional diseases and by the data on the medical geography of unfavourable geochemical regions. Since the last century Selie`s theory of the total adaptive syndrome has gained recognition.

According to it human response on any external effect is considered as a cascade of intricately organized physiological reactions showing up in phases. By adaptation standard is meant functional stability of an organism, by homeostasis – the mechanism of maintenance of physiological variables within a certain fluctuation range. Live organisms are considered as open dynamic self-organizing systems. The notions of the ecological capacity of landscapes, ponds and the biota have been introduced.

Phenomenological models have been involved in mathematical simulation which analyze common trends, give a group characteristic of an object of ecological monitoring, estimate the entry-exit relationships of vital functions of live systems without analyzing mechanisms of their transformation. Medico-ecological safety of areas is evaluated from integral indices demonstrating the absolute deficit of population in dangerous areas (the degree of life-span reduction, the level of the early maternal and infantile mortality).

The development of information technologies in the twenty first century (I would not use such abbreviations; it is a very minor disturbance though; do not pay to much attention to it, although I dislike it. Latin is not a modern language anymore, although I have myself a "Latin" education, fortunately!). has shown a limitation of this approach. The analysis of the demographic structure in Russia, the advances in adaptology, informative medicine and population genetics have given a new view of the problem of medico-ecological zoning of territories. The main problem of ecologically unfavourable territories is not the absolute but the comparative deficit of labour power, competition inability and low adaptability of the population. Maintenance of the functional plasticity of the biota and Man in particular has become the problem of the day.

The notion of "standard" is associated now with the number of invariant programs of adaptation accessible to an individual, "homeostasis" - with the ability of a live system to quickly pass from one level of physiological functioning to another, different in mechanisms of self-regulation. In mathematical simulation the more and more attention is paid to non-liner relationships, to on-line and deductive models which describe mechanisms and regularities of transformation of live systems.

3.   Anthropic environmental factors as catalysts of natural selection and accelerators of human technogenic evolution

An active influence of Man on the nature significantly changes the character of his connection with the environment, requires additional adaptive abilities of an organism. Global technogenic catastrophes and industrial emissions make adaptation to technogenic pollution a necessary requirement of life for a tremendous amount of people. The character of adaptation against the background of a prolong stress may both quantitatively and qualitatively differ from adaptive responses to an acute stress.

The situation requires predicting the cost of adaptation – its physiological, psychological and social value for an individual and his descendants. For Man, external stress factors may be natural or artificial (technogenic). As a biological system Man has an evolutionary fixed algorithm of an early adaptation to the first stress factors, e.g. to the seasonal weather change. Technogenic stress factors are extreme ones, the algorithm of adaptation to which being formed just now, in parallel with the development of technogenesis.

As catalysts of human evolution technogenic environmental conditions are peculiar for their frequency component which is often in synchrony with an individual's own biorhythms; therefore, they have an informative effect and are able to quickly and significantly change the individual short-term and long-term programs of human adaptation.

The possibility of rapid human evolution under anthropically changed environmental factors was theoretically proved by N.V. Timofeev-Ressovsky and I.I.Schmalhausen. N.V. Timofeev-Ressovsky predicted it for the first ten years of the XXI century. He isolated 3 fundamental principles in the theoretical biology which create the theoretical premises for the prediction: 1) selection of stable specimens in a species under the influence of environmental factors, 2) inheritance of favourable parents' characters, 3) divergence of trends of progressive evolution, the biota developing simultaneously several parallel ways.

In our research we tried to analyze biosocial programs of human adaptation. The psycho-social component of the complex behavioural response was traced using Szondy's psychoanalytical method. The method allowed evaluating geno- and phenotypical manifestations of unconscious human adaptation. We analyzed 763 case histories. We studied daily, 7-day, seasonal, 7-year fluctuations of physiological functions; frequency components of the cardiac action and electro-skin conductivity during various pressure tests. 6.000 cases were analyzed to estimate the rate of adaptive responses in conditions of chronic stress. The results of the investigation are given in the corresponding chapter.

4. External geoecological factors modifying the human long-term adaptation program

Geoecological varieties of human adaptation models were studied in the following comparisons: 1) biosocial adaptation in residents of the western and eastern Ural geozones, 2) adaptive responses in individuals affected by low radiation doses compared to natives of the same latitudes subjected to a greater geomagnetic effect in the Far North, 3) the role of geoecological variance in the development of electrophysiological manifestation of social stress in young Uralians (soldiers from Chechnya). We obtained the following results.

1) The process of adaptation of Man and other live systems to conditions of technogenesis in the Urals cannot be considered completed. (300 years of mining industry, 50 years of nuclear enterprises). The technogenic pressure does not promote intraspecific unification of the biota components; on the contrary, it activates the process of bifurcation of populations in subgroups less sensible to the effect of dangerous factors. This results in a rapid biosocial differentiation of the society.

2) The structure of heart-vascular diseases differs significantly between western and eastern geozones in Sverdlovsk region. In the western geozone disorders of regulatory systems and failures of adaptive processes like acute vascular cases predominate. In the eastern geozone – disorder of executive structures with predomination of chronical pathology at the cellular level. The number of heart-vascular cases in the south-western part is more than in the north-eastern one.

3) Comparison of patients with the same heart diseases but with various ecological life has shown that according to the degree of the response of regulatory systems and the level of activation of metabolic processes in the myocardium the patients are arranged in the following decreasing order: the liquidators of the Chernobyl accident consequences, residents of middle latitudes, migrants from the TransUrals. Compared to the Ural residents the liquidators is a dissipative adaptation model, while migrants from the TransUrals – an assipative one.

4) Electrophysiological manifestations of stress in soldiers from Chechnya were to zone extent unified; this group differed from other groups of people affected by chronic stress in extreme situations – from the Afghan soldiers and from the liquidators of the Chernobyl accident consequences.

5) For the young Uralians, the biophysical (electrophysiological) manifestation of the desadaptation syndrome was more frequent than the psychological or clinical ones.

6) The natural creative potential was more frequently demonstrated inn teenagers, whose phenotype contained more heterozygote signs.

For the geological conditions in the Urals, the south-western and north-eastern parts differ in geomorphic, crystallographic, geochemical characteristics as well as in the geologically predetermined and industrially increased gradient of the redox-potential. This suggests that the basic electrophysiological indices in the tested individuals determined by geoecological conditions of their residents are a more significant component of the adaptive syndrome than their social modification caused by participation in war.

During the investigation general regularities, individual and group variations of biosocial desadaptation were revealed.

It was found that individuals of all groups were comparable in the basic inherited biosocial requirement: they all needed recognition, love and respect. In all groups this requirement was blocked and reduced to the level of need of individual love and confidence relations in dear ones.

Moreover, almost half of the tested lost the need of even individual love.

It was established that individual clinico-psychological varieties of the desadaptation syndrome depended on the algorithm disturbed. Thus, in patients with an inherent passive strategy of adaptations (search for recognition by demonstrating attractiveness, loyalty, fear of disturbing traditions) somatic diseases with injuries at the cellular level were observed (myocardium infarction, oncopathology). In individuals with an inherent active behaviour the personality destruction was observed:

phenotypically manifested intra- or extraversion, pathopsychological development (liquidators of the Chernobyl accident consequences, war participants, narcomen).

The disbalance between the geno- and phenotypical programs of biosocial adaptation increased in the younger individuals: in children with disturbed heart rhythm both tendencies of self-destruction, the somatic and the psychological ones, were observed.

We have isolated 3 groups of the liquidators of the Chernobyl accident consequences with various degrees of resistance to the combined effect of psychological stress and low radiation doses. We described the psychological markers of an unconscious individual which can predict particular forms of psychosomatic pathology in the liquidators for the postponed period; in particular, to isolate groups of individuals predisposed to the development of local of system forms of the heart-vascular diseases or not.

Psychological criteria of desadaptation are predictive markers of early cardiac deaths in conditions of chronic social-ecological stress. During the first 10 years after the Chernobyl accident cases of a sudden cardiac death were significantly more often registered among the liquidators with neuro-circular distony and systematic disturbances of haemodynamics than among those with local circulation disturbances like ishemic heart diseases.

We compared the psychoanalytical portraits of the liquidators with pathology of brain and heart vessels. Representatives of the first group had bimodal strategy of adaptation: they tried to preserve their property at the cost of disturbance of personnel contacts and moral-ethical principles. Individuals of the second groups had unimodal strategy of adaptation: they tried to preserve moral-ethical principles even at the cost of their property and their social status.

Thus, 5-7 years after the Chernobyl accident the same group of the liquidators showed the tendency to bifurcation of the evolutionary process and several trends of biosocial adaptation of various biomedical and socio-psychological value.

5. Summary

Our 20-year observations on human adaptation to technogenic conditions support the following idea.

In contemporary conditions processes of biosocial adaptation of Man depend rather on the individual and populational specifics of his biocybernetic state than on his physiological and biochemical peculiarities. This is the principle distinction between the models of human adaptation to acute and chronic stress. The first model has been studied and described by H.Selye as the concept of non-specific adaptation syndrome. The second one should be studied in detail within the framework of geoecology. The first one is based on fading harmonic fluctuations, the second one on the quantum transitions.

We suggest a hypothesis of various (geoecologically and psychologically caused) ways of long-term models of human biosocial adaptation; variations of the models are the basis of the rapid evolutions of residents of technogenic areas.

Heterogeneity of long-term adaptation programs of the Ural residents caused by geoecological heterogeneous conditions of their residence and by different psychological predisposition to the socio-ecological stress is the morphological basis of modification of the Uralians. It gives them a chance to survive in conditions of technogenesis, to begin a new subspecies of Homo-sapiens, Homo technogenicus and

Homo adapting, i.e. a Man who is more stable to the effect of unfavourable ecological factors of geophysical and geochemical nature. The degree of adaptive heterogeneity of live forms in modern ecology is an impotent criterion of medico-ecological zonation of areas; it allows to quantitatively estimate the potential ability of the residents to technogenic transformation and creative activity in new ecological conditions.

We suppose that a detailed study of the key mechanisms of adaptogenesis of the Uralians may be help to find methods effectively increase their stress-resistance, viability, professional competitive ability and social mobility.

PHYTOVIRUSES AS INDICATORS OF ENVIRONMENT

A. BOYKO
National Taras Shevchenko University of Kyiv, Biological Faculty,
64 Volodymyrskaya St., 01033 Kyiv, UKRAINE

Up-to-date ecological researches testify that different stressful influences on the biosphere lead to changing of virus properties. It means that ecological problems are not only environmental problems; they create a very difficult situation that is concerned with quick changing of pathogen viruses properties. All this testifies that people must be more attentive to probable outbreaks of a viral infection during an ecological disbalance. Virologists, geneticists, biochemists and radio-biologists are solving the problem of environmental mutagens that is related to an influence of some viruses on the genes and chromosomes of cells. In a number of cases specialists of virology, physics, mathematics and other branches of science can solve the problems using methods of biological system investigation [1].

Being a part of ecological system of our planet, viruses can fulfill some informational functions which take place at the genetic level of the biosphere. Some viruses, for example TMV, can be indicator systems of a pollution of the environment reacting to different processes by its mutability, which induces proper symptoms. These symptoms are formed as a result of interactions of viral infection and environmental factors.

Viruses have original properties that give an opportunity to use them in model experiments of genetic engineering, biochemistry, ecology and other branches of biological sciences. As vector systems viruses are used for receiving transgenic organisms.

Ubiquity of viruses points out their original properties. But the main penchant of researchers to viruses consists of their contamination of human, animal, plant and microorganism. Viruses of influenza, herpes, leucosis and immunodeficiency lead indeed to big epidemic outbreaks.

We discovered that rod-shaped and isometric viruses have diamagnetic properties. They can orient in the magnetic field and create original structures.

Viruses of plants have been studied intensively in many laboratories in different countries. Viral infections can lead to a wide spreading over some ecological factors. Phytoviruses infect fruits (12-85%), hops (28-100%), wheat (10-70%), potatoes (10-65%) vegetables (8-75%), citrus plants (16-84%), and mushrooms (5-62%) according to our experimental data. There are reductions of yield and quality. We investigated that TMV keeps its structural and infecting properties in different tobacco products. It is confirmed that herbs, for example plantain, stramonium, chamomile, nettle, viburnum and others are contaminated by viruses at different environmental conditions. These pathogens can reduce officinal properties of herbs. Viruses can stop the production of antibiotics, which are produced by microscopic fungi in the active environment.

At the same time it is investigated if the plant organism can carry human and animal viruses. During watering of tomatoes by sewage enterovirus accesses fruits in 5-6 days through the rootage. In this case the virions keep their biological properties and can infect mammals when appropriate conditions are around.

It is a peculiarity of all viruses that they have only one kind of nucleic acid, which can be presented by RNA or DNA in different forms. This can be one or more

F. Brechignac and G. Desmet (eds.), Equidosimetry, 57–64.

fragments, single-stranded or double-stranded. When viral RNA is positive, it serve as mRNA and can induce the synthesis of new virions (TMV, Tobamovirus). The genome of some viruses can be presented by a negative RNA. In this case enzyme polymerase is required for synthetizing of positive strand of RNA. We do not know why the process of evolution has done so. Interesting is that such a situation is characteristic for many viruses of human, plants and animals. Structures of genome, lipids, proteins and envelopes of these viruses are very similar. Rhabdoviruses are very active at the places of radiation accidents.

Phytoviruses induce some symptoms on infected plants: mosaic formation, yellowing, phytoblastema, dwarf growth, necrosis and others. It is important that we have different methods for virus diagnostics and identification. One of them is remote diagnostics of infections on a large agrocenosis. It can be done both directly close to the ground and from space with the help of special equipment.

Activity of viral spreading depends on various vectors that are spread in a biocenosis.

For persistent viruses it is typical to have a latent period which take place as vectors. Some viruses can reproduce as vectors during survival. Phytoviruses can be transmitted by horticultural sundryagricultural equipments or seed material. It is interesting that vesicular stomatitis virus, alfalfa wound swelling virus, wheat virus, sazan rubella virus and lettuce necrotic yellowing virus can be propagated in insect cell culture. It is proposed that the concentration of the virus in tumor tissue is 10^{11} per 1 gram. The virus has system spreading and some stress situation pushes the quick appearance of tumors that can be observed visually at farms near Kiev. Many plants viruses can be localized on various soil types.

It is needed to point out that accumulation of viruses in an infected plant material is different. From 1 liter of fresh tobacco leaves sup we can get 1,9-2,8 grams of TMV, leaves and fruits of tomato – 1,0-2,0 g., and from cucumber leaves and fruits – 0,6-2,3 grams of Cucumber Mosaic Virus (Cucumovirus). It is interesting that this virus is similar to virus of poliomyelitis and some others viruses of human.

Taking into account a wide spread of phytoviruses on various species of plant it is necessary to point out that they can be the first drive to weakening of plant to other pathogens such as bacteria, fungi, micoplasmas and viroids. These viruses infect these pathogens.

Decreasing of nitrogen fixing can be observed in infected plants in case of disbalance of ecological niches. Various kinds of fungi are activated on the fruits, which are contaminated by viruses. The fruit becomes dangerous to people and it is necessary to prohibit its using for any production.

It is important that unidentified structures are formed in cells of nightshade during some viral infections and breach of ecological niches. They inhibit the plant, and the fruit becomes more sensitive to infection by fungi and bacteria. These structures are typical to ecological niches with increased content of radionuclide.

As it is been reported, phytoviruses have a wide host range. For example, TMV contaminates tomato, tobacco, rose, cherry, sunflower, cucumber, and hops. There is rather difficult situation with tomato and cucumber in greenhouses. In fact, these plants and their fruits are contaminated almost up to 32-100%. The plants are infected not only by viruses but also by other phytopathogens.

Rod-shaped virus can be obtained from ascocarp and mycelium of field mushrooms. This virus reduces yield and quality of production. During rabbit immunization there were visible symptoms of virus systemic infection (oppression, secreting of tears and rotting of the ears). The virus is related with TMV, which infects plants. Rod-shaped virus of field mushrooms can infect mushrooms together with spherical viruses. Their role in a human organism wasn't investigated. But previous researches show that these viruses have very interesting properties in a cell culture of mammals. The cell infected by this virus becomes destructed. It has been demonstrated that TMV RNA obtained from aromatic rose can induce the assembling of new virions in HeLa cells. The cells disintegrated, stunted and differed from controls. Next researches show that the virus is very similar to initial strains of TMV. It is pointed out that similar results can be obtained in other cells of mammals.

So, phytoviruses induce great losses to national economics. They have original properties in non-host organisms and can even induce some pathological processes in mammals organisms. These populations and their influence on the environment should be studied carefully.

All factors that have an influence on viruses and other biological objects can be divided in: in vitro and in vivo factors. Extracellular factors are still not studied properly. But some recent scientific works, appeared lately, describe the migration of phytoviruses by water, soil and animal faeces. The virions mainain their biological properties during some time.

Rise and spreading of disease depends on many factors that are in a complex structure of biocenosis and reflect changes during interactions between various organisms at complicated flows of energy. Scientists that study a wide disease spreading situation do not dispose of proper data about the influence of environmental conditions on viruses. Despite its importance for understanding of theoretical and practical problems of general virology, ecology, biotechnology and physiology it may be due to the rule that TMV infects hundreds of various plants.

Phytoviruses are those objects, which are used to solve many virological and other biological problems by many researchers now. Some viruses are used as indicator system in projection of living quartersalive systems, estimation of permissible standards of various kinds of measurements, estimation of sterilizer chamber and even during space researches in conditions of imponderability. Viruses can be launched into atmosphere by thunderstorm activity and available for sorption. At the same time viruses yield a great loss to national economy under some ecological conditions, and that is why most works are madding made on plants from agrocenoses, what are executable tasks [4].

Various strains of some virus can be obtained from biocenoses of Ukraine and other regions, and they have characteristic properties. Some of them can induce lateent infection, others cause serious pathological changes. The degree of lesions depends on virulence of virus, stability of plants and some other ecological factors [6]. Besides, some viruses infect only some species of plants.[1]

A very difficult question of interactions and viruses' circulation at ecological niches is the research of complex factors of the biotic and abiotic environment, which can change viral properties. For example, some TMV strains have unequal nucleotide sequences and alterations in the envelope protein. There are viruses that infect plants using vectors and require a helper virus. We suggest that the infected plants have some factor of transmission. These results are very interesting for understanding known mechanisms of transmission, and for investigation of new ways of virus circulation in biocenoses and agrocenoses. Some plants are sensitive in different ways to various vectors, such as hookworm, leafhopper and others. Transmission of a virus by vector is a very difficult process, which has some interactions with plant and environmental factors. Some phytoviruses are reproduced in the vector organism and induce some pathology. There is similar behavior of mammal viruses, which are transmitted by vectors. Most of them are investigated and some scientistscan forecast a virus infection using computers.

Phytoviruses do not often collide with abiotic factors. Viruses can be affected only through the organism of vector or plant. Many viruses are known to stay in different kinds of soil, water, and faeces for a long period of time.

When penetrating the cell, viruses use cell resources for its reproduction. Essential parasitism of viruses is determined by the cell conditions of and its metabolism. The vector of phytoviruses is the only way of contribution. Mutability of plant viruses depends on many factors of the environment including a vector. For example beet soil-borne virus is transmitted by Polymyxa betae. The only way for preventing of an infection is the application of strong hygiene and introduction of new stable cultivars.

Passage of some virus strains on proper plants can induce adaptation changes to a vector. A wide spread of phytoviruses depends on a wide range of host plants. There may be some ecological factors influencing mutability of virus, too: temperature, humidity, conditions of plant growing and interference among various strains and virus adaptation during viral infection.

On the basis of special researches we found out that viruses infect all floras are infected by viruses in different countries (Ukraine, Russia, Vietnam, USA, Poland). And there are pestholes of disease which appear by the action of some environmental factors.

Viruses in biocenoses are characterized by their circulation and adaptive specialization in some plant species. Along with aphids, leafhoppers can exist in biocenoses. Most of plant viruses are propagated inside their organs. Transovial transmission is characteristic for some of them. Viruses transmitted by leafhoppers are pathogen to the vector, and these viruses are similar to mammal viruses in their replication, physical structure and biochemical properties. It leads to some interactions between these viruses and vectors in biocenoses and to passing through a difficult evolution mechanism.

Investigations have shown that there are some pathogens which are transmitted by leafhoppers. At first they were identified as viruses but it has been found that they are mycoplasmas. A very difficult life cycle was shown to be typical for them. Interactions with the cell showed that they have $\acute{\alpha}$-amanitine which is sensitive to DNA-dependent RNA-polymerase being similar to the one of plants.

There are also viruses, fungi, viroids and Rickettsia-like organisms. Tomato, which has symptoms of stunt, shows mixed infection (TMV and mycoplasmas) and there are rod-shaped viruses and mycoplasmas at hops with dwarf symptoms. It is important that there is mixed infection in biocenosis. Now it is proved that some viruses infect mycoplasmas.

The problems examined do not cover the whole difficult situation of biocenosis and agrocenosis. There are other vectors in ecological niches, such as ticks, soil microorganisms, and microscopic fungi. A feature of many viruses is their localization in a seed. Some unknown phenomenon takes place during seed infection.Seed infection gives steadfast source of maintenance of circulation. It is supposed that viruses that are not transmitted through seed cannot infect haploid cells.

Interactions of viruses with environmental factors can be described by complex processes which appear, under influence of ecological conditions, on organisms infected by viruses. The mentioned data give us an opportunity to underline that these and others factors provide proper circulation of viruses in biocenosis and their reproduction in the organism. Cell factors can provide an opportunity for viruses to induce a necrotic reaction in a system. It is supposed that viral RNA is a universal form of original transport of viral infection into plant cells.

Only some of all these factors induce the development of both heterogenic and homogeneous strains. There is some homology between some fragments of viral genome.

An estimation of artificial factors that have an influence on various strains is very important for comprehension of interesting properties of virus and other biological objects. Tens of phytoviruses are constructed very simply and have proper biochemical and physics properties. Activity of viruses becomes apparent through complex chemical components. It has been discovered that in some cases viruses change their properties not only in vitro, but also in vivo. They lose their infectious properties or show new virulence. Changes may occur at a range of temperatures, radiation treatment and influence of various compounds. There are such diagnostic criteria as temperature inactivation, titre and others. An interaction between a virus and cell components can be disbalanced in plant organisms own to different factors [1].

Many phytoviruses can save their infectious properties for years in various plant fossils, or when lyophilized. Inhibitors of viral infection can stop viral infection and increase yield. Some changes of biochemical processes are observed during their action. Antiviral activity of preparations depends on the condition of the plant, the way of injection and other factors. Single preparations have an original influence on different characteristics of replication of virus during plant ontogenesis. At early stages of infection TMV supposedly can produce some molecules that control the synthesis of new virions. Antiviral activity of some preparations depends on sensitivity of plants and indirect capacity of action on the virus through the plant. There are however losses of infectious properties of viruses in vivo as well.

A direct influence of some substances on a virus can induce a change of buoyant density of virions and their stability in solutions. These and other researches are very important for understanding of molecular reactions, regularities of morphogenetic processes during viral infection.

Changes of properties of virion under the influence of various factors can be discovered on plants. Temperature-stable mutants, forming of different cell inclusions are among them.

The investigation of influence of physical factors on virus and infected plant is also very important. There are not too many such researches due to many problems. It is important not only to know biological properties, but also to use them in the fight against viral infection. Therefore, there was a necessity of carrying out the experiments in vivo and in vitro. There were various methods for of solving this problem. Various organisms and viruses have different sensitivity to radiation treatment. Lacings are created during an action of ultraviolet. Capsomers play the most important role in RNA damage. The capsomers of TMV have a protective function for NA. Such radiation treatment has an activating role for increasing poliovirus sensitivity. Inactivation of TMV RNA can take place by photodynamic effects, which can be used in some experiments.

Some changes of virion properties can occur during α-irradiation of infected plants. The reaction of potato sensitive to Potato X-Virus can vary in a wide range in biocenosis during permanent α-irradiation.

Disbalance of an appropriate virus reproduction in infected plants can lead to some changes in enzyme activity. This process becomes irregular, when there was *irradiation*. This irradiation can induce infraction of virions structure and of the structure of the infected cell. Some doses of irradiation stimulate a growth and development of perennial plants chronically infected by viruses. Small doses of irradiation do not lead to great changes in biochemistry and metabolism of infected plants. In this case there is increase of some substances synthesis, which is stabilized for individual species of plants. Increasing of radiation can however affect contaminated plants. It can also lead to ecological breakdown.

Table 1. Reduction of atmospheric nitrogen fixing by legumes infected by viruses

Region	Reduction of atmospheric nitrogen fixing at contaminated legumes
Chernobyl	53.0
Vyshgorod	40.6
Kyiv	33.0
Vasylkov	62.2
Bila Tserkva	51.0
Zhashkiv	60.1
Uman	43.0
Odessa	37.7
Vietnam	37.0
USA	30.0-72.0
Poland	44.0

Though we approached to understanding of mechanisms of post-radiation, and further investigations on the cell level are required.

One of the important problems is studying influence of magnetic fields on virus and infected plant. Studying only viruses can answer a range of questions, which are related to magnetobiology. It is proved that among biological molecules, only proteins and nucleic acids demonstrate anisotropy. Magneto-biological effects and primary mechanisms of influence of magnetic fields on biostructures are explained.

Microorganisms are known to have a high sensitivity to permanent magnetic fields (PMF). The fields effects depend on the condition of the object, and some characteristics of the field. These and other data are concerning the mutability under the influence of the fields. There are also opposite views.

It is investigated that PMF have stimulating effects in various plant species. We showed such stimulation for the first time [1, 2].

Important model experiments were carried out on RNA and DNA of different origin, which were in a solution and have a little influence of PMF. PMF can induce well-defined changes in chemical and physical properties of NA, and determine the orientation of genomes in PMF. The orientation was obtained for TMV and isometric viruses, which infect rose plants. Some ranges of PMF can induce disturbance of organelles structure that were contaminated by viruses.

A heat treatment is considered as a factor of influence on plants during vegetation in phytotrons. There are also methods like meristem culture, previous irradiation-irradiation and PMF meristem or garden material treatment. Interesting results were obtained about space heliophysical factors at department of virology of Kiev National University [5, 6]. It is shown that phytoviruses have some regularity in their reproduction, and change of microgravity and factors of space flight shows interesting interactions of plants with infection. It is important that clinostating of infected plants leads to some changes, for an example in reproduction of wheat viruses[1].

All presented data underline very strong changes, which can take place during viral infection at some environmental conditions.

There is an opportunity to look at the conditions into the cell as factors of the environment, which can influence on replication and realization of genetic material. From our point of view, very interesting is the complementary character of interactions between virus and host cell organelles. Our researches , show that only the complementary character can help to understand some positions of virus evolution, their assembling and application in biotechnological researches and genetic works[1].

Many factors can be mentioned which influence plant viruses and induce some changes in virus properties. These investigations open a way to solving theoretical and practical tasks in virus ecology. It should be pointed out that fast creation of up-to-date biotechnologies, carrying out tasks of genetic engineering with using of viruses requires quality control of virus contamination in various ecological niches. Carlavirus, tobamovirus and ilarvirus were studied.

The authors create the mathematical description [4] of process of infringement of virus particles morphology at influence gamma-irradiation in vitro, with help of which it is possible to investigate connections between change of the morphological and infectious characteristics of a virus (for example that the speed of infection decrease is more, than speed of fall of viruses structure (number of virus particles of pattern group).

Mathematical models of the influence of a constant magnetic field (CMF) on the development of a plant and on infectious properties of a virus according to a hypothesis about complex interaction of influences CMF simultaneously on a virus and on a plant are constructed. They confirm, that either for of an attacked plant to function or for a virus infection to function there is an extreme size of CMF (0,2 tesla), determining the maximal oppression of a virus infection.

A mathematical model of spasmodic reduction of the level of tobacco mosaic virus infection in vitro under influence of increased temperature, is constructed. The model enables to investigate an initial stage of an oppression of a virus and its complete loss of infectious properties. The concrete values of parameters of the model for given supervisions are found.

According to our researches, for some viruses it is possible to cause original crystallization. The mathematical formula for the description of the dynamics of the occurrence of virus inclusions is given in publications [5, 6].

There is constructed an formulation for the evaluation of nodule nitrogen fixing process breaking in infected leguminous plants is constructed. The mathematical models for the analysis of the quantitative and qualitative characteristics of hop at a virus infection are offered. Besides the dynamic nonlinear model of the hop plant growth at a virus infection is given.

During development of discrete dynamic models of the plants processes at a virus infection on the basis of biological laws it is taken into account such biostructure of the model which determines requirement of elements for the certain substances (potential intensivities), and the real value of substances which can be used by these elements (real intensivities of processes) is calculated. With the help of such approach to construction of mathematical models of a plant development it is possible to carry out researches of the models of virus reproduction influence on a general plant's condition in various ecological situations. Here the virus reproduction process can be considered as a "superfluous" element of the biostructure, on which the plant model uses its resources.

The string Dean plant development and nodule nitrogen fixing at a virus infection model is developed on basis of the developed; biostructure. The model enables to investigate processes at various changes of ecological conditions.

On the first sight the mathematical modeling in virology does not essentially differ from mathematical modeling of other biological systems. However viruses are original genetic populations, which have not the synthetic system, and use environment of the organism-owner, changing, first of all, its information code. Therefore researches on a joint of mathematics, cybernetics, virology and ecology can spill light on the questions about nature of viruses. Today the virus concept is wider then frameworks of biology. To take into consideration for example computer viruses, the damage from which becomes more significant in conditions of scientific and technical progress, and soon their influence will be not less essential, than influence of biological ones [1-6].

References

1. L. Boyko. Ecology of plant viruses, Kyiv: Higher School, 1990, 167 pp. (Russian)
2. Yuri V. Zagorodni, V. Voytenco, A. Boyko. Studying of an influence on plants' organisms in condition of ecological instability and its estimating by mathematical modeling in CEIS system. // In book: Process Modeling, Cottbus. BTU, -, Berlin, Springer, 1999.– p.86-98.
3. Yuri Zagorodni, Anatoliy Boyko. Mathematical modeling in plant viruses researches, Kyiv, EcsOb, 2001, 153 pp.
4. Polischuk V. P., Budzanivska I. G., Rizhuk C. M., Patyka V. P., Boyko A. L. Monitoring of plant viral infections in biocenoses of Ukraine, Kyiv: Fitosociocenter, 2001, 220 pp. (Ukrainian)
5. Yu. Zagorodni, V. Voytenco, A. Boyko Studying Of The Influence Of Phytoviruses On Plant's Organism In Conditions Of Ecological Instability And Its Estimation By Mathematical Modelling in CEIS System/ Papers of Processes Modelling, Cottbus, BTU. – Springer. 1999. –P. 86–98
6. Yu. Zagorodni, A. Boyko, I. Beiko Skrygun Constuction Of Computer Stimulation Of Critical Plants State Under Influence Of Phytoviruses Infection And Ecological Unstability.// Papers of 15th IMACS World Congress On Scientific Computation, Modelling And Applied Mathematics, Berlin/Germany, August 1997

PROTEIN AND RADIOACTIVITY LEVELS OF *PATELLA COERULEA* LINNAEUS AROUND DARDANELLES

Mustafa ALPARSLAN
Çamakkale Onsekiz Mart University, Faculty of Fisheries, 17100,
Çanakkale, TURKEY.

Mehmet N. KUMRU
Ege University, Institute of Nuclear Sciences, 35100, Bornova, Ýzmir,
TURKEY

1. Introduction

Marine potential food sources are not sufficiently known and evaluated yet. Additionally, the other food sources are exposed to hazardous environment pollution such as domestic and industrial pollutants. These pollutants may be of microbiological and physical-chemical nature like radiation. However, new food sources are needed and they have to be produced under hygienic conditions for ever-growing populations. *Patella coerulea* has a high protein content and it is possible to produce this gastropod under laboratory conditions. These marine products are presently being collected from mediolittoral rocks and stones. Some suspicious barrels dumped from vessels or ships around the Dardanelles can be detected. Unfortunately, not any publication on this subject was found in our region.

Weidman et al. [2] studied relations between mollusk (bivalves) and geochemistry. Corals have been used previously to reconstruct high resolution geochemical records of the surface of subtropical oceans, but no comparable tools have been developed for the colder, higher latitude oceans. It is reported that the application of accelerator mass spectrometry microsampling techniques allows it now. The carbonate shell of the long lived (approx. 200 years) mollusk (bivalves) at Arctic islands can be used to fulfill this role for the mid and high latitude North Atlantic Ocean, together with the sampling methods used to produce the first history of bomb ^{14}C in the Northern Atlantic Ocean from Georges Bank (41° N, 67° W.)

Alpaslan et al. [1] (1995) did research on the seawater in relation to microbiological and physico-chemical parameters (salinity, ph, temperature, radium, total beta activity, silicate, nitrate, nitrite, orthophosphate, ammonium, anionic, detergent).

Mayr [3] studied the capacity of a bivalv mollusk, Anomalocardia brasiliana, which lives in the local bottom sediment, to remobilize ^{60}Co previously sorbed in the sediment. Processing is made by raising the temperature to the boiling point (350°C - 400°C) and adding additionally K_2SO_4 and Cu, Se (such as Hg catalyser). Under these circumstances organic nitrogen is converted to $(NH_4)SO_4$. Excessive NaOH is added and free NH_3 is distilled into balanced acid solution by water vapor. The remaining acid is refiltered.

2. Material and methods

2.1 BIOLOGICAL SAMPLING AND ANALYSIS

The Species-*Patella coerulea* L. belongs to:
Phylum-Mollusca

65

F. Brechignac and G. Desmet (eds.), Equidosimetry, 65–72.

Classis-Gastropoda
Ordo-Prosobranchia
Subordo-Arcaeogastropoda
Family-Patellidae

Patella coerulea L. (Mediterranean Limpet) has a low conical shell up to 46 mm long (3rd station - Çardak); exterior of shell is with coarse radial corrugations leading to a wavy margin and many fine unequal radial ribs. The colour outside is grey-brown-red with white spots and radial marks; concentric growth rings may show as contrasting colour bands; inside coloration of fresh shell with dark stripes leading to the apex which is blue mother-of-pearl. The appearance of this limpet is very variable: colours fade with age an after dark. Habitat is firmly detached to the rocks, usually in the horizontal plane on the shore (similar species P. lusitanica) [4, 5].

Samples were collected from supralittoral areas and taken to the laboratory immediately. Shell and the meat were separated from each other. Total weight, length and width were determined.

We investigated 79 samples. One of these samples was 46 mm length in Çardak. These organisms were reported from Bosphorus and around the Bay of Yzmir [6]. Chosen stations are from Kilitbahir to Kepez such as Kilitbahir (1), Lapseki (2), Çardak (3), Gelibolu (4) and Kepez (5) (Fig.1). All the samples were taken from supralittoral zones of the stations and taken to the laboratories as soon as possible. These samples were analyzed for protein (total Nitrogen) and radioactivity. The gastrointestinal system was discarded from the meat and homogenized in the blender. The protein analyses (Nitrogen determination) were performed by means of the Kjeldahl method.

2.2. CHEMICAL AND RADIOCHEMICAL ANALYSIS

Indicators: Toshira indicator (as volume ± 5 %, in 50° of Alcohol saturated methyl red solution and in 25.50 % ethanol, 0.025 % 25 methylene blue mixed and phenolphthalein. Samples (homogenized will be settled to the Kjeldahl balloon, 2 ml H_2SO_4, 1 gr. K_2SO_4 and spatula $CuSO_4.5H_2O$ and boiling stone is added. Firstly, to be heated for 10 min. slowly and then later heat be increased for approximately 30 min. untill solution turns colourless.

The time for dissolving changes from protein to protein, but is generally completed within two hours. The balloon is to be cooled till room temperature and approximately 10ml of distilled water be added. The vapour balloon is heated in the parnas-wager tool and water vapour passed into the system for 1-2 min, when the cooling is stoppedand some short heating is allowed. Thus accumulated material in the distillation balloon is reabsorbed. After this, it will open and continue to heat [8].

Method: The main principle depends on the weight of the glass balloon and the sample weight 10 gr., on cartridge. First of all, the fat is extracted with ether, the ether is then evaporated and and the remaining fat weighed [9].

(1) Total fat % = Last weight-tare x 100 / 10 gr. sample weight.
(2) Dampness % = (M1-M2) x 100 / (M1 + M0)
M0 = Tare
M1 = Plate weight (before heating and drying with the sample process)(gr.)
M2 = Plate weight with the sample after (after heating and drying process)(gr.)

(1) Samples were mixed with sand and ethanol completely and we began the drying process in the water bath, and dried the sample in 103±2 °C till it reached stable weight.

(2) The principle is to first disconnect the parts of the lipids, cover the sample with HCl and strain the reminder. Extraction is made by n-hexane method or petroleum ether.

Sample Preparation: Samples were stored under sterile conditions until analysis. The samples were ashed at about 350°C to minimize the loss of ^{137}Cs during ashing. After ashing, the samples were transferred to an air tight plastic container of 2.8 cm diameter and 8.0 cm height and were stored for a minimum period of one month so as to make sure that radon and thoron daughter products could attain radioactive equilibrium with their respective parent [226]Ra and [228]Th [9].

The standard source has been prepared using calibrated solutions of various isotopes and mixing them in a known weight of Analar sodium oxalate to approximate the ash matrix of the sample. The thorium source was prepared by mixing thorium nitrate powder with sodium oxalate. The radium source was prepared by mixing uranium ore with sodium oxalate. The source for [40]K was 20gr of Analar potassium chloride. The sources of [226]Ra and [228]Th were also stored in airtight plastic containers for a minimum period of one month.

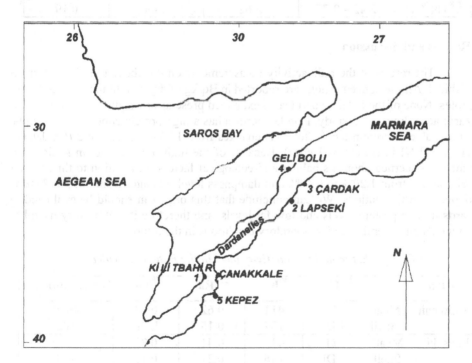

Fig.1. Distribution of Stations.

The thorium nitrate and uranium ore used for the source preparation is more than 15 years old and hence ensures that [226]Ra and [228]Th has attained radioactive equilibrium with its daughters. The ^{137}Cs source was prepared by mixing ^{137}Cs standard solution with sodium oxalate.

Samples and standards were counted in identical containers, and generally in 3"X3" well type NaI (Tl) integral line assembly housed in a 15cm thick lead shield lined with cadmium and copper [10]. The counting time for each sample was 1000 min. Each of the samples was counted three times giving a standard error 2-3 %; all activity values given in the following section are means of three measurements. Table 1 gives the energy region used for the estimation of various radionuclides, the conversion factors (concentration of the isotope/cpm due to the isotope in its γ energy region) as estimated from the standard source end the minimum detectable activities of the nuclides. The energies used for the estimation of ^{228}Th (2.62MeV), ^{226}Ra(1.76MeV), ^{40}K(1.46), ^{137}Cs(0.66MeV) and gross gamma (0.40-3.00MeV).

Table 1. Energy region used for the estimation of various isotopes, conversion factor and the minimum detectable activity of the system used.

Nuclides Analyzed	Energy region (MeV)	Energy used (MeV)	Concentration Per cpm (Bq)	Minimum detectable Level (Bq)
^{137}Cs	0.59 – 0.73	0.66	4.03	0.12
^{40}K	1.38 – 1.58	1.46	3.62	0.57
^{226}Ra(^{214}Bi)	1.62 – 1.92	1.76	1.37	0.15
^{228}Th(^{204}Tl)	2.42 – 2.77	2.62	2.49	0.19

3. Results and discussion

The results of the radioactivity measurements on the above samples are given in Table 2. All the concentrations are reported in Bq/kg of dry weight for meat and shell samples. None of the Patella coerulea L. examined presents any significant radiological hazard; according to our study, Patella coerulea has a high protein content. As it can be seen in Table 5. protein labels (total N) changes from 17.68% in station 2 (Çardak) to 13.10 (total N) in station 5 (Kepez). Because of the high protein level in station two might be concerned about its habitat and ecological factors. In addition to this, the fat level chances from 1.20% to 0.40% and dampness levels changes from 78% to 74% as it depends on the stations. We can conclude that this organism should be evaluated in regards to high protein levels and low fat levels, and therefore it will be very useful to produce Patella coerulea under laboratory conditions in the future.

Table 2. Estimated radioactivity level in the Patella coerulea L.

Stations		^{137}Cs	^{40}K	^{226}Ra	^{228}Th	Gross Counts
Kilitbahir	Meat	19	411	0.62	0.14	4000
	Shell	LDL	137	0.15	LDL	100
Lapseki	Meat	11	415	0.41	0.22	4600
	Shell	LDL	146	0.21	0.12	200
Çardak	Meat	10	354	0.35	0.17	4200
	Shell	LDL	129	0.18	0.12	150
Gelibolu	Meat	17	436	0.42	0.28	5000
	Shell	LDL	154	0.23	0.21	300
Kepez	Meat	13	489	0.53	0.37	2800
	Shell	LDL	128	0.19	0.19	200

(LDL: Lower Detection Limit)

Table 3. Estimated shell length, total meat, total fat dampness and total N levels in Kilitbahir Station.

Shell/mm	Total Meat/gr.	Meat/gr. (without gastrointestinal system)
44	4.17	2.83
42	6.13	4.20
34	2.74	2.00
35	2.81	2.82
40	3.78	2.82
36	2.22	1.68
38	2.23	1.73
42	3.04	2.333

(Total fat: 1.1 0. %   Dampness: 77.4%   Total N: 16.42%)

Table 4. Estimated shell length, total meat, total fat dampness and total N levels in Lapseki Station.

Shell/mm	Total Meat/gr.	Meat/gr. (without gastrointestinal system)
39	3.68	3.04
35	1.80	1.35
40	2.19	1.51
38	1.83	1.45
36	2.62	1.95
36	2.64	1.96
37	1.80	1.45
33	1.74	1.36
34	1.84	1.58
32	1.69	1.32
30	1.71	1.30
33	1.70	1.27
34	1.73	1.35
36	1.75	1.38
37	1.83	1.45
41	2.24	1.71
39	3.83	2.83
40	3.82	2.81
40	3.40	2.35

(Total fat: 0.5 0. %   Dampness: 77.8%   Total N: 14.90%)

Table 5. Estimated shell length, total meat, total fat dampness and total N levels in Çardak Station.

Shell/mm	Total Meat/gr.	Meat/gr. (without gastrointestinal system)
44	4.18	2.89
46	6.03	4.43
36	2.72	2.05
39	2.86	2.29
41	3.89	2.83
38	2.23	1.70
36	1.95	1.44
44	3.76	2.64
36	2.24	1.59
40	2.87	2.37

(Total fat: 1.0 0. %   Dampness: 78.0 %   Total N: 17.68%)

Table 6. Estimated shell length, total meat, total fat dampness and total N levels in Gelibolu Station.

Shell/mm	Total Meat/gr.	Meat/gr. (without gastrointestinal system)
40	3.70	3.05
38	1.87	1.34
40	2.18	1.50
38	1.83	1.44
36	2.63	1.95
38	2.72	2.15
37	2.00	1.42
36	2.33	1.84
31	1.23	0.92
32	1.83	1.55
33	1.72	1.36
30	1.66	1.29
35	1.27	1.02
30	1.63	1.00
32	1.31	0.93
35	1.81	1.41
33	1.46	1.12
34	1.91	1.42
30	1.62	1.22
30	1.40	0.99
33	0.79	0.56

(Total fat: 0.4 0. %   Dampness: 77.8%   Total N: 13.68%)

Table 7. Estimated shell length, total meat, total fat dampness and total N levels in Kepez Station.

Shell/mm	Total Meat/gr.	Meat/gr. (without gastrointestinal system)
30	0.77	0.57
30	1.41	0.90
32	0.79	1.22
35	1.92	1.42
34	1.47	1.14
35	1.81	1.42
32	1.30	0.94
30	1.64	1.10
35	1.26	1.04
30	1.67	1.30
33	1.73	1.39
35	1.84	1.58
30	1.23	0.95
36	2.33	1.83
36	2.20	1.44
37	2.73	2.16
36	2.64	1.98
38	1.84	1.45
40	2.18	1.53
37	1.88	1.39
40	3.70	3.10

(Total fat: 0.7 0. %   Dampness: 74.0 %   Total N: 13.10 %)

Low values of ^{60}Co and ^{137}Cs, BF$_5$ do not allow to classify *A. brasiliana* as a good biological indicator for pollution by their radionuclides.

Chen-Shunhua [3] made research on accumulation, distribution and excretion of low level ^{65}Zn and ^{134}Cs in some marine mollusks. It was indicative that there was no significant difference between the absorption rate of Gastropoda and Bivalve for ^{137}Cs, being much lower than the one of ^{65}Zn.

4. References

1- Alpaslan, M., Kumru, M.N., Yaramaz, Ö., Sunlu, U., 1995, Researches made during tourism season around Urla related to microbiological and physico-chemical Fresenius Envi. Bull, 4:545-549 Basel (Switzerland)
2- Weidman, C. and Jones, G., 1993. Development of The Mollusc Arctica Islantica As a Paleoceonographic Tool for Reconstruction Annual and Seasonal Records of delta ^{14}C and delta ^{18}O in the mid-to-high, latitude North Atlantic Ocean. Isotope Techniques in the Study of Past and Current Environmental Change in the Hydrosphere and the Atmosphere. Proc. Of an International Symposium Held in Vienna. 19-23 April, Vienna (Austria). IAEA 623 pp. 461-470
3- Mayr, L.M., 1984. Bioaccumulation and Elimination of ^{60}Co and ^{137}Cs by Anomalocardia Brasiliana (Gmelin, 1791) (Mollusca Bivalvia). Remobilization of ^{60}Co, Retained in Marine Sediment by Microbial Activity. Thesis (M.Sc.) University Federal, Rio de Janeiro, RJ. (Brasil), Ins. De Bioficica
4- Chen-Shunhua,Xu Lisheng, Zhao-Xiaokui, Zhong-Chuangguang, Liu-Zhensheng, LiZaofa., 1993. Accumulation, Distribution and Excretion of Low Level of ^{65}Zn and ^{134}Cs in Some Marine Mollusc. Acta-Agriculture-Nucleatae-Sinica. V.7(1). pp.45-51

72

5-   Campbell, A.C., The flora and fauna of the Mediteeanean sea. Sant Vicene dels Horts, Barcelona. D.L.B. pp.128-129 (1982)

6-   Campbell, A.C., The hamlyn guide to the flora sea, Sant Vicene dels Horts, Barcelona. D.L.B. pp.29286p. 128 (1982)

7-   Hartmann, L. Techniques modernes de laboratoire et explorations fenetionelles. Teme I L'explansion scientifique Française, pp.83-85 (1983)

8-   Türk standartlarý, Meat and meat products determination of moisture, Türk standartlarý Enstitüsü TS 1743/ November pp. 32-33 (1974)

9-   Bakac, M. And Kumru, M.N., Radiactivity Measurements In The Çine Steam For Uranium Reconnaissance. Turkish Journal of Nuclear Sciences. Vol. 20 No:1 june 1993, pp.23-32

10-   Küçüktaþ, E. And Kumru, M.N., 1993. Natural Radioactivity Around The Muðla YATAÐAN Region. Turkish Journal of Nuclear Sciences. Vol. 20 No: 202 December 1993, pp. 13-20

EQUIDOSIMETRIC COMPARISON OF EFFICIENCY OF EFFECTS OF GAMMA-IRRADIATION AND CHEMICAL TOXIC AGENTS (COPPER AND PHENOL) ON THE RED ALGAE OF THE BLACK SEA

N. TERESTCHENKO,
The A.O. Kovalevsky Institute of Biology of the Southern Seas, NASU,
Nakchimova prosp.2, Sevastopol, 99011,
UKRAINE

V. VLADIMIROV
The A.O. Kovalevsky Institute of Biology of the Southern Seas, NASU,
Nakchimova prosp.2, Sevastopol, 99011,
UKRAINE

1. Introduction

The present environmental situation is characterized bythe presence of many deleterious anthropogenic factors in ecosystems. Often pressure of natural factors is increased to such high levels, that they exert injurious effects and even death of separate individuals, species and degradation of whole ecosystems in different arias. This problem is topical concerning aquatic ecosystems including the Black Sea ecosystems [1-3].

It is important to develop the new approach of ecology – equidosimetry and to use it for decision with regard to environmental problems. The subject of equidosimetry is a comparative evaluation of effect of different chemical toxicants and various ecological deleterious factors, their equivalent effective doses of acting on biological organisms and their systems. Equidosimetry is the base for evaluation of environmental risk for biota and whole ecosystems, for the choice of an environmentally permissible burden on ecosystems which are created through different factors.

2. Problem of a common approach to equidosimetric evaluation of the effect of various deleterious factors

There are a lot of various deleterious factors in nowadays ecosystems; therefore it is necessary to have a complete concept of evaluation of ecological effects using a common dosimetric system. This common basic system can serve as a generic unit of expression of equidosimetric evaluation of ecological factors . This is a highly complex question as the origin of factors and mechanisms of their influence on biota are different and also the responses of biological organisms are compound and diverse . In spite of the high complexity of some investigators [3-9] try to solve this problem by using the effect of ionizing radiation [5-8]. as a standard of comparison It is proposed to use the units of ionizing radiation i.e. equivalent dose as the base for the common unit of equidosimetric evaluation of the damaging effect of various factors i.e. anecological Gy-eq. [5-6]. This approach was used for ecological evaluation of equivalency of the effect of a toxic agent on a whole natural ecosystem as well as on artificial experimental microcosm [5, 8, 9].

F. Brechignac and G. Desmet (eds.), Equidosimetry, 73–78.

3. Effect acute gamma-irradiation, phenol and copper on the red algae

3.1. OBJECTS OF INVESTIGATION

This article studies of the effect of acute gamma-irradiation and chemical molismants on the Black Sea red algae and makes the comparison of the equivalent effective doses of these factors that cause a similar effect on the algae.

The investigation was made on multicellular red algae. They are not only the natural value of diversity of the marine flora, but they are valuable representing good biological raw material for economic needs. Four species of the Black Sea red algae were investigated: *Callithamnion corymbosum* (J. E. Smith) Lyngb., *Ceramium rubrum* (Huds.) J.Ag., *Grateloupia dichotome* J.Ag., *Gracilaria verrucosa* (Huds.) Papent..

3.2. THE RANGE OF STUDIED FACTORS

The following factors were studied : pressure of acute gamma-irradiation (^{137}Cs, at a dose rate 0,5 Gy.sec^{-1}) at doses 7, 47, 407, 1207 and 2407 Gy and as chemical molismants: phenol (for the concentration in water from 1 to 10 000 mkmol .l^{-1}), copper (Cu^{+2} in form of sulfate, for the concentration in water from 0,01 to 100 ikmol.l^{-1}) on the red algae. These experiments were considered as a model of an accident or a planned release of waste from various technologies in the sea environment and as an approach to evaluate the potential or real danger these technologies can present toan ecological permissible state of aquatic ecosystems.

3.3. CYTOFLUORIMETRICAL METHOD OF REGISTRATION OF PHYSIOLOGICAL STATE OF RED ALGAE

We used a cytofluorimetrical method for the registration of the physiological state of the red algae in the study on the action of toxic factors on hydrobionts [10]. Under the pressure of different deleterious factors a destruction of algae pigment complex takes place. The quantum *issue* of the lightcollecting pigment (phycoeritrin) intensively increases as a result of pigment complex destruction. Phycoeritrin luminescence is located in the yellow part of spectrum. Destruction of the algae pigment system leads to the death of the algae cell. Damaged cells have bright-yellow luminescence under with the luminescent microscope. The number of damaged cells in percentageof whole number of cells in area of observation characterizes the degree of damage of whole organism [10]. Measurements of number of damaged algae cells were carried out after 14 days of cultivating algae in experimental water.

3.4. COMPARISON OF VARIOUS AGENT EFFECT ON THE RED ALGAE

The study shows that the efficiency of toxic agents was determined by the specific properties species of the algae as well as by the origin of the factor. The investigated molysmants demonstrated the highest efficiency of damaging effect on Callithamnion. corymbosum, the lowest on Callithamnion. verrucosa. Callithamnion. rubrum takes the second and Callithamnion. dichotome takes the third position concerning to efficiency of the damaging effect of toxic agents (Table1).

Table 1. Influence the various deleterious agents on cells of four species of red algae

Species of red Black Sea algae	Part of deleterious cells, %	Intensity of factors		
		Dose of gamma-irradiation, Gy	Concentration of phenol, μmol /l	Concentration of Cu^{+2}, μmol /l
Callithamnion corimbosum	0	0	1	0,01
	0,1	7	10	-
	1,5	47	100	0,1
	5	407	-	-
	18	1207	1000	1
	100	2407	10000	10 – 100
Ceramium rubrum	0	7	1 - 10	0,01 – 0,1
	0,3	47	100	-
	6	407	1000	1
	12	1207	-	-
	30	-	10000	10
	50	2407	10000	-
	100	-	-	100
Grateloupia dichotom	0	7	1 - 10	0,01 – 0,1
	0,1	47	100	-
	1,5	407	-	1
	5,0	1207	1000	10
	32	-	10000	-
	40	2407	-	-
	80	-	-	100
Gracilaria verrucosa	0	7	1 - 10	0,01 – 0,1
	0,1	47	100	
	0,5	407	-	1
	3	-	1000	10
	10	1207	-	-
	15	-	10000	-
	20	2407	-	-
	50	-	-	100

The comparison of intensity of the factors that cause the same effect of damage in hydrobionts allowed to making up lines of equivalent doses for each species. For example for *C. corymbosum* the action dose of 47 Gy of gamma-irradiation causes the same damage in algae like 100 μmol/l of phenol and like 0.1 μmol.l^{-1} of copper; 1207 Gy of gamma-irradiation is equivalent to the action of 1000 μmol.l^{-1} of phenol and 1 μmol.l^{-1} of copper; in the same time 2407 Gy of gamma-irradiation causes the same effect as 10000 μmol.l^{-1} of phenol and 10 μmol.l^{-1} of copper. The results presented in Table 1 allow to put together equidosimetric ranges for all studied red algae species.

4. Comparative equidosimetric evaluation of effective equivalent doses of different deleterious agents

The quantitative correlation of factor intensity at every equidosimetric line is determined with peculiarities of characteristic dependencies "dose-effect" for every factor. In the same time the undertaking to keep the full (0-100% effect) dependencies "dose-effect" was not specially attempted in this study. The range of chemical toxicant concentrations and of doses of gamma-irradiation was chosen on the base of well-known data [5, 8, 11, 12].

We used the concentrations of chemical agents and doses of gamma-irradiation ranging between ecologically safe levels to damaged levels. In this way a very wide range of intensity of factors was investigated, especially in the range of low doses. The results are represented in a "dose-effect" coordinate (although not giving the full picture of dependencies "dose-effect") to reflecting the species specific sensitivity character of algae in relation to to the toxic factors. They characterize the level of intensity of each factor that causes a same damaging effect (Fig. 1).

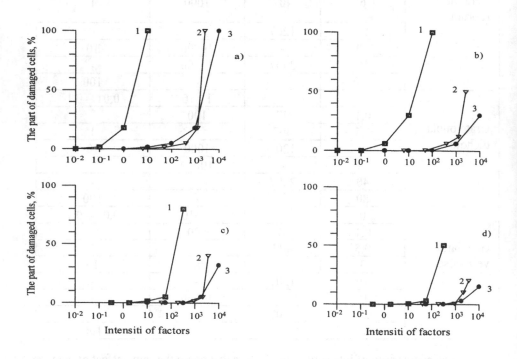

Fig 1. Dependences of «dose-effect» for red algae under pressure of toxic agents, where: species of red algae: a) - Callithamnion corymbosum, b) - Ceramium rubrum, c) - Grateloupia dichotome, d) - Gracilaria verrucosa; deleterious factors: 1 – phenol, 2 – gamma-irradiation, 3 – copper (Cu^{+2}); intensity of factors – phenol, coppermkmol /l, gamma irradiation – Gy.

In spite of the species specific character of the effect of the toxic agents it is possible to line up the factors as a whole according to the increase of their damaging effect efficiency on red algae: phenol → gamma-irradiation → copper (Cu^{+2}).

If we consider the obtained data only with respect to 100% damaging effects to algae cells it is possible to use the unit, that was proposed early [5,6] i.e. ecological Grey-equivalent, for the comparative equidosimetric evaluation of equivalent effective doses of chemical toxicants . It is expressed for chemical agents as Gy/mkmol /l. In this case for C. corymbosum, we reached the following meaning of ecological Grey-equivalent in connection to the studied chemical toxicants:

gamma-irradiation/phenol – 0,24 Gy/µmol /l;
gamma-irradiation/copper (Cu^{+2}) – 240 Gy/µmol /l.

Our investigation on red algae shows that the ecological Grey-equivalent for copper is considerably higher than for phenol. Probably it is connected to the origin of the phenols and their specificity of interaction with macroalgae [13, 14].As a result they have a lesser toxic effect on algae than copper or gamma-irradiation.

5. Conclusions

Analysis of data of magnitudes of ecological Grey-equivalent for chemical toxicants and their "dose-effect" dependencies allow to conclude that in case the efficiency of toxic effect of chemical agents is less than efficiency of damaging effect of gamma-irradiation these toxicants lead to an equidosimetric evaluation represented at ecological Grey-equivalent of less than 1 (as in case with phenol). Toxic chemical agents with a toxic effect efficiency higher than the damaging effect of gamma-irradiation have accordingly an ecological Grey-equivalent higher than 1 (as in the case of copper).

In this article the equidosimetric evaluation of effect nuclear and chemical factors (in ecological Grey-equivalent) on cells of red algae was given. This evaluation allows to make prognosis of the comparative ecological danger of the studied toxic agents for red Black Sea algae.

6. References

1. Polikarpov G.G., Zaitsev Y.P. Horizons and strategy of search in marine biology. Report at Session of Presidium of Academy of Science Ukr.SSR 18 May 1968. Kiev, Naukova dumka, 1969, 31 pp. (in Russian).
2. Polikarpov G.G., Zaitsev Y.P., Zats V.I., Radchenko L.A. Pollution of the Black Sea (Levels and sources). in: Proceedings of the Black Sea Symposium, Ecological Problems and Economical Prospects, 1991, Publ. The Black Sea Foundation for Education, Culture and Protection of Nature, Istanbul, 1994, P. 15-42.
3. Polikarpov G.G. Marine radioecology in Academy of Science of Russian and Ukraine. Biologiya morya V. 25, ¹ 6, 1999. pp 475-479. (in Russian).
4. .Polikarpov G.G. Conceptual model of responses of organisms, populations and ecosystems to all possible dose rates of ionizing radiation in the environment. Radiation Protection Dosimetry, V.75, 1998, pp 181-185.
5. Polikarpov G.G. Effects of nuclear and non-nuclear pollutants on marine ecosystems. Marine Pollution. in: Proceedings of a Symposium Held in Monaco, 5-9 October 1998, IAEA-TECDOC-1094, Vienna, 1999, pp 38-43.
6. Polikarpov G.G. Biological aspect of radioecology: objective and perspective. Comparative Evaluation of Environmental Toxicants. in: Proceedings of the International Workshop on

Comparative Evaluation of Health Effects of Environmental Toxicants Derived from Advanced Technologies, Chiba, January 28-30, 1998. Tokyo, Kodansha Scientific LTD, 1998, pp 3-15.

7. EgorovV.N., Policarpov G.G., Terestchenko N.N. et al.Investigation and evaluation of pollution< damage and state of ecosystems on the Black Sea shelf. in: Ecological safety of coastal and shelf zone and complex investigation of resources of shelf. Sevastopol, 2001, pp 111-127. (in Russian).

8. Fuma S., Miyamoto K., Takeda H., YanagisawaK. et al., Ecological effects of radiation and other environmental stress on aquatic microcosm. in: International Workshop on Comparative Effects of Health Effects on Environmental Toxicants Derived from Advanced Technologies. National Institute on Radiological Sciences: Abstracts, 1998, pp 21-22.

9. Polikarpov G.G., Tsytsugina V.G. Ñomparision cytogenetic and ecosystemic efficiency of effect radioactive and chemical mutagens in hydrosphere. Dokladi of NAN Ukraine, 1999, ¹ 6, pp 199-202. (in Russian).

10. Vladimirov V.B., Terestchenko N.N. Way of evaluation physiological state of red algae. Author's certificate. SU 111190234 A, 07.11.85, 1985, Bull., ¹ 41.

11. Bernhard M., Zattera A. Major pollutants in the marine environment. in: Marine Pollution and Marine Waste Disposal, Pergamon Press, Oxford and New York, 1975, pp 195 - 300.

12. Woodhead D.S. Dosimetry and the assessment of environmental effects of radiation exposure. in: Radioecology after Chernobyl: Biogeochemical Pathways of Artificial Radionuclides, SCOPE 50. Eds Sir F. Warner and R.M. Harrison, 1993, pp 291 – 306.

13. Glushko J. M.Toxic orgenic agents in economic waters. Lenungrad, Chimiya, 1976, pp 214. (in Russian).

14. Stom D.I. Meaning of oxidation and detoxication of phenols with aquatic plants. in: Self-purification of water and migration of pollution through trophic chain. Ì., Nauka, 1984, pp 91-97. (in Russian).

MACROPHYTES AS BIOINDICATORS OF RADIONUCLIDE CONTAMINATION IN ECOSYSTEMS OF DIFFERENT AQUATIC BODIES IN CHERNOBYL EXCLUSION ZONE

A. KAGLYAN, V. KLENUS, M. KUZ'MENKO, V. BELYAEV,
Yu. NABYVANETS, D. GUDKOV.
Department of Radioecology, Institute of Hydrobiology, NASU, Geroiv Stalingradu Ave.12, Kyiv, UA-04210, UKRAINE

Studied water bodies of left-bank flood-lands of Prypyat' river are located in the 10-km zone of Chornobyl NPP. They are among the most heavily contaminated aquatic objects in the exclusion zone. Higher aquatic plants represent the majority of hydrobionts in those water bodies and are characterized by high productivity and ability to actively absorb radionuclides . Occupying the dominating place among other components of freshwater biocenosis according to their biomass, higher aquatic plants play an essential role in processes of radionuclides distribution in water ecosystems. They serve as a natural "biofilter" that accumulates radionuclides and by that way removes them temporarily from the water ecosystem's matter turnover. Diversity of aquatic plants species results in a wide range of radonuclides uptake.

Maximal values of radionuclides content in highest aquatic plants were observed for Glyboke Lake. Apparently, those phenomena occurred due to high concentrations of radionuclides in water and bottom deposits. The lake is covered with macrophytes. They grow intensively and occupy almost half of the lake's water table. About 25% of water plants consist of air-water species and about 75% include water species. A major part of highest water plants grow in the dam area and on the lake's perimeter with reed mace (*Typha angustifolia*) and glyceria (*Glyceria maxima*) as dominants. Quite frequently the sedge (*Carex sp.*) was also identified on the lake's perimeter. The main stretch of the lake was covered with yellow pond lily (*Nuphar lutea L. Smith*) and *Stratiotes aloides L.* in a ratio of 1:2 [1]. Groups of *Stratiotes aloides L.* were quite powerful and as a rule with hornwort (*Ceratophyllum demersum L.*) in lower tier. All radionuclides concentrations are given in Bq/kg of air-dry weight.

In 1998 concentrations of ^{90}Sr in plants varied from 7.2 to 58.2 kBq/kg, and ^{137}Cs from 2.3 to 411 kBq/kg. Values of the upper limit of ^{90}Sr content in plants from other water bodies that are located in flood-lands, varied from 21.6 to 43.2 kBq/kg. Concentrations of ^{137}Cs in water plants from water objects were found to be within following limits: Daleke Lake – 2.2 – 96.4 kBq/kg; Krasnyans'ky Branch before the dam and in cold section of Chornobyl NPP cooling pond – 0.68 – 53.8 kBq/kg; warm section of Chornobyl NPP cooling pond and Krasnyans'ky Branch behind the dam – 1.2 – 27.4 kBq/kg. Minimal values of concentrations were observed in aquatic plants from Prypyat' river (Chornobyl town). Concentrations of ^{90}Sr varied from 0.03 to 2.9 and ^{137}Cs from 0.06 to 1.36 kBq/kg.

In Fig. 1-9 the dynamics of ^{90}Sr and ^{137}Cs content in specific aquatic plant species that were available for systematic sampling for the period of 1989-2000 is presented. Gradual decreasing of ^{90}Sr and ^{137}Cs concentrations in aquatic plants of both sections of Chornobyl NPP cooling pond is clearly seen (fig. 1 and 2).

F. Brechignac and G. Desmet (eds.), Equidosimetry, 79–86.

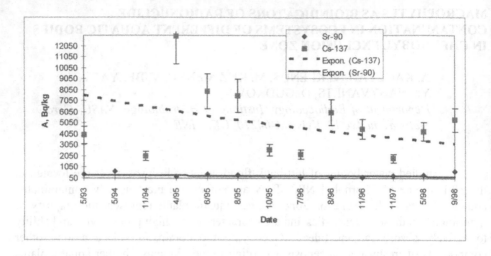

Fig. 1. Dynamics of the radionuclides content in Phragmites australis Trin. from warm section of cooling pond.

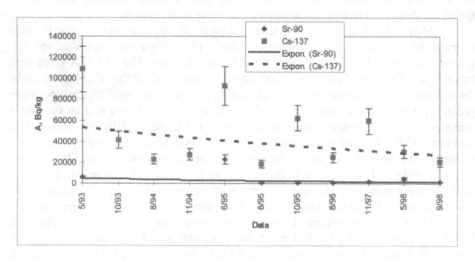

Fig. 2. Dynamics of the radionuclides content in Ceratophyllum demersum L˙ from cold section of cooling pond.

The dynamics of the radionuclides content in higher aquatic plants from Glyboke Lake was quite interesting (Fig. 3-5). As one can see, the content of radiocaesium in macrophytes slightly decreased while ^{90}Sr concentrations exhibited tendency of increasing in all collected water plant species. It probably could be explained by radionuclides leaching from "hot particles"; their amount in lake is still extremely high. That process led to strong binding of radiocaesium with bottom deposits and transformation of ^{90}Sr to more mobile soluble forms. At the same time, for other studied water objects a trend of ^{137}Cs and ^{90}Sr decreasing content in higher water plants was identified (Figs. 6-9).

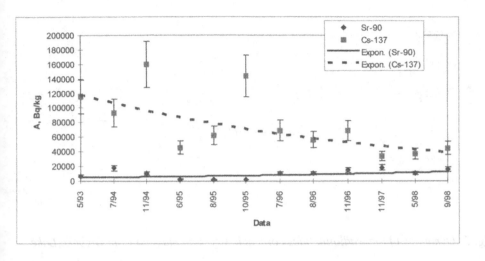

Fig. 3. Dynamics of the radionuclides content in Nyphar lutea L. from Glyboke Lake.

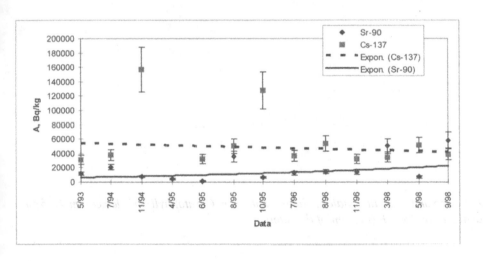

Fig. 4. Dynamics of the radionuclides content in Stratiotes aloides L. from Glyboke Lake.

It is obvious from Fig.10-12, that macrophytes from Glyboke Lake were most heavily contaminated with ^{90}Sr. Other water objects from the point of view of radionuclide concentrations in aquatic plants could be ranged in following decreasing order: Daleke Lake and Krasnyans'ky Branch – Chornobyl NPP cooling pond – Prypyat' river. Concerning decreasing of ^{137}Cs concentrations that rank looked like Glyboke Lake – Daleke Lake – Krasnyans'ky Branch (except hornwort from cold section of cooling pond) – Chornobyl NPP cooling pond – Prypyat' river.

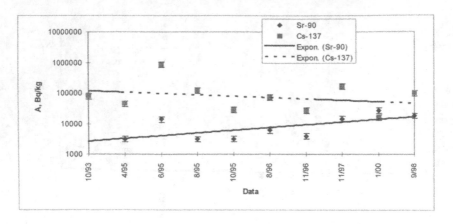

Fig. 5. Dynamics of the radionuclides content in Clyceria maxima from Glyboke Lake.

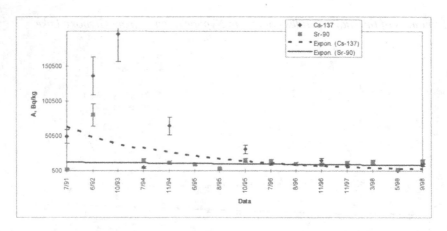

Fig. 6. Dynamics of the radionuclides content in Ceratophyllum demersum L. from Krasnyans'ky Branch (in front of the dam).

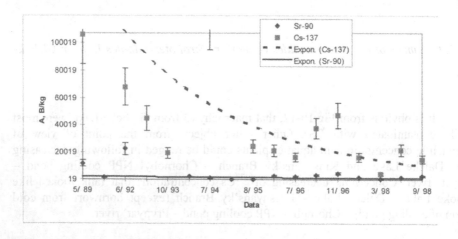

Fig. 7. Dynamics of the radionuclides content in Phragmites austral is Trin. from Daleke Lake.

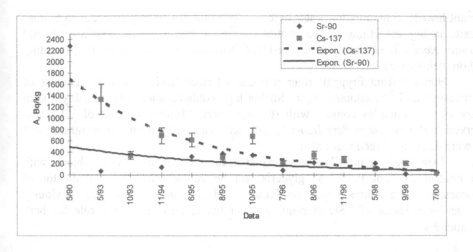

Fig. 8. Dynamics of the radionuclides content in Typha latifolia L. from Prypyat' river (Chornobyl town).

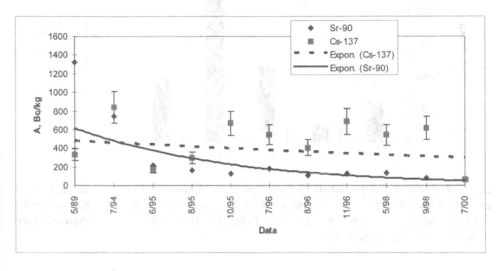

Fig. 9. Dynamics of the radionuclides content in Ðotamogeton perfoliatus L. from Prypyat' river (Chornobyl town).

Accumulation of radionuclides by different species of highest aquatic plants growing in water bodies of different types is affected by various factors. There is vegetation period, water pH etc. For example, while considering highest aquatic plants collected in August-September, it can be seen, that in isolated and stagnant water objects like Glyboke Lake and Krasnyans'ky Branch the maximal concentrations of ^{137}Cs were observed in glyceria and the smallest ones were observed in reed mace. Maximal concentrations of ^{90}Sr were found in *Potamogeton Perfoliatus L.* and *Stratiotes aloides L.* while minimal radionuclide concentrations were determined in sedge. High radiocaesium concentrations in glyceria could be explained with its growing in the water-bank line where according to our findings for isolated and stagnant heavily contaminated water objects concentrations of radionuclides in bottom deposits are highest. Highest concentrations of ^{90}Sr in *Potamogeton Perfoliatus L.* and *Stratiotes aloides L.* obviously were a result of formation of metal carbonates (including ^{90}Sr) on their surface.

Horwort from Prypyat' river was characterized having the highest levels of radiocaesium and ^{90}Sr contamination. Such a high contamination was probably due to uptake of radionuclides coming with running water. Minimal content of ^{137}Cs was observed in *Potamogeton Perfoliatus L.* and reed mace and minimal concentrations of ^{90}Sr were detected in reed mace also.

Therefore, for heavily contaminated isolated or stagnant water objects with high content of "hot particles" glyceria can be considered as a bioindicator of radiocaesium while *Potamogeton Perfoliatus L.* and *Stratiotes aloides L.* can obviously serve as bioindicator of ^{90}Sr. In running water horwort can play that role for both radionuclides.

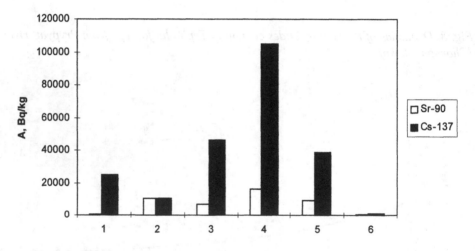

Fig. 10. Radionuclides content in Ceratophyllum demersum L.: 1 - cold section of cooling pond ChNPP, 2- Krasnyans'ky Branch (in front of the dam), 3- Krasnyans'ky Branch (behind the dam), 4 - Glyboke Lake, 5 - Daleke Lake, 6 - Prypyat' river (Chornobyl town), August 1996.

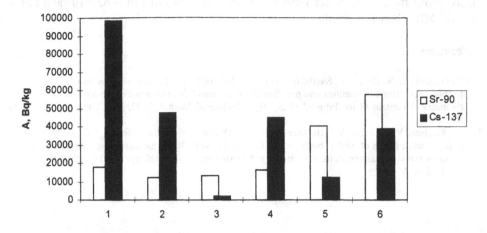

Fig. 11. Radionuclides content in higher aquatic plants of the Glyboke Lake
(September, 1998): 1.- Glyceria maxima, 2 - Carex sp., 3 - Typha angustifolia L., 4 -
Nyphar lutea L., 5 - Ceratophyllum demersum L., 6 - Stratiotes aloides L.

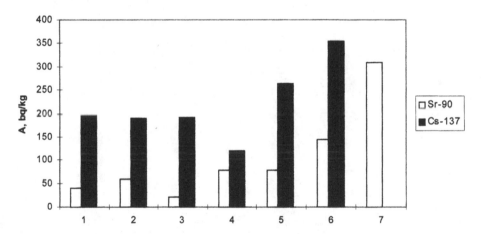

Fig. 12. Radionuclides content in aquatic plants of the Prypyat' river (Chornobyl
town), September, 1998): 1.- Glyceria maxima, 2 - Carex sp., 3 - Typha Latifolia L., 4
-Ðotamogeton perfoliatus L., 5 -Ðotamogeton Pectinatus L., 6 - Ceratophyllum
demersum L., 7 - Cladophora sp.

Based on data concerning phytomass stock in water objects of the exclusion
zone [1] and on our own data regarding radionuclides content in higher water plants we
have calculated the strontium-90 and ceasium-137 deposition in different species of
macrophytes at the end of the vegetation season of 1998.

The content of ^{137}Cs and ^{90}Sr in ther highest aquatic plants (September 1998)
was as follows: Glyboke Lake – $(4.7 – 8.4)*10^3$ and $(3.6 – 4.0)*10^3$ MBq; Daleke Lake
– 30-248 and 12-30 MBq; Krasnyans'ky Branch (behind the dam) – $(21.8 – 53.2)*10^3$
and $(19.9 – 54.1)*10^3$ MBq correspondingly. Thus, the stock of ^{137}Cs and ^{90}Sr in

macrophytes from these water objects was in the limits of $(26 - 62)*10^3$ and $(24 - 58)*10^3$ MBq correspondingly.

References

1. D.I. Gudkov, L. N. Zub, A.L. Savitskiy and others. Macrophytes of exclusive zone of Chernobyl NPP: formation of plant communities end peculiarities of radionuclide contamination under conditions of left riverbank floodplain of the Prypiat' river. Hydrobiological Journal V.37, N 6, 2001, pp.64 -81 (in Russian).
2. A.E. Kaglyan, V.G.Klenus, V.V.Belyaev and others. Dynamic content of ^{90}Sr and ^{137}Cs in hydrobionts: higher aquatic plants of water bodies of the Chernobyl' NPP 30km exclusive zone. Naukovi zapysky Ternopil'skogo peduniversytetu. Ser.: Biology, Special issue: Hydroecology. N4 (15), 2001, pp.14 -17 (in Ukrainian)

EFFECTS OF RADIOACTIVE AND CHEMICAL POLLUTION ON PLANT VIRUS FREQUENCY DISTRIBUTION

V.P. POLISCHUK, T.P.SHEVCHENKO*, I.G. BUDZANIVSKA,
A.V. SHEVCHENKO, F.P. DEMYANENKO, A.L. BOYKO
Virology Dpt., Taras Shevchenko' Kyiv national university, 64 Volodimirska st., Kyiv 01033, UKRAINE
*Institute of Agroecology and Biotechnology of UAAS, 15 Metrologichna st., Kyiv, UKRAINE
e-mail: virus@biocc.univ.kiev.ua

1. Introduction

Remediation and exploration of soils in Chornobyl estrangement area also suppose growing of agricultural crops in the region. However, plants in this area are permanently under effect of the constant radioactive pressure [2]. Nowadays, it must be taken into account that stress might simultaneously have an abiotic (radioactive, chemical) nature and a biotic one as well. Virus infection can be an inducement of biotic stresses in plants. As we demonstrated in previous works, ionizing radiation can lead to mutations in plant organisms and in plant viruses, too. Owing to mutations' accumulation, there is a possibility of new more pathogenic virus strains' appearing. Arisen in Chornobyl region, these highly virulent strains can be spread over a large territory affecting agricultural crops' yield [2, 3].

Similarly, increasing of heavy metals' concentration as a factor of chemical pollution may lead to common, low-specific physiological and biochemical changes in plants, the most common from them could be membrane damages, changes in enzyme activity, inhibition of root growth, etc [1, 5]. The result of these interactions may be almost the same.

Research on the course of virus infections under the effects of radioactive and chemical pollution and monitoring for plant viruses circulation may reveal plants that are genetically resistant to virus infections, may help to explain the nature of plant virus strain diversity and allow to develop efficient protective measures.

2. Materials and methods

In our work we tested some local areas in Chornobyl region: Kopachy, Novo- and Staro-Shepelichi (10-km zone), Chornobyl surroundings, Opatchichi, arid areas in Shepelichi and Buryakivka villages. Plants from the ecologically clean place in Shatsk National park were taken as controls . Gamma-radiation dose rates for regions to being compared were 300-520 µR/h and 5-10 µR/h for Chornobyl NPP contaminated area and ecologically clean territory of Shatsk National park, correspondingly. Plants were sampled on 15 points.

Also, we collected plant and soil samples for heavy metal content investigation using atomic absorbance spectroscopy (AAS) from two regions of Ukraine (Boguslav area in Kyiv region, and area of Zmiyivsky Power Plant in Kharkiv region) by the same procedure.

All samples were tested in indirect ELISA with polyclonal rabbit antiserums for the following viruses: TMV, PVX, PVY, PVM, PVS, PVF, PLRV, BYMV, AMV and

F. Brechignac and G. Desmet (eds.), Equidosimetry, 87–92.

CMV by the common method. Goat anti-rabbit IgGs conjugated with horseradish peroxidase were used as second antibodies, and H_2O_2 as a substrate [4, 9]. To stop reaction, we used 2M H_2SO_4. Results were counted at the wavelength of 492 nm.

A spectrum of typical indicator plants was used to determine biological properties of viruses as follows: *Chenopodium amaranticolor*, *Datura stramonium*, *Nicotiana tabacum* cv. Samsun, *N. glutinosa*, *N. tabacum* cv. trapeson, *N. debney³*, *N. sylvestris* and *Plantago major*.

In order to study virus morphology, a JEM-1200 electronic transmission microscope was used.

3. Results and discussion

Our work was focused on the assessment of some plant virus distribution (TMV, PVX, PVY, etc.) in wild and cultural flora of Chornobyl NPP area in order to develop recommendations concerning the most efficient way of remediation of contaminated soils in this estrangement area.

Analysing some specimen of *Elytrigia repens*, *Taraxacum officinale*, *Plantago major* and *Cirsium arvense*, we could reveal serious differences between samples from Chornobyl and from other areas. Investigations in 2 fields near to the Opatchichi settlement showed that the plant virus frequency distribution was much higher in the case of Chornobyl samples comparing to control plants from Shatsk National park. It may be due to a decrease in protection reactions of plants growing under conditions of radioactive pollution, and to increase of virus infectivity as well. The most grim values were obtained for such pathogenic viruses as PVX, TMV and PVY which proves a need in careful monitoring for virus circulation to avoid their spreading over the agrocenoses of Ukraine.

Plant species diversity was much less abundant on the arid area comparing to normal forest. It was demonstrated that, similarly to Opatchichi area, plant virus frequency distribution was much higher in the case of Chornobyl NPP area samples comparing to control plants, which can reflect a level of radionuclide pollution or a seriously reduced quantity of plant species presented (Fig.1). It should be stated that previous sample analysis from different areas of the Chornobyl region (Novo-Shepelichi, Kopachi, Yanov, Buryakivka) proves the general tendency toward the increase of epiphytothic background (i.e. increased ability of viruses to infect plants) in agrocenoses and biocenoses as well. Moreover, viruses extracted from plants in this region developed non-typical symptoms on the corresponding indicator plants.

Frequency distribution was much higher for the majority of viruses investigated in Chornobyl samples even comparing to unexploited areas of other regions including Kharkiv with serious level of chemical pollution, especially with heavy metals pollution. The purpose was to check the dependence between increased heavy metals content in soil and its influence on the course of virus infection in plants growing on Ukrainian agrocenoses [6].

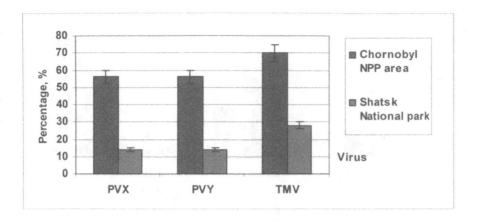

Fig. 1. Virus antigens' presence in regions differing in radioactivity pollution.

Experiments conducted in relation with heavy metal (cadmium, lead, copper, zinc and nickel) soil content have shown their different concentrations in the regions investigated (Fig.2). Contrary to Kyiv region and control area of Shatsk National park, Kharkiv region had remarkably high level of pollution with these heavy metals regardless of the almost similar absorbing capacity of soils of these regions [10]. Obviously, increased heavy metal concentration in the area of Zmiyivsky Power Plant (Kharkiv region) mainly is due to the activity of the power plant.

Much more frequent revealing of TMV, BMV, CMV and PVY antigens was typical for this region (Fig.3). Together with heavy metals, adsorption of virus particles also depends on the soil features such as mechanical structure, pH, etc [8]. The ecologically clean area of Shatsk National park was comparatively free from virus antigens possibly owing to differences in soil structure and to the less obvious anthropogenic pressure. Although, comparing distribution of virus antigens in Kyiv and Kharkiv regions we revealed their higher concentration in Kharkiv despite the same soil structure and its absorbing capacity [11].

Modeling experiments using TMV-*Nicotiana tabacum* system were conducted in order to determine the effect of virus infection on plants grown in soil with higher content of heavy metals (copper, lead and zinc soluble salts), which corresponded to the estimated level of Kharkiv region [11] and was higher in comparison to maximum permissible concentrations determined for these metals in Ukrainian soils [7, 12]. Visually, control plants grown in sterile non-polluted soil looked absolutely normal, whereas plants grown in soil with heavy metals added developed stunting (Fig.4). Virus-infected plants grown in non-polluted soil demonstrated the development of typical disease symptoms (leaves deformation, enations, yellowing) in two weeks post infection. Infected plants grown in the polluted soil clearly showed a disappearing of virus infection symptoms in 10 days post infection, and died in 20 days post infection (dpi).

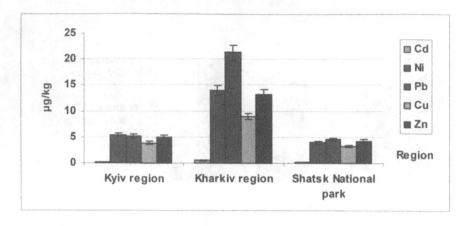

Fig. 2. Heavy metal content in soil probed for virus antigens' presence.

We assume that the results of these model experiments show increased virus contamination of plants grown in the chemically polluted environment, probably owing to the inhibition of general physiological activity (immune state) [12] that makes virus infection development possible [13, 14], and secondary soil contamination by viruses as well.

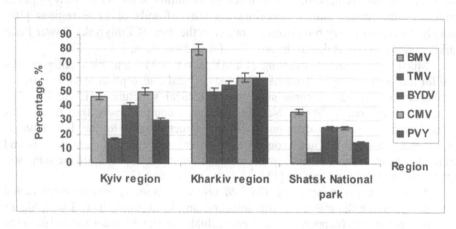

Fig. 3. Virus frequency distribution in regions differing in pollution level.

Virus interactions and circulation in agro- and biocenoses, where there are a lot of external biotic and abiotic factors influencing their ratio, infectivity, strain diversity, is one of the hardest problems for plant virology. Studies of viruses circulating in the area of radioactive pollution are important not only from the ecological point of view, but for controlling strain pathogenicity.

Fig. 4. Combined effect of two stressors on experiment Nicotiana tabacum plants.
1 – intact control plant; 2 – virus-infected plant grown in non-polluted soil; 3 – non-infected plant grown in heavy metal polluted soil; 4 – virus-infected plant grown in heavy metal polluted soil.

Following assessment of plant virus distribution in human-polluted areas will allow to construct a map of plant virus and sanitary state for areas exposed to biological risk, and then to develop a program prognosing the arising of plant diseases of important agricultural crops. In case of remediation of contaminated area soils it could be a serious problem due to the absence of any phytovirus control on this territory for 12 years. Therefore, viruses can be used as an indicative model for evaluation of general changes in the environment.

4. Conclusions

1. Level of virus contamination of plants correlates with the level of pollution of the same area. This fact does not depend on the kind of pollution (radionuclide or heavy metal).

2. Combined effect of viruses and heavy metals provides more serious plant disease development and growth inhibition than in case of every single factor's influence (synergy).

3. Viruses can be used as an indicative model for evaluation of general changes in the environment.

5. References

1. Barcelo J., Poschenrieder Ch. Plant water relations as affected by heavy metal stress: a review // J. Plant Nutr. -1990. -V.13, N.1. -pp.1-37.
2. Boyko A.L. Ecology of plant viruses / Kyiv. –Vischa Shcola. -1990. –167p. (*In Russian*).
3. Boyko A.L., Stepanyuk S.A., Garifulin O.M. Obtaining and analysis of provirus cDNA for C-fragment of TMV CP for plants isolates from radioactively polluted Ukrainian region // Biopolymers and Cell. -1996. –V.12, N.6. –pp.112-115 (*In Russian*).
4. Clark M.F., Adams A.N. Characteristics of the microplate method of enzyme-linked immunosorbent assay for the detection of plant viruses // J. of Gen. Virol. -1977. -V.34. -pp.475-483.

5. Foy C.D., Chaney R.L., White M.C. The physiology of metal toxicity in plants // Ann. Rev. Plant Physiol. -1978. –V.29. –pp.511-566.
6. Garmash N.U. Effect of increased doses of heavy metals on their accumulation in wheat and beans in ontogenesis // Physiology and biochem. of agricultural plants. -1989. – V.21, N.2. -pp.141-146 (*In Russian*).
7. Kabbata-Pendias À., Pendias H. Trace elements in soils and plants / CRC press, Ink.- Boca Raton, Florida. -1986. -349p.
8. Kegler H., Spaar D. et al. Viruses in soil and ground water (review) // Arch. Phytopath. Pflanz. - 1995. - N.29. -pp.349-371.
9. Methods in molecular biology. ELISA: theory and practice. Ed. by Crowther J.R. / Humana Press. -1995. -223p.
10. Peterson P.J. Adaptation to toxic metals // Metals and micronutrients: uptake and utilisation by plants / Ed. D.A.Robb, W.S.Pierpoint. –New York: Academic Press. -1983. –pp.51-69.
11. Polischuk V.P., Senchugova N.A., Budzanivska I.G., Holovenko O.L., Boyko A.L. Monitoring and strategy in virus diseases' prognosing for agricultural plants in various agrocenoses of Ukraine // DAN. -1998. –N.2. –pp.184-187 (*In Ukrainian*).
12. Rauser W.E., Dumbroff E.B. Effects of excess cobalt, nickel and zinc on the water relations of *Phaseolus vulgaris* // Env. Exp. Bot. -1981. –V.21, N.2. –pp.249-255.
13. Stiborova M., Doubravova M., Brezinova A., Friedrich A. Effect of heavy metal ions on growth and biochemical characteristics of photosynthesis of barley (*Hordeum vulgare*) // Photosynthetica. -1986. –V.20, N.4. –pp.418-425.
14. Weckx J., Clijsters H. Heavy metals induce oxidative stress // Physiol. Plant. -1992. –V.85, N.3, Pt.2. –p.73.

SPECIES DIVERSITY AS A FACTOR OF RADIOSTABILITY OF ALGAL CENOSIS

N.L. SHEVTSOVA, L.I. YABLONSKAY

Institute of Hydrobiology, NASU, Prosp. Geroyev Stalingrada, 12, Kyiv, UA-04210, UKRAINE

1. Introduction

Anthropogenic contamination of the environment became the determinant factor of destabilization of both terrestrial and aquatic ecosystems. The problem of resistance and flexibility, not only of biosystems of different levels of organization and evolutionary development, but also of ecological alignments under deteriorative conditions of the environment is one of the important issues in modern ecology.

Resistance as a whole, and radioresistance particularly – the main integral index of structural and functional status of biosystems or ecosystems (cenoses, populations), characterizes their ability to withstand to external extreme factors to the habitat.

Owing to structural hierarchy of vital systems each level of organization is interconnected. A higher level is apt to regulate deviations of a lower one by feedback, and vice-versa. Alterations at cellular level can lead to alterations in a population, cenoses etc levels and vice versa. Therefore it is logical to assume that species diversity may be one of the ecological protections (preservatives is something that is used for preventing to making babies) for stability.

2. Role of species diversity and richness in model algaecenoses stability under gamma-irradiation

An appropriate model for testing such an assumption can be the model of cenosis of fresh-water planctonic algae, which offers some advantages.

- The possibility of maintaining of the algal collection in laboratory conditions for a rather long time, practically without damage to specific and qualitative structure;
- easy access and determination of structural cenosis indexes – cell number abundance and species richness (number of species, their dynamics ratio, other different specific indexes);
- well studied radiosensitivity of fresh-water planctonic algae in culture

Algal samples were collected from the Dnieper river near Kiev at different seasonal periods. The mixes of different species numbers were selected (Table 1) for series of experiments. In all cases predominant were the Protococcales *Scenedesmus quadricauda, Sc. quadricauda* v. *vesiculosus, Sc. quadricauda* v. *africanus, Sc. acuminatus, Sc. acuminatus* v. *tetradesmoides, Sc. acuminatus* v. *biseriatus, Sc. acuminatus* v. *elongatus, Sc. bijugatus, Sc. obliquus, Sc. protuberans, Sc. costatus, Sc. apiculatus, Coelastrum microporum, C. sphaericum, Tetrastrum punctatum, T. glabrum, T. staurogeniforne, Tetraedon incus, Pediastrum duplex, P. duplex* v. *cornutum, P. boryanum, P. kawraiskyi, Dictyosphaerium pulchellum, Didymocystis planctonica, Ankistrodesmus acicularis, Francia tenuispina, Ankistrodesmus pseudomirabilus* v. *spiralis, Golenkinia radiata,* and *Kirchneriella irregularis.*

F. Brechignac and G. Desmet (eds.), Equidosimetry, 93–97.

Table 1. Outline characteristic of model algaecenosis structure

Total number of species	Dominant species (nominator, in abundance; decrease order)	Main groups ratio 1/2/3, %
25	Sc.quadricauda Sc.acuminatus Coelastrum microporum	100/0/0
30	Sc.quadricauda Sc.acuminatus Didymocystis planctonica	100/0/0
53	Sc.quadricauda Sc.acuminatus Coelastrum microporum	85/11/4
68	Sc.quadricauda Sc.acuminatus Ankistrodesmus pseudomorabilis	80/17/3
77	Sc.quadricauda Sc.acuminatus Coelastrum microporum	82/13/5

Note: Main groups of algae (divisions) 1 – Chlorophyta, 2 – Bacilariophyta, 3 – Cyanophyta.

The algae were cultivated in Fitzgerald medium No. 11 [1]. Irradiation was produced by gamma rays of ^{60}Co in a UK-250000 apparatus at absorbed doses between 1 and 10^5 Gy at different exposure rates (Table 2).

Table 2. Dose and power range in experiment

Radiation absorbed dose, Gy	1	2.5	5	10	50	250	500	1000	5000	10000	20 000
Exposure rate, mGy/sec[1]	1.5	1.5	1.5	1.5	15	153	153	326	326	326	326

A 5-day-old synchronized suspension mixed cultures was irradiated. Synchronization was induced by means of a succession of dark and light phases [2]. After irradiation, the culture was inoculated into fresh liquid medium in 1:5 proportions (density before inoculation $1 \cdot 10^6$ cells/ml) and exposed to light and dark phases of 8 and 16 h respectively. The cultivation temperature was 20-22°C. The survival rate was determined as the ratio of the number of cells in the irradiated version to the number in the control, which was taken as 100%. In both cases, the cells were counted simultaneously in a Goryayev chamber [1]; numbers of species were also determined.

Index of species richness was calculated by formula:

$$d_i = S_i/N_i \ [3] \ ,$$

here d_i - index of species richness; S_i - species number in i-variant; N_i – medium cells numerous in i-variant

Diversity index – $S_\% = (S_i/S_n) \bullet 100\% \ [4]$

S_i – species amount in i-variant; S_n - species amount in basic polyculture; $S_\%$ - index of diversity

The survival curves of the protococcal algae (Fig. 1) show a pronounced, consistent plateau at doses between 1 and 50 Gy and decline at doses between 50 and 20,000 Gy. The presence of these segments indicates that postirradiation recovery occurs in the model cenosis and that some of the cells have elevated radioresistance. The survival rate corresponding to the beginning of the radioresistant segment of the curve depends on the state of development of modal algal cenosis. Maximum radioactive injury was observed at 15-20 days – logarithmic stage of the algaecenosis development. It was expected relevant number of species corresponded to medium injury range (Fig.1).

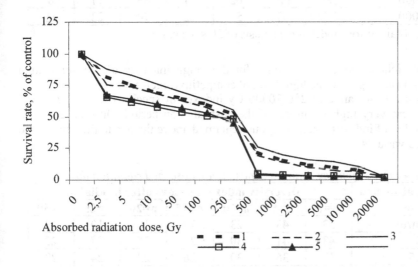

Fig.1 Survival rate of cells of model algaecenosis with different number of species after gamma-irradiation as a function of dose; Number of species in model cenoses:1 25; 2 - 30; 3-53; 4 –68; 5 – 77.

This was a model algaecenosis with medium numbers of species – 30 - among the tested ones. Due to high species level competitive exclusion the maximum injury was observed for model algaecenosis with 50-70 number of species.

Changes in the dominant species complex were observed. Over the range of absorbed dose of gamma-radiation 1-50 Gy *Scenedesmus quadricauda* - the dominant species of basic polyculture and control stay the monodominant. But under the doses of 250, 500, 1000, 2000 such subdominant as *acuminatus var .biseriatus* appeared. Under the highest doses as 5.000, 10.000, 20.000 Gy *Senedesmus quadricauda* was the monodominant species.

Over the range of absorbed radiation dose below or equal to LD_{50} (= 250 Gy for all examine model cenosis) an increase of the indexes of species richness was observed (Tabl.3).

Table 3. Species richness index changes depend upon absorbed radiation dose

Radiation absorbed dose	Species richness index on n-day after irridiation, %						
	n=1	n=5	n=10	n=15	n=20	n=25	n=30
Control	193	53	10	2	1	1	1
1	241	41	30	3	3	2	2
2.5	169	63	21	4	2	3	2
5	135	28	15	4	1	1	1
10	46	25	14	4	3	1	1
50	330	54	11	3	2	1	1
250	161	56	38	9	10	8	6
500	218	59	64	116	39	15	10
1000	304	84	81	45	75	20	15
5000	232	84	77	60	60	19	9
10 000	280	110	62	61	58	11	6
20 000	179	97	56	62	74	23	15

(Note: data obtained for model algaecenosis of 30 species)

Orientation of richness index is evidence for the high intensity of the domination process in this dose range and the high level of competitive exclusion.

Within a dose range of 250-20000 Gy indexes of species richness of model algaecenosis were very high for most of the postradiation period. Only on 25th and 30th day the value of indexes decreased, but remained above the one in the control and other irradiated variants.

Table 4.Diversity index changes depend upon absorbed radiation dose

Radiation absorbed dose	Diversity index on n-days after irridiation						
	n=1	n=5	n=10	n=15	n=20	n=25	n=30
Control	53	43	23	10	10	10	13
1	43	41	30	27	25	27	22
2.5	37	36	30	26	22	24	22
5	43	38	35	34	31	21	21
10	42	35	24	24	23	21	21
50	33	20	17	23	17	13	17
250	27	23	20	10	10	13	13
500	30	20	17	20	13	10	10
1000	30	27	20	17	17	13	13
5000	27	13	10	10	10	10	10
10 000	20	13	10	10	11	10	7
20 000	13	13	13	12	12	11	9

(Note: data obtained for model algaecenosis of 30 species)

By confrontation of the values from the tables 3 and 4 and from survival valuesit becomes understandable, that such a picture is labeled by a considerable decrease of mean number of cells of all species, while the number of species stay constant, above or at the level of the control. Such rather exotic situation has allowed to assume that the increase of species richness and diversity is somehow brought aboutto lower (to reset) the oppressing effect of gamma-radiation and to increase the stability of a system. The hypothesis that species diversity increases stability of a system, was supported by Margalef (1968) " the Ecologist have seen the possibility of construction of feedback systems under a species diversity, measured any way" [3] McNaughton has suggested, that at a level of primary producers species diversity is a means of functional stability of the collection. We consider that the data of our experience testify for the promotion of the given hypothesis.

References

1. 1975. Metody fiziologo-biokhimicheskogo issledovaniya vodorosley v gidrobiologicheskoy praktike (Methods of Physiological and Biochemical Investigation of Algae in Hydrobiology). Naukova dumka Press, Kiev.

2. Lorenzen, H. 1957. Synchrone Zeilteilung von *Chlorella* bein verschiedenen Licht-Dunkel-Wechzeln. Flora, **144**, No. 4, pp. 473-496.[Rus.]

3. Odum E. P. 1986 Basic ecology, **2** Moscow "Mir"[Rus.]

4. Margalef R. 1968 Perspectives in Ecological Theory, Chicago, University of Chicago Press. 122 pp.

5. McNaughton S.J.,1978. Stability and diversity in grassland communities, Nature, **279**, Pp.351-352

Part 3.

Radioecology and Ecotoxicology in General Ecology

MODERN PROBLEMS OF ECOTOXICOLOGY

G. ARAPIS
Agricultural University of Athens
Iera Odos 75, Athens 11855, GREECE

1. Introduction

With the interest that scientists and the general public show today towards ecology and environmental protection, a number of new ecological branches are emerging, including ecotoxicology.

During the 1960s, 1970s, and 1980s scientists have acknowledged that nature is much more complex than previously thought. Not everything could be reduced to an experiment in the laboratory. Wolfram talked about irreducible systems, to which most biological systems belong, but required a synthesis of many laboratory experiments and/or observations in situ [1, 2].

Ecotoxicology belongs to one of the new sciences which emerged as a consequence of the adverse effects of pollution on complex natural systems. It has been acknowledged in this scientific discipline, that natural (and other) systems are so complex that it is impossible to reach an understanding of all the details of these systems. We have to accept that in the ecological sciences our description of natural systems and their processes will inevitably have a certain degree of uncertainty due to their enormous complexity.

Nevertheless, ecotoxicological research started in the 1960s attempting to reveal as many details as possible of the processing of toxic substances in the environment. This research proved to be extremely valuable as it was able to point towards some general rules and classification of the behaviour of toxic substances. It became, however, clear that the ultimate goals of ecotoxicological research, to determine all processes in nature for all toxic subtonics and pollutants of interest (there are probably about 100,000 chemical compounds used in such an amount that they could threaten the environment), would be comparable to the fate of Sisyphus.

Nowadays, the ecosystem as a whole starts is to be considered as a living, evolving and dynamic entity, and not simply a conglomeration of physical and biotic components. Appropriate examples are drawn from various species, populations, communities, and ecosystems to emphasise and explain the role of ecological factors and phenomena. Thus, at the level of organisms the effects and the way they adapt for example to temperature, moisture, light, photoperiod, ionising radiation, salinity, pH and toxicants are taken into account. At the population level, parameters such as growth, reproduction, mortality, spatial pattern, dispersal, migration and communication are important. At the community level, additional attributes such as diversity, competition, parasitism, predation, etc. are given. Finally, at the ecosystem level, the concepts of trophic levels and webs, nutrient cycles, maturity, succession, niche, stability, homeostasis, etc., are also taken into consideration.

F. Brechignac and G. Desmet (eds.), Equidosimetry, 101–109.

2. A framework for ecotoxicology

Ecotoxicology can be simplified to the understanding of only three functions: Firstly, there is the interaction of the introduced toxicant, xenobiotic, with the environment. This interaction controls the amount of toxicant or the dose available to the biota. Secondly, the xenobiotic interacts with its site of action. The site of action is the particular protein or other biological molecule that interacts with the toxicant. Thirdly, the interaction of the xenobiotic with a site of action at the molecular level produces effects at higher levels of biological organisation. If ecotoxicologists could write appropriate functions that would describe the transfer of an effect from its interaction with a specific receptor molecule to the effects seen at the community level, it would be possible to accurately predict the effects of pollutants in the environment. We are far from a suitable understanding of these functions. Unfortunately, we do not clearly understand how the impacts seen at the population and community levels are propagated from molecular interactions [3].

Nevertheless, techniques have been derived to evaluate effects at each step from the introduction of a xenobiotic to the biosphere to the final series of effects. These techniques are not uniform for each class of toxicant, and mixtures are even more difficult to evaluate. Given this background however, it is possible to outline the current levels of biological interaction with a xenobiotic: Chemical and physicochemical characteristics, bioaccumulation/biotransformation/biodegrade-tion, site of action, biochemical monitoring, physiological and behavioural, population parameters, community parameters and ecosystem effects [3, 4].

Each level of organisation can be observed and examined at various degrees of resolution. The factors falling under each level are illustrated in the Figure 1. Examples of these factors at each level of biological organisation are given below.

2.1. CHEMICAL AND PHYSICOCHEMICAL CHARACTERISTICS

The contribution of the physicochemical characteristics of a compound to the observed toxicity is called quantitative structure activity relationships (QSAR).

It must be remembered that in most cases the interaction at a molecular level with a xenobiotic is happenstance. Often this interaction is a by-product of the usual physiological function of the particular biological site with some other low molecular weight compound that occurs in the normal metabolism of the organism. Xenobiotics often mimic these naturally occurring organisms, causing degradation and detoxification in some cases, and in others toxicity.

2.2. BIOACCUMULATION/BIOTRANSFORMATION/BIODEGRADATION

Bioaccumulation, which is the increase in concentration of a toxic substance in a tissue compared to the environment, often occurs with materials that are more soluble in lipid and organics (lipophilic) than in water (hydrophilic). Biotransformation is the process when compounds are transformed into other materials by the various metabolic systems that reduce or alter the toxicity of materials introduced to the body. Finally, biodegradation is the process that breaks down a xenobiotic into a simpler form.

Ultimately, the biodegradation of organics results in the release of CO_2 and H_2O to the environment.

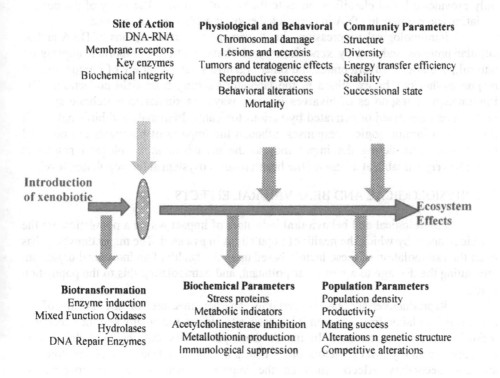

Site of Action	Physiological and Behavioral	Community Parameters
DNA-RNA	Chromosomal damage	Structure
Membrane receptors	Lesions and necrosis	Diversity
Key enzymes	Tumors and teratogenic effects	Energy transfer efficiency
Biochemical integrity	Reproductive success	Stability
	Behavioral alterations	Successional state
	Mortality	

Introduction of xenobiotic → Ecosystem Effects

Biotransformation	Biochemical Parameters	Population Parameters
Enzyme induction	Stress proteins	Population density
Mixed Function Oxidases	Metabolic indicators	Productivity
Hydrolases	Acetylcholinesterase inhibition	Mating success
DNA Repair Enzymes	Metallothionin production	Alterations n genetic structure
	Immunological suppression	Competitive alterations

Figure 1. Parameters and indications of the interaction of a xenobiotic with the ecosystem. The parameters given are only an example of what needs to be comprehended for the explanation of the effects of xenobiotic upon an ecosystem. However, biological systems are organized within a hierarchy and that is how ecotoxicology must face environmental problems.

2.3. RECEPTOR AND THE MODE OF ACTION

The site at which the xenobiotic interacts with the organism at the molecular level is particularly important. This receptor molecule or site of action may be the nucleic acids, specific proteins within nerve synapses or even present within the cellular membrane. It is also possible for it to be very non-specific.

2.4. BIOCHEMICAL AND MOLECULAR EFFECTS

DNA adducts and stand breakages are indicators of genotoxic materials, compounds that affect or alter the transmission of genetic material. One advantage is that the active site can be examined for a variety of organisms. The methodologies are proven and can be used virtually regardless of species. However, damage to the DNA only provides a broad classification as to the type of toxicant. The study of the normal variation and damage to DNA in unpolluted environments was just initiated.

Immunological suppression, due to damage to certain regions of DNA and to cellular proteins, induced by xenobiotics could have subtle but important impacts on natural populations. Invertebrates and other organisms have a variety of immunological responses that can be examined in the laboratory setting from field collections. The immunological responses of bivalves in some ways are similar to vertebrate systems and can be suppressed or activated by various toxicants. Mammals and birds have well documented immunological responses although the impacts of pollutants are not well understood. Considering the importance to the organism, immunological responses could be very valuable at assessing the health of an ecosystem at the population level.

2.5. PHYSIOLOGICAL AND BEHAVIOURAL EFFECTS

Physiological and behavioural indicators of impact within a population are the classical means by which the health of populations in assessed. The major drawback has been the extrapolation of these factors based upon the health of an individual organism, attributing the damage to a particular pollutant, and extrapolating this to the population level.

Reproductive success is certainly another measure of the health of an organism. In a laboratory situation it is possible to measure fecundity and the success of offspring in their maturation. In nature these parameters may be very difficult to measure accurately. Many factors other than pollution can lead to poor reproductive success. Secondary effects, such as the impact of habitat loss on zooplankton populations will be seen in the depression or elimination of the young age classes.

Mortality is certainly easy to assay on the individual organism. Macroinvertebrates, such as bivalves and cnideria, can be examined and since they are relatively sessile, the mortality can be attributed to a factor in the immediate environment. Fish, being mobile, can die due to exposure kilometres away or because of multiple intoxications during their migrations. By the time the fish are dying, the other levels of the ecosystem are in a sad state.

2.6. POPULATION PARAMETERS

Population numbers or density have been widely used for plant, animal, and microbial populations in spite of the problems in mark recapture and other sampling strategies. Since younger life stages are considered to be more sensitive to a variety of pollutants, shifts in age structure to an older population may indicate stress. In addition, cycles in age structure and population size occur due to the inherent properties of the age structure of the population and predator-pray interactions.

The determination of alterations in genetic structure, i.e. the frequency of certain marker alleles, has become increasingly popular. The technique of gel electrophoresis has made this procedure seem quite simple and easy. Population geneticists have long used this method to observe alterations in gene frequencies in populations of bacteria, protozoa, plants, various vertebrates and the famous Drosophila. The largest drawback to this method is ascribing differential sensitivities to the genotypes in question. Usually, a marker is used that demonstrates heterogeneity within a particular species. Toxicity tests can be performed to provide relative sensitivities. However, the genes that have been examined up to now are not genes controlling the xenobiotic metabolism. These genes have some other physiological function and act as a marker for the remainder of the genes within a particular linkage group.

Pollution can be indicated by alterations in the competitive abilities of organisms. Obviously, bacteria that can use a xenobiotic as a carbon or other nutrient source and bacteria that can detoxify a material have a competitive advantage, with all the remaining factors being equal. Xenobiotics may also enhance species diversity if a particularly competitive species is more sensitive to a particular toxicant. These effects may lead to an increase in plant or algal diversity after the application of a toxicant.

2.7. COMMUNITY EFFECTS

The structure of biological communities has always been a vastly used indicator of stress in a biological community. Early studies on cultural eutrophication emphasised the impacts of pollution as they altered the species composition and energy flow of aquatic ecosystems. Various biological indices have been developed to judge the health of ecosystems by measuring aspects of the invertebrate, fish, or plant populations. Perhaps the largest drawback is the effort necessary to determine the structure of ecosystems and to understand pollution-induced effects from normal successional changes. One of the most widely used indexes of community structure has been species diversity. Many measures for diversity are used, from such elementary forms as species number to measures based on information theory. A decrease in species diversity is usually taken as an indication of stress or impact upon a particular ecosystem. Diversity indexes, however, hide the dynamic nature of the system and the effects of island biogeography and seasonal state. Related to diversity is the notion of static and dynamic stability in ecosystems. Traditional dogma stated that diverse ecosystems were more stable and therefore healthier than less rich ecosystems. May's work in the early 1970s did much to question these almost unquestionable assumptions about properties of ecosystems [5]. We certainly do not doubt the importance of biological diversity, but diversity itself may indicate the longevity and size of the habitat rather than the inherent properties of the ecosystem. Rarely are basic principles such as island biogeography incorporated into comparisons of species diversity when assessments of community health are made. Diversity should be examined closely as to its worth in determining xenobiotic impacts upon biological communities.

Currently it is difficult to pick a parameter that describes the health of a biological community and have that form a basis of prediction. A single variable or magic number may not even be possible. In addition, what are often termed biological communities is based upon human constructs. The members of the marine benthic invertebrate community interact with many other types of organisms, micro-organisms, vertebrates, and protists that in many ways determine the diversity and persistence of an organism. Communities can also be defined as functional groups, such as the intertidal community or alpine forest community that may more accurately describe functional groupings of organisms.

2.8. ECOSYSTEM EFFECTS

Alterations in the species composition and metabolism of an ecosystem are the most important impacts that can be observed. Acid precipitation has been documented to cause dramatic alterations in both aquatic and terrestrial ecosystems. Introduction of nutrients certainly increases the rate of eutrophication. Effects can occur that alter the landscape pattern of the ecosystem. Changes in global temperatures have had dramatic effects upon species distributions. Combinations of nutrient inputs, utilisation, and toxicants can significantly alter aquatic ecosystems [3].

3. Ecotoxicology and modelling

The application of ecotoxicological models in environmental management of emissions of toxic substances quickly became a powerful tool. Two equally applicable approaches were developed from the early 1970s: a more ecological approach and a more chemical approach. The ecological approach was mainly inspired by the general application of ecological models in environmental management. It focused on the processes in ecosystems influenced by toxic substances and the distribution of the toxic substances in the ecosystem as a result of these processes. Jorgensen gives several illustrative examples of this approach [6]. The chemical approach focuses on the distribution of the chemical of interest in the environment in accordance with the properties of the chemical.

The two approaches complements each other. The chemical approach allows us to compare easily the fate and thereby the effect of two or more chemicals emitted into the environment. This type of model does not yield a very accurate estimate of the concentrations and the related effects of the focal chemical(s). However, it does indicate which part of the environment is threatened and which chemicals we can or cannot allow to be emitted into the environment and at which concentrations. These models are, for instance, extremely useful in our effort to substitute environmentally unacceptable chemicals with other more acceptable ones. The ecological approach, in contrast to the chemically oriented models, yields more accurate values of chemical concentrations in the environment, which is of importance in the management decisions associated with the individual ecosystems and with risk assessments in the use of chemicals in certain contexts. These models may be used to determine with reasonable accuracy the amounts of emissions which are permissible from individual sources, but they cannot judge which chemical among a range of possibilities is most safe to use, in contrast to chemical models.

Today, a wide range of models has been developed. They comprise the experience of many case studies where models have been an integrated part of the environmental management decision on the use and emission of toxic substances [7]. The range of available ecotoxicological models is very wide with respect to chemical compounds, environmental problems, and the ecological components and processes involved. It is likely that more sophisticated models will be developed in very near future to the benefit of environmental management. Furthermore the two types of ecotoxicological models tend to approach each other in the sense that the chemical models include more and more ecological compartments and processes and the ecological models place more emphasis on the general chemical properties of the compound of interest. Another development that took place, particularly in the 1980s, was the translation of the model results in a management context. The uncertainty of the model was to a certain extent reflected in the quantification and interpretation of the model results.

4. Estimation of ecotoxicological properties

Ecotoxicology, as emphasised above, deals with very complex systems and therefore it is not possible to know all the details of all the ecological components and processes involved in a focal ecotoxicological problem. That is the reason, we have to use models to provide us with the overview needed to consider at least the most important components and processes. This requires, however, that the properties of the ecological components along with the processes and their interactions with the toxic substance of interest are known. This implies that we have to know the pertinent properties of the huge amount of compounds that we are using in our modern society today and the properties related to the interactions of these substances with the wide range of ecological components. The result is that we cannot assess the full environmental consequences for more than a handful of chemicals today.

Research on estimation methods has therefore been intensified during the last decade. Due to this research it is now possible to make reasonable estimations for a wide range of pertinent properties of the chemicals for which measurements are not yet available. There are, as for the models, two methods: a chemical method and an ecological method.

The first method builds on the principle that chemicals of similar structure and with related formulae have similar properties. These are the so-called QSAR and SAR methods. Details of these methods can be found, for instance, in [8] and [9]. These methods are under constant development, and the introduction of the concept of molecular connectivity has contributed during the last few years to the significant reduction in uncertainty of parameters estimated by these methods.

The ecological method is based on allometric principles, i.e. that the interaction between an organism and its environment is related to the surface area of the organism, which makes it possible to relate the size of the organism to such important parameters as uptake and excretion rates, biological concentration factor, and ecological magnification factor. It implies that when these rates are known for one organism, they may be estimated for other organisms provided that the sizes of the organisms are known [10].

5. Ecosystem considerations

Ecotoxicology focuses on the effects of toxic substances not only at the organism and population level, but also increasingly at the ecosystem level. During the last decade, generally there has been an increasing effort to understand ecosystems at the system level [11, 12, 13]. Through the research in this field during the last years it has been possible to reach to an understanding of the hierarchical organisation of ecosystems, the importance of the network that binds the ecosystem components together, and the cycling of mass, energy and information.

Biomarkers have been developed to improve the estimation of exposure – including sublethal exposure – of populations of critical species in ecosystems [14]. They provide increased accuracy in the estimation of impacts from chronic exposures to defined toxicants in an environment. Whatever their usefulness might be, these methods cannot be used to assess the impacts of toxic substances and even chemical mixtures at ecosystem level. It must not, in this context, be forgotten that the properties of an ecosystem cannot be equated to the sum of the properties of its individual components. First, many detrimental effects, e.g. impairment of reproductive performance and reduction of growth potential, may occur at concentrations well below those causing lethality. Second, even if perfectly understood, the toxicity of a chemical for a specific population is of little value in characterising the toxicity that may be manifested in many ecosystems. Therefore, the current approach must be replaced with examinations of the toxicity of toxicants throughout several ecosystems, which will require a strong emphasis on basic ecological research.

Risk assessments of widely used toxicants are often based on more or less complex models. It is necessary to expand these risk assessments to encompass the risk of: a) reductions in population size and density, b) reduction in diversity and species richness, c) effects on frequency distribution of species, and d) effects on the ecological structure of the ecosystem, particularly on a long-term basis [15].

This expansion of the risk assessment concept to a much wider ecosystem level has not yet provoked much research. New approaches, new concepts, and creative ideas are probably needed before a breakthrough in this direction will occur. The concepts of ecosystem health and ecosystem integrity are probably the best tools developed up to now.

6. References

1. Wolfram, S. 1984. Cellular Automata as Models of Complexity, Nature, 311, pp. 419-424.
2. Wolfram, S. 1984. Computer Software in Science and Mathematics, Sci. Am., 251, pp. 140-151.
3. Landis, G. W. and M.-H. Yu. 1995. Introduction to Environmental Toxicology. Impacts of Chemicals Upon Ecological Systems. Lewis Publishers, pp. 1-328.
4. Suter, G. W. and L. W. Barnthouse. 1993. Assessment Concepts. In Ecological Risk Assessment. G.W. Suter, II. Ed., Lewis Publishers, Boca Raton, pp. 21-47.
5. May, R. M. 1973. Stability and Complexity in Model Ecosystems. Second Edition, Princeton University Press, Princeton, New Jersey.
6. Jorgensen, S. E. 1983. Modelling the Distribution and Effect of Toxic Substances in Aquatic Ecosystems, in Application of Ecological Modelling in Environmental Management, Part A (S. E. Jorgensen, ed.), Elsevier, Amsterdam, The Netherlands.
7. Jorgensen, S. E., B. Halling-Sorensen and S. N. Nielsen, Eds. 1995. Handbook of Environmental and Ecological Modelling, CRC Lewis Publishers, Boca Raton, USA.
8. Jorgensen, S. E. 1990. Modelling in Ecotoxicology, Elsevier, Amsterdam, The Netherlands.
9. Lyman, W. J. and D. H. Rosenblat. 1990. Handbook of Chemical Property Estimation Methods: Environmental Behavior of Organic Compounds, American Chemical Society, Washington, DC, USA.
10. Jorgensen, S. E. 1994. Fundamentals of Ecological Modelling, Developments in Environmental Modelling, 19, 2nd Edition, Elsevier, Amsterdam, The Netherlands.

11. Jorgensen, S. E. 1992. An Introduction of Ecosystem Theories: A Pattern, Kluwer Academic Publishing, Dordrecht, The Netherlands.
12. Jorgensen, S. E. 1997. An Introduction of Ecosystem Theories. A Pattern, 2nd Edition, Kluwer Academic Publishing, Dordrecht, The Netherlands.
13. Hall, C. A. S., ed. 1995. Maximum Power: The Ideas and Applications of H. T. Odum, University Press of Colorado, USA.
14. Peakall, D. B. and J. R. Bart. 1983. Impacts of Aerial Applications of Insecticides on Forest Birds, CRC Crit. Rev. in Environ. Control, 13, pp. 117-165.
15. Jorgensen, S. E. 1998. Ecotoxicological Research-Historical Development and Perspectives. In: Ecotoxicology-Ecological Fundamentals, Chemical Exposure and Biological Effects. Ed. G. Schuurmann and B. Markert. Wiley-Spectrum. New York-Heidelberg.

ECOSYSTEM GIS-MODELLING IN ECOTOXICOLOGY

V. DAVYDCHUK
Institute of Geography, NASU, Volodymyrska St.,44, Kyiv, 01034, UKRAINE
chornob@geogr.freenet.kiev.ua

1. Introduction

Out-door ecotoxicological evaluations of the well-being/vulnerability, stability, capacity and other parameters of the ecosystems under co-influence of the technogenic pollutants of both radioactive and chemical character, via estimation of their biomass stock and specie number [1] or by any other characteristics and parameters, related to the ecosystems, requires at least integral assessments of the study area in terms of the diversity and structure of ecosystems. Identification, structuration and localization of the ecosystems examined in the real environment by their components, elements and relations are very useful for the ecotoxicological studies by the reasons:
- experimental - formal passportization the conditions of the field data collection;
- methodological – data analysis;
- and practical - further extrapolations and applications.

To characterize the ecosystem spatial and component structure and pollution, a number of the relative cartographic information layers is to be created and used. The Geographic Information System (GIS) is an effective tool for creation of such a layers for cartographic modelling of the whole ecosystem, it's elements, and pollution field reconstruction, as well as for the further manipulation with the relevant cartographic information layers in solution of the ecotoxicological problems.

2. Structure of the ecosystems and ecotoxicology

At least at the level of the experimental data collection, extrapolation and application, ecotoxicological studies operate with a diversity of the ecosystems in the real environment. To consider this, the GIS create at the base of topographic map and satellite imagery. A complex landscape approach is proposed as a structural base for the description, identification, classification and parameterisation of the ecosystems and further characterisation of their structural elements, components and relations between them.

Landscape as a natural terrestrial system consists of the biomatic and geomatic natural components, as well as relief, surface deposits, air, water, soil, plant and animal associations. The landscape structure of the territory causes a local grooving conditions and proportions between washing-off and infiltration, and therefore determines configuration and intensity of a geochemical fluxes/barriers and influences a number of parameters, which are sufficient in the ecotoxicological context.

The legend of the landscape map is based on the principle of synthesis. For the best field identification of the landscape units (ecosystems), the legend includes a number of the co-ordinated physiognomic characteristics of the natural components and

F. Brechignac and G. Desmet (eds.), Equidosimetry, 111–118.

elements. An example of the landscape map for the Polissky district of Kyiv province (Ukraine) and a fragment of its legend is give on the Fig. 1.

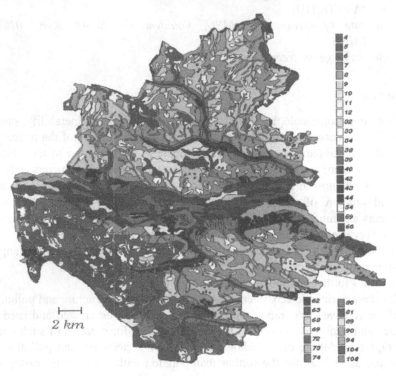

Fig. 1. Polissky district. Landscape map.

Legend (fragment):

Fluvioglacial, moraine-fluvioglacial and limnoglacial plains
2. Fluvioglacial ripply plains, composed by fluvioglacial dusty sand (thickness >2 m), with thin (1,5-5 cm) interlayers of loamy sand on the depth 0,8-1,5 m, with sod-podzolic dusty-

sandy soil, covered by oak-pine grassy forests and long-fellows
4. High moraine-fluvioglacial ripply plains, composed by fluvioglacial dusty sand, with thin (1,5-5 cm) interlayers of loamy sand on the depth 0,8-1,5 m, bedded by moraine boulder loam on the depth 2,5-3,5 m, with sod-

podzolic dusty-sandy soil, covered by oak-hornbeam-pine forests, cultivated pine forests and long-fellows

7. Extreme-moraine ridges, composed by fluvioglacial dusty sand, bedded by moraine boulder sandy loam on the depth 0,4-1,0 m, with sod-podzolic dusty-sandy soil, covered by pine-hornbeam-oak forests and long-fellows

9. Lowered moraine-fluvioglacial plains, composed by fluvioglacial sand (thickness 0,5-1,5 m), bedded by moraine boulder loam and derivates of neogenic bunter clays, with sod-podzolic sandy soil, covered by humid birch-pine bilberry-mossy forests and long-fellows

10. High moraine-fluvioglacial ripply plains, composed by fluvioglacial sandy loam, bedded by fluvioglacial sand (thickness 0,5-1,5 m) and moraine boulder loam, with sod-podzolic sandy-loamy soil, covered by hornbeam-oak forests and long-fellow grass associations

12. Flat limno-glacial lowlands, composed by fluvioglacial dusty sand, bedded by limno-glacial dusty loam on the depth 0,3-0,7 m, with sod-podzolic gley dusty-sandy soil (pH 4,2-4,5, humus 2,5-2,8%), covered by humid ash-oak forests

33. Watershed catchments at the tops of the dells, with flat bottoms, composed by deluvial dusty sand, with sod-podzolic gley dusty-sandy soil, covered by humid pine-birch mossy forests

34. Watershed catchments at the tops of the dells, with flat bottoms, composed by deluvial sandy loam, with sod-podzolic semi-gley sandy-loamy soil, covered by semi-humid ash-oak forests

River terraces
38. Smoothed terraces, composed by alluvial sand, with sod-podzolic sandy soil, covered by pine mossy-lichenous forests

54. Rear lowered flat parts of the terraces, composed by eutrophic peat (thickness 0,5-2,5 m), with peat bog soil (organic matter >75%), covered with alder forests and sedge-reedous bog coenoses

Flood plains
60. Low flat flood plains, composed by eutrophic peat (0,5-2,5 m), with alluvial peat bog soil (organic matter 60-70%), covered by alder forests and sedge-reedous bog coenoses

62. High flat flood plains, composed by alluvial dusty sand, with alluvial dusty sandy soil, covered by forb-grassy meadows

69. High segment-crest flood plains, composed by alluvial layered sand and loam, with alluvial dusty-sandy soil (pH 4,5-5,0, humus 1-2%), covered by forb-grassy meadows on the crests (60%) and with alluvial gley loamy soil, covered by forb-sedge-hygrophytous meadows in the depressions (40%)

Erosion network
80. Balkas (ravines) with steep slopes (>12°) and flat bottoms composed by sandy loam, with sod gley dusty-sandy soil, covered by humid ash-oak forests

81. Balkas (ravines) with steep slopes (up to 12°) and flat bottoms composed by eutrophic peat (0,3-0,5 m), with peat bog soil, covered by humid ash-alder forests and sedge-hygrophytous meadows

90. Dells with concave bottoms, composed by sandy loam, with sod semi-gleic dusty-sandy soil (pH 5,0-5,5, humus 1,5-2,0%), covered by humid ash-oak forests

94. Dells with flat bottoms, composed by eutrophic peat (thickness 0,3-0,5 m), with peat soil, covered by alder forests and sedge-reedous bog coenoses

114

Depressions

104. Depressions with 2-4° slopes and flat bottoms, composed by eutrophic peat (thickness 0,3-0,5 m), with peat-gleic soil (organic matter 60-80%), covered by alder forests, willowish shrubs and sedge-reedous bog coenoses

Dunes

108. Dunes with steep slopes (up to 20°), composed by eolic sand, with sod-podzolic sandy soil, covered with pine lichenous forests and xerophytous grass coenoses

At the base of legend of the landscape map, the ecosystem can be classified directly at least by the following signs, which are of the ecotoxicological importance:
- genesis;
- form of relief and surface inclination;
- lithology of the surface deposits;
- soil type;
-grooving conditions;
- plant associations and land use structure.

By reclassification of the synthetic legend, a number of maps of the appropriate structural elements and components of the landscape complexes can be generated, as a separate information layers. Among them map of the soil types (Figure 2) and many others.

2 km

Fig. 2. Polissky district. The soil types.

Reclassification can be extended in details on the ecotoxicological significant parameters and attributes of the components, which are derived from the field sampling data, materials of the field investigations or additional thematic cartographic materials. Adopting this additional information into the frame legend of the landscape map, we can extend the classification list for each component or element of ecosystems with a corresponding parameters or characteristics, for example:

- soil acidity, chemical composition, humus content and forms
- clay minerals and their percentage by the soil horizons;
- depth of the gleyic horizon and ground water level;
- linear and surface soil erosion;
- zones of the washing-off, transit and accumulation;
- geochemical barriers: sorptional, gleic, mechanical, alcaline;
- dominant plant species, ediphicator species, ground cover/indicators;
- etc.

By this way, the basic landscape map can be used for the best extrapolation the experimental data mentioned above, therefore for the ecotoxicological cartographic parameterisation of the ecosystems.

Furthermore, in frame of this approach, additional thematic maps (information layers) can be adopted. For example, at the woodland area, if the maps of the forest types and the woodstand age groups are available, we can classify the forest ecosystems by the phytomass stock – total (Table 1) and by the separate fractions, after that corresponding maps can be derived.

Table 1. Stock of the phytomass for the typical woodstands of Polissia by the age groups (ton / hectare, dry matter)

Forest types	Age groups			
	Very young	Young	Average	Mature
Pinetum cladinosum	19,0	139,5	181,6	207,5
Pinetum vaccinoso-polytrichosum	-	162,2	257,2	280,5
Querceto-pinetum□herbosum	80,4	249,9	341,3	361,9
Querceto-carpineto-pinetum herbosum	35,0	191,5	308,8	398,1
Carpineto-quercetum carioso-aegopodiosum	-	141,5	285,3	387,1
Fraxineto-carpineto-quercetum spiraeto-geumosum	-	215,7	415,7	541,3
Populeto-betuletum vaccinoso-polytrichosum	-	115,1	155,5	203,7
Alnetum hygrophyto-herbosum	-	107,8	279,2	280,7

Manipulating with the forest types and the woodstand age groups information layers, an optimal regional specie nomenclature, corresponding to the certain growing conditions, can be defined for the forests, meadows and bog coenoses, which is of importance for evaluation the ecosystem well-being.

3. Landscape-based GIS

To manipulate effectively with a number of the cartographic information layers, which are to be arranged around the landscape map, some elements of the GIS technologies can be used. Following an approach mentioned above, a landscape map and its legend are proposed as a structural base of the radioecological/ecotoxicological GIS applications.

The landscape based radioecological/ecotoxicological GIS provides the possibility for more extensive modelling structure/parameterisation, contamination and well-being of the ecosystems. It allows use the procedures both reclassification (creation maps of the separate elements and components from complex maps) and overlay of the different information layers. As an example, concerning to the task of the cartographic modelling of the forest phytomass contamination by Cs-137, which is an important step to evaluating the well-being of ecosystems of the experimental site mentioned below, this procedure includes following steps, which are presented at Fig. 3.

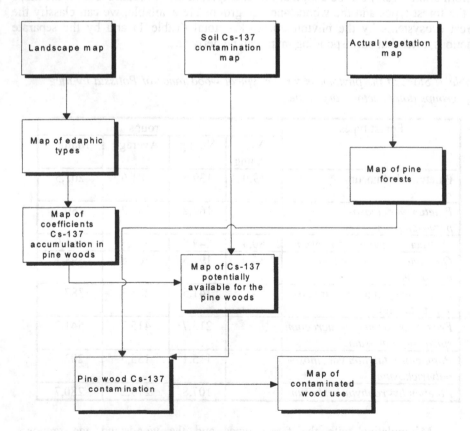

Fig. 3. Cartographic modelling of the phytomass contamination.

4. Experimental site description

Polissky district, which is proposed as an experimental site, situated in the Western periphery of the Chornobyl zone.

Climatic conditions. Climate of Polesie is temperate semi-continental under influence of cyclonic activity. It is characterised by moderately cold winter and moderately warm summer. Average temperature of January, which is a coldest month, -5-6°C. Average temperature of February is higher, but this is a month of the temperature extremes (-22-25°C, even -33-35°C). July is characterised with both maximal average temperature (17-19°C), and highest temperature extremes (35-36°C).

The season of vegetation begins in the mid April and continues to the late October. Its duration is about 220 days. Sum of the active temperatures (above 10°C) is about 2660°.

Average annual precipitation is about 580 mm, with variations between 400-850 mm. Up to 70% of the annual precipitation is related to the warm period of year.

Landscapes and ecosystems. The low and plain relief (110-145 m above sea level) of the fluvioglacial sandy plain was formed at the North-Eastern slope of the Ukrainian Crystalline Shield within the area of the Dnieper stage of an ancient quaternary glaciation, which corresponds to the Zaal stage of the Western Europe.

Among the most typical landscapes of the region there are: river flood plains and terraces, extreme-moraine ridges, moraine-fluvioglacial and limnoglacial plains. This part of Ukraine is notable for the considerable swampiness. The bogs account for 22,5% of the district.

Its northern part presents landscape of low, ripple moraine-fluvioglacial plain, which lies 15-35 m lower, then extreme-moraine ridge. This level of the relief is composed by fluvioglacial sand, bedded by morainic loam at depth 0,8-1,0 m. The concentration of the dust particles less 0,05 mm is very low (4-8%). This explains poorness of soddy-podzolic sandy soil formed here (pH 4,2-4,5, humus 0,8-1,5%). The landscape of low moraine-fluvioglacial plain is covered by pine forests.

A number of isolated depressions of 0,3-0,8 km diameter scattered on the surface of the landscape. Swamps with alder forests or sedge-reedous bog coenoses occupy most of them.

Landscapes of the river terraces are very typical for Chernobyl zone. Flat surfaces of terraces are composed of alluvial sands with deficiency of the clay and dust particles (4-6%). Thus sod-podzolic sandy soil formed here is dry and poor (0,4-0,6% of humus). The terraces are cowered mainly by different-age pine forests of artificial origin. Back, lowered parts of the river terraces, which are situated close to terrace joint, are occupied by eutrophic bogs with peat bog soil (thickness of the peat is about 0,5-2,5 m) and forests of *Alnus glutinosa (L.) Gaertn.* or by sedge-reedous bog coenoses.

Mostly typical forest ecosystems of the region are *Pinetum cladinosum* and *Pinetum phodococco-dicranosum*, with the soil cover of *Koeleria glauca (Sorend.) DC, Festuca sulcata (Hack.) Num.p.p., Antennaria dioica (L.) Gaertn., Phodococcum vitis-idaea (L.) Avror., Dicranum scoparium Hedv., Cladonia silvatica (L.) Hoffm.* and *C. rangiferina (L.) Webb.* Under influence of the regular forest farming, the wood stands of the territory belong mainly to the young and average age groups.

The landscapes of flood plains of Uzh river and its tributaries represent the lowest level of relief of the test site. Flat and segment-crest surfaces of the frontal and middle parts of the flood plains with a number of the meander lakes and bypasses are composed by layered alluvial sands and loams and covered by the shrubs of *Salix acutifolia Willd.* with grassy- xerophytous soil cover, forb-grassy and sedge-hygrophytous meadows on alluvial soils. Flat and segment-crest surface of the frontal and middle parts of the flood plains with a number of the meander lakes and bypasses are composed by layered alluvial sands and loams and covered by the xerophytous shrubs, forb-grassy and sedge-hygrophytous meadows on the alluvial soils.

Polissia is notable for the considerable swampiness of the river flood plains and terraces. In the accident zone bogs account for 22,5% of the territory. Mostly typical swamps occupy rear lowered flat parts of the river terraces and flood plains, composed by a eutrophic peat (thickness 0,5-2,5 m), with a peat bog soil (pH 5,0-5,5, organic matter up to 75%), covered by alder forests and sedge-reedous bog coenoses. Two thirds of the bogs and wetlands were drained and ameliorated during last decades before the accident, to increase acreage of the arable areas. After the accident the drainage systems are degrade, and natural humidity of the drained wetlands is restore.

5. Conclusion

The landscape based GIS is an effective tool for the structuration and classification of the ecosystems, for the field experimental data collection, analysis and extrapolation, as well as the ecotoxicological parameterisation of the ecosystems and solution the regional and local ecotoxicological problems.

6. References

1.   Kutlakhmedov Y., Polikarpov G., Kutlakhmedova-Vyshnyakova V.Yu. Radiocapacity of Different Types of Natural Ecosystems and their Ecological Standardization Principles // J. Radioecol. – 1997. – V.6 (2). – p. 15-21.

MAPPING TRANSFER PARAMETERS OF RADIONUCLIDES IN TERRESTRIAL ENVIRONMENTS

S. DENYS, G. ECHEVARRIA, J.L. MOREL

Laboratoire Sols et Environnement, UMR 1120 ENSAIA-INPL/INRA, 2 av. de la Forêt de Haye, BP 172, F- 54 505 VANDOEUVRE les NANCY, FRANCE

E. LECLERC-CESSAC

Agence nationale pour la gestion des déchets radioactifs (Andra), 1-7 rue Jean MONNET, - 92 298 CHATENAY MALABRY cedex, FRANCE

Abstract

Safety assessment models for potential sites selected for underground repositories of high level and long-lived radioactive waste requires the prediction of phytoavailability of such radionuclides at the regional scale. In this context, the areas in which the radionuclides may be highly mobile or accumulate have to be well known, as they will contribute to a maximal dose for Man. The parameters controlling the phytoavailability of the radionuclides are mostly defined by experiments based on sieved-soil sample and a methodology is needed to extrapolate these parameters to the regional scale by taking into account the variability of the soil properties within the landscape. A mapping of three radionuclides phytoavailability (^{63}Ni, ^{99}Tc and ^{238}U) was conducted here using the MapInfo mapping software. The distribution frequency of the phytoavailability parameters was represented over the 186 km^2 area of the French laboratory for the study of deep underground nuclear waste disposal, in Bure. Isotopically exchangeable pool of Ni, E_t and pH of the soil were the two phytoavailability parameters chosen respectively for ^{63}Ni and for ^{238}U and were both measured on sieved samples coming from the soil units defined at 1 : 50 000. The redox potential, E_h was the parameter used for ^{99}Tc and was measured in the field. For each radionuclide, class of soils were built according to their properties in term of radionuclides phytoavailability and maps of phytoavailability were drawn at 1: 25 000.

Results allowed a prediction of the phytoavailability of the ^{63}Ni, ^{99}Tc and ^{238}U at a regional scale, based on the superimposition of the laboratory measurement of the parameters significantly controlling the mobility and the phytoavailability of the three radionuclides and knowledge of soils over the area. Critical areas were also determined where either accumulation in soils may be highest or accumulation in plants may be highest. They also allowed defining the most likely transfer parameters for the three radionuclides in the area of Bure.

F. Brechignac and G. Desmet (eds.), Equidosimetry, 119–129.

1. Introduction

The transfer of radionuclides in the biosphere is a major issue in the calculation of the final dose to man in safety assessments of underground nuclear waste disposals. These safety assessments rely on a method which often selects critical groups, which are the populations located in places where the dose is maximal. This critical group is defined by a specific scenario (e.g. self-sustainable populations) and is situated in those areas in which radionuclides may accumulate or be highly mobile. The soil is the medium that supports terrestrial ecosystems and exerts a significant control on both the transfer of radionuclides to the food chain and the radiation exposure to living organisms. Soil properties may vary considerably across the landscape according to topography, geology, climate and history of land use. Therefore, the variability of soil characteristics at the local scale may subsequently introduce a large variability in transfer of radionuclides in the food chain and the ecosystems. Mobility and transfer of radionuclides in the soil-plant system are specific to soil type and land use (crop, farming practices,...). The distribution of radionuclides transfer parameters in soils is required to determine both the extreme values and their frequency in specific sites. The Bure site (French laboratory for the study of deep underground nuclear waste disposal) displays a wide variety of soil types and land uses. The soil and landscape approach is therefore essential to select best estimates of transfer parameters values of radionuclides.

Among the mechanisms governing plant uptake of radionuclides, the ability of the soils solid phase to supply the soil solution is one of the main processes involved. This ability can be estimated through the K_d, i.e. the soil: solution distribution coefficient of the radionuclide in the soil. Phytoavailability of radionuclides is primarily based on this property and is commonly estimated at the sieved-soil sample level by measuring the uptake of the radionuclide by plants after a homogeneous contamination of the soil sample. Transfer Factor, TF (i.e. soil-to-plant concentration ratio of the radionuclide) or Effective Uptake, EU (i.e. fraction of the radionuclide transferred to the plant) are two parameters used to assess the soil-to-plant transfer of radionuclides and thus their phytoavailability (1, 2, 3, 4). However, those measurements do not take into account the high heterogeneity of soil profiles and their actual functioning including hydrological properties and vertical migration processes. Therefore, a methodology is needed to extrapolate the prediction of the phytoavailability of the radionuclide obtained at the soil sample level to a regional scale, in agreement with the site specificity of soil physical and chemical properties.

The objective of this work was to assess the spatial distribution of transfer properties of radionuclides in soils at the Bure site following a three-step approach. The first step was to measure the mobility of the three radionuclides in soil samples taken from the site and from other situations and their transfer to plants in pot experiments. This allowed for the identification of soil characteristics that controlled these parameters. The second step was to verify that the mobility and the leaching of radionuclides did not strongly influence their transfer to plants in soil cores. Finally, the third step was to extend transfer values to the regional scale according to the spatial distribution of soils and their properties in the Bure site after having measured the parameters that control radionuclides mobility and transfer in each of the soil units present. Three radionuclides were chosen to illustrate this approach: nickel-63, uranium-238 and technetium-99. They were chosen as model radionuclides in terms of mobility, presence in natural environments, and importance for the biosphere safety assessment of nuclear fuel disposal.

2. Materials and Methods

2.1 SITE SELECTED FOR THE STUDY

The site for this study is a 186 km^2 area next to the underground Andra research laboratory around the Bure municipality in Meuse, and bordering the Meuse and Haute Marne departments (France). It is located on the eastern side of the Sedimentary Basin of Paris and is dominated by a Jurassic limestone plateau and carbonated marl outcrops in the slopes. This area is also affected by iron-rich sandy deposits of the early Cretaceous period.

The pedological map of the area was previously established at the 1: 50 000 scale. The pedological survey allowed the identification of 9 typological soil units (TSU) over the entire area (table 1). The most widespread units (FAO classification) on the plateau were Rendzic Leptosols; alcaric Cambisols and Calcisols. In lowlands, hydromorphic soils were found (Fluvic Gleyic Cambisols). Dystric Cambisol and Luvisol were located on sandy and loamy deposits and were mostly located in forested areas. These soils need to be considered as pH is a significant factor influencing Ni mobility in soils through dissolution of iron and manganese hydroxides (5, 6). pH also influences strongly U mobility (7). Variation of redox potential between soils will affect also ^{99}Tc mobility.

2.2. SOIL SAMPLING

For each soil unit from the 1:50 000 survey, the A_p horizon of a typical soil profile was sampled to measure the phytoavailability of the three radionuclides on a 5 mm sieved soil sample. The variability of soil properties was taken into account by study at a finest scale of a plot of the plateau area ("Glandenoie plot").

2.3. ^{63}Ni PHYTOAVAILABILITY MAPPING

Isotopic exchange is the main mechanisms that will influence the behaviour of ^{63}Ni in the environment (1, 8). Stable Ni is present worldwide at varying concentrations. However its bioavailability depends on total concentration and on soil physico-chemical conditions. There is a negative significant relationship between E.U. of Ni by plants and the labile pool of stable Ni measure by isotopic exchange kinetics (IEK). Measuring the labile pool of Ni will provide information on the probability of transfer of ^{63}Ni to plants and will therefore permit the classification of soils relative to this parameter (1). Moreover, experiments in undisturbed cores on the three major soil units of the Bure area showed that this approach was also verified when the water regime of the soils (evapotranspiration and drainage) on which plants were grown was close to real conditions (9). This confirmed that the approach at the level of the soil sample was fully appropriate to assess the potential of [63]Ni availability in soils.

The phytoavailability of stable Ni in the soil samples was estimated by E_{90d} after performing isotopic exchange kinetics (IEK) on five replicates of each sample. The method was carefully described in detail in previous works (1, 8). The E_{90d} pool represents the reservoir in which [63]Ni is isotopically diluted during an average period corresponding roughly to the time of a growing season (8).

Prior to mapping, classes of [63]Ni phytoavailability were defined from E_{90d} measurement by ANOVA test ($\alpha=5\%$) using the STATBOX Pro software. Definition of classes was based on the homogeneous group of the Neuwman-Keuls test. In addition, a subjective classification was possibly adopted to isolate the E_{90d} values which were found in more than one group. According to the dilution of [63]Ni in the isotopically exchangeable pool of Ni, the higher the E_{90d} the lower the phytoavailability of [63]Ni. The Map-Info 4.0 mapping software was used to give a polygonal representation of each class of [63]Ni phytoavailability superimposed on the pedological maps of either the area either the plots. Each topographic point of the maps was identified by its Lambert II coordinates and a layer was associated to each class of phytoavailability. Each layer corresponded to a polygon, outlining the zones within which the E_{90d} values ranged from the lower to the higher value of the associated class.

2.4. [99]Tc PHYTOAVAILABILITY MAPPING

Literature regarding ^{99}Tc mobility and phytoavailability in the environment points to the determining role of redox potential in soils on the fate of ^{99}Tc in soils (2, 4, 10, 11, 12). Reduction of [99]TcO_4^- seems irreversible (12) and therefore its mobility over the long term might be significantly affected even by occasional reducing conditions in soils. If [99]TcO_4^- is to be expected in soils then K_d values will be comprised between 0 and 0.25 and TF will depend on the dilution level of the radionuclide in soil pore-water and the transpiration capacity of crops. Experiments of ^{99}Tc addition on undisturbed soil cores sampled on the three main soil units of the Bure area, showed that in the case of contamination of soils through irrigation water (i.e. during the growing season), evapotranspiration was high enough to retain the water and practically inhibit downward transfers of mobile [99]TcO_4^- in the soil profile (13). Under such conditions, TF was only influenced by the dilution of the radionuclide in the pool of easily available water in the soil. When comparing TFs to crops (wheat, maize) soils were ranked according to their water holding capacity. Reduction of [99]TcO_4^- in soil reducing sites might have a much higher influence on ^{99}Tc transfer parameters than leaching in such a scenario. Transfer factors in soil cores are in general 10 times lower than in pot experiments as a consequence of this dilution effect (13).

Therefore, the map of mobility and transfer of ^{99}Tc in the area is based on the redox potential map. A portable Platinum electrode (Schott Gerätte PT 737 A) coupled with a millivoltmeter (WTW pH 330/SET) was used to measure in situ E_h values in the month of May 2001. For each of the 9 soil units, three different measurements were realised to determine E_h. Again, mapping of ^{99}Tc potential mobility was realised using the same procedure as for ^{63}Ni. Classes of soils were built on the basis of their redox potential and therefore of their potential to reduce ^{99}TcO$_4^-$.

2.5. ^{238}U MOBILITY MAPPING

The mobility of ^{238}U in soils is strongly regulated by soil pH (7). In the Bure area soil pH varies within the same range (5.5 to 8.5) to these reported by Echevarria et al., 2001. It is therefore possible to calculate K_d values of ^{238}U for each soil unit using the relationship:

$$Log\ K_d = -1.29\ pH + 11.1 \qquad\qquad Eq.\ [1]$$

For each soil unit, soil pH was determined in triplicate on 5 sampling points. pH was determined on a 1:2.5 soil-solution ratio after a 4h equilibration time. K_d of ^{238}U was then mapped in the same way as for ^{63}Ni on the basis of different classes of soils according to their mean K_d value.

3. Results and Discussion

3.1. ^{63}Ni PHYTOAVAILABILITY

The E_{90d}-values varied among the 9 TSU identified (Fig. 1). They ranged between 0.2 mg Ni kg^{-1} DS on the Calcic Cambisols (soil 9) and 22.4 mg Ni kg^{-1} DS on the Colluvial Calcic Cambisols (soil 1). Four groups were defined from the Neuwman-Keul test (α = 5%): group a including only TSU 1, group b including TSU 2 and 3, group c including TSU 3, 4 and 5 and group d including TSU 4, 5, 6, 7, 8 and 9. According to E_{90d} values obtained from each TSU, 4 classes of ^{63}Ni phytoavailability were defined (Table 1).

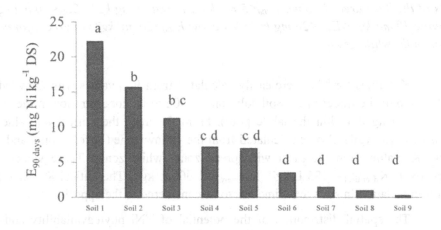

Fig.1 E_{90d} values of 9 soil samples representative of each TSU. Letters identify homogeneous groups according to the ANOVA Neuwman-Keuls test (α = 5%).

The corresponding map of ^{63}Ni phytoavailability was established by differentiating the four classes obtained from the soil map (Figure 2). Most of the Bure area is classified as class II and III, respectively as moderate and high transfer of ^{63}Ni.

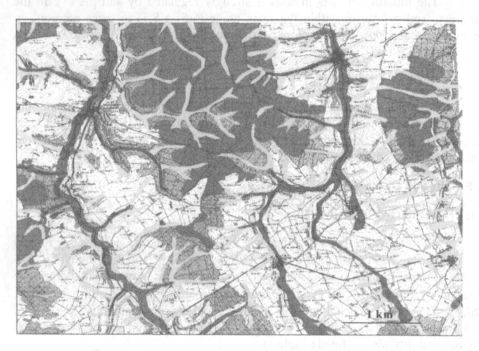

Fig.2 Map of ^{63}Ni phytoavailability potential according to the labile pool of Ni in soils of the Bure area. In blue: $E_{90d}<5$ mg kg^{-1}; in red: 5 mg kg$^{-1}<E_{90d}<10$ mg kg^{-1}; in white 10 mg kg$^{-1}<E_{90d}<20$ mg kg^{-1}; in yellow $E_{90d}>20$ mg kg^{-1}; in green: forested areas in the white group.

K_d values for ^{63}Ni were easily calculated from E_{90d} values with knowledge of the ratio of the tracer in the soil solution t = 90d to its concentration in the solid phase (exchanged within the labile pool). From the map, the critical K_d value in terms of phytoavailability is calculated from the yellow zone (lowest value) and the likely K_d value from the most widespread zone (white zone). These values are respectively $K_{d \text{ critical}} = 185$ L kg^{-1}; $K_{d \text{ likely}} = 1\ 600$ L kg^{-1}. The critical K_d value for ^{63}Ni accumulation in soils over time has to be considered as the opposite.

The spatial distribution of the potential of ^{63}Ni phytoavailability and its corresponding soil K_d allowed localisation of critical groups or ecosystems. In the case of safety assessment in the biosphere of radioactive waste disposal, we consider soil K_d as a parameter responsible for the accumulation of the radionuclide over the long term in soils following application through irrigation water. K_d values allow for the expression of loss through leaching (the other losses being plant uptake and radioactive decay). In this case, the higher the K_d, the greater the radiation and dust inhalation exposures for organisms living on the soil. The critical ecosystem regarding this aspect is the yellow area and therefore consists of the colluvial soils occurring in dry valleys. These soils are very widespread across the Bure area and are sometimes located close to villages. If we consider ^{63}Ni phytoavailability, the lower the E_{90d} value, the higher the accumulation of the radionuclide in plants and therefore the intake by man and other domesticated or wild animals. Critical ecosystems in this case are located in the alluvial soils of the Bure area. This is an important factor as in this karstic limestone plateau, all villages are situated in the valleys where streams flow. Therefore, the entire population of the area dwells in the most critical place in terms of intake exposure. Garden soils are mostly situated in this area, which is of great concern to human populations.

3.2. ^{99}Tc PHYTOAVAILABILITY

As $^{99}TcO_4^-$ inputs are supposed to occur during the growing season, most of the added radionuclide will be retained in the upper layer of soils while being taken up by crops (approximately 50% of the input). Therefore, the reducing conditions at this depth will probably affect the fate of ^{99}Tc more than at any other depth. Consequently, measurement of the redox potential concerns solely the surface horizon of the profile. The E_h values ranged from −70 to 350 mV, with the lowest values being found in grassland soils located in wetland areas close to streams (-70 mV) or situated in alluvial zones (-20 to -4 mV). At the surface of such soil profiles between 0 and 15 cm, iron oxidation spots are visible which are due to gleying. However, the majority of the soils of the area present positive E_h values at the soil surface. In agricultural soils (Rendzic Leptosols, Calcisols, Fluvic and Dystric cambisols), E_h ranged between 260 and 315 mV. Slightly lower values can be found at the surface of soils developed on poor draining marls (250 mV). E_h variation can be found with respect to topography. Hence, in the Glandenoie plot, E_h values on luvisols (Luvisols) are as high as 300 mV on the upper part of slopes and reach +68 mV further down the slope, where water stagnates. The highest E_h value was found in Dystric Cambisols from forested areas.

These results allowed us to draw two classes of soils according to their reducing properties in surface horizons (Figure 3), namely:

Class I: $220 < E_h < 350$ mV

Class II: $-70 < E_h < 220$ mV

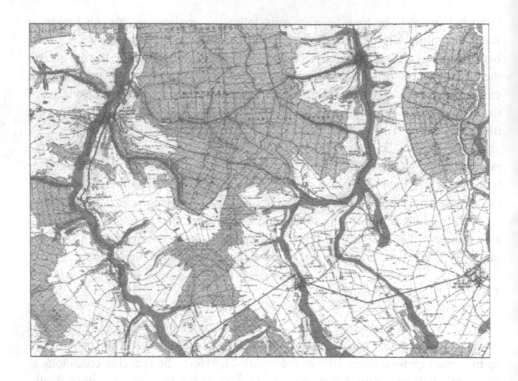

Fig 3. Map of E_h values in soils of the Bure area (measured in May 2001).
In white: 220 mV<E_h<450 mV; in blue: -70 mV<E_h<220 mV; in green: forested
areas in the white group.

To each class, a level of soil plant transfer of ^{99}Tc may be associated a level of soil plant transfer of ^{99}Tc. The highest transfer is obtained for the highest redox potential. If a decrease of the transfer of ^{99}Tc occurs, it is most likely to affect the soils from class II. The critical area in terms of ^{99}Tc accumulation in soils over the long term corresponds to the soils of class II. In this area long-term K_d values of ^{99}Tc may differ significantly from 0 (typical values in rendzinas) and reach 1.5 due to temporarily reducing conditions (2). In terms of plant transfer, the critical area in which transfer factors should reach the order of 10^1-10^2 (depending on the dilution rate in soil water and evapotranspiration) covers approximately 95 % of the territory, *i.e.* class I.

3.3. ^{238}U MOBILITY

Calculation of ^{238}U K_d values according to Eq. [1] showed large variations over the range of soils in the Bure area. They varied from 2.6 to 1297 L kg^{-1}. Highest values were found in acid soils (Dystric Cambisols and Luvisols), both cultivated and forested. Lowest values were found in the Rendzic Leptosols and the Calcaric Cambisols on marls. This leads to a quite complex pattern of K_d

values in such a small area where there may be a vast variability within 500 m. In the case of U, mapping of K_d values is therefore essentially needed to locate surface of critical areas. In this respect, four classes of soils were established according to their mean K_d values and they were reported on figure 4. Classes were built on the basis of orders of magnitude between soils and not from statistical tests as for [63]Ni. However, results from statistical differentiation did not differ much, but they were less discriminating for the map:

Class I:	$K_d < 10 \, l \, kg^{-1}$	pH>8
Class II:	$10 \, l \, kg^{-1} < K_d < 100 \, l \, kg^{-1}$	7.1>pH>8
Class III:	$100 \, l \, kg^{-1} < K_d < 1000 \, l \, kg^{-1}$	6.3>pH>7.1
Class IV:	$K_d > 1000 \, l \, kg^{-1}$	pH<6.3

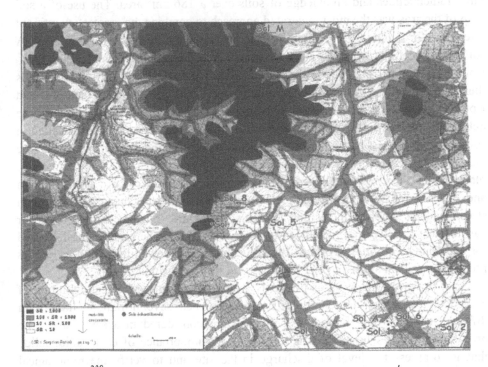

Fig 4. Map of $^{238}U \, K_d$ in soils of the Bure area. In white: $K_d<10 \, L \, kg^{-1}$; in pink: 10 L $kg^{-1} < K_d < 100 \, L \, kg^{-1}$; in blue $100 \, L \, kg^{-1} < K_d < 1000 \, L \, kg^{-1}$; in purple $K_d > 1000 \, L \, kg^{-1}$; in green: forested areas in the white group.

The location of the critical group in terms of accumulation of ^{238}U and other radioisotopes of uranium in soils is the area covered by class IV. These soils are mainly forest soils or soils at the boundary between fields and forests. Any change in land use, especially cultivation of former forest soils, would lead to an increase in the critical area (irrigation should not occur in forests in the current state). Some agricultural practices such as soil liming that increase soil pH can also greatly influence this pattern. Soil-to-plant transfer is generally enhanced by low K_d values. For uranium, this is not always the case as complexes such as uranyl carbonate are not as readily taken up by plants as the uranyl ion (7, 14). However, the highest mobility of uranium should lead to highest plant uptake. Consequently, the critical area in terms of U transfer to crops is, as for ^{99}Tc, the area covered by soils from class I, *i.e.* the limestone plateau.

4. Conclusion

A prediction of the phytoavailability of the [63]Ni, ^{99}Tc and [238]U at a regional scale was based on the superimposition of the laboratory measurement of the parameters significantly controlling the mobility and the phytoavailability of the three radionuclides and knowledge of soils over a 186 km^2 area. The use of a soil map of the area was the most integrated approach to combine the greatest quantity of relevant information as to the mobility and phytoavailability of the radionuclides. A soil type is defined from the material from which the soil has developed and also from the biological, chemical and physical processes that led to its development. It is therefore possible with limited experimental approach, to extrapolate data obtained from a given type of soil to the whole area it covers. Results allowed the determination, for the three radionuclides, of critical areas in which accumulation in soils may be highest and other critical areas where accumulation in plants may be highest. They also allowed definition of the most likely values of K_d and transfer to plants of the three radionuclides in the area of Bure. In the case of [63]Ni, inhalation and external radiation exposure to man and other living organisms is highest in stream valleys where villages and gardening areas are situated. Critical areas for soil-to-plant transfer are situated nearby in dry colluvial lowlands, also possibly used for gardening activities. For the two other radionuclides, information is not as precise but critical areas may not be situated in similar location to [63]Ni.

However, this work is primarily a preliminary approach. The definition of a critical area is an extremely difficult task when several radionuclides with different chemical and biological behaviour are considered in a same area. The only way to solve the problem is to classify the importance of each radionuclide that is to assess the level of discharge in the site and to verify the radiological impacts of each radionuclide and the sensitivity of organisms to this impact. The scope of equidosimetry also takes into account the interactions with other pollutants derived from other human activities or natural processes. Sensitivity of organisms to pollutants will probably show synergistic or antagonistic effects which should be taken into account. Geographic Information Systems (GIS) present one option to deal with the complexity of the data that needs to be handled, firstly to specialize them and to define critical areas and secondly to link data about the behaviour of different pollutants and their effect on a same location. It is also possible to include models in the GIS approach, which despite being simple, may be very helpful in helping to resolve the problem of equidosimetry in a specific site. The GIS approach is also essential to take into account the time dimension (evolution of soils, human scenarios,...) in the monitoring of the contamination of ecosystems.

Acknowledgements

Author wishes to thank Aurélie Pingot, Parissa Ghazi-Nouri and Louis Florentin for their useful assistance in the project. They are also grateful to the French agency for radioactive waste management (Andra) for funding this research.

References

1. Denys S., G. Echevarria, E. Leclerc-Cessac, S. Massoura and J.L. Morel (2002) Assessment of plant uptake of radioactive nickel from soils. *Journal of Environmental Radioactivity.* **In press.**
2. Echevarria, G., P. C. Vong, et E. Leclerc-Cessac and J.L. Morel (1997). Bioavailability of ^{99}Tc as affected by plant species and growth, application form, and soil application. *J. Environ. Qual* **26**: 947-956.
3. Ng, Y.C. (1982). A review of transfer factors for assessing the dose from radionuclides in agricultural products. *Nuclear Safety,* **23** : 57-71.
4. Sheppard, S. C., M. I. Sheppard, et W. G. Evenden (1990). A novel method used to examine variation in Tc sorption among 34 soils, aerated and anoxic. *Journal of Environmental Radioactivity* **11**: 215-233.
5. Andersen P.R and T.H. Christensen (1988) Distribution coefficient of Cd, Co, Ni and Zn in soils. *Journal of Soil Science* 39 : 15-22.
6. Quantin C., T. Becquer and J. Berthelin (2002) Mn-oxide : a major source of easily mobilisable Co and Ni under reducing conditions in New Caledonia Ferrasols. *C.R. Geoscience,* **334**, 273-278.
7. Echevarria, G., M. I. Sheppard and J. L. Morel (2001). Effect of pH on the sorption of uranium in soils. *Journal of Environmental Radioactivity* **53** : 257-264.
8. Echevarria G., J.L. Morel, J.C. Fardeau and E. Leclerc-Cessac (1998) Assessment of phytoavailability of nickel in soils. *Journal of Environmental Quality.* 27, 1064-1070.
9. Denys S., G. Echevarria, E. Leclerc-Cessac and J.L. Morel (2001) Fate of ^{63}Ni in undisturbed soil cores. *In ICOBTE Proceedings, Guelph, Ontario, Canada. [CD-Rom].*
10. Landa, E. R., L. H. Thorvigt, et R. G. Gast (1977). Effect of selective dissolution, electrolytes, aeration, and sterilization on ^{99}Tc sorption by soils. *J Environ. Qual.* **6** (2) : 181-187.
11. Echevarria G. (1996) Contribution à la prévision des transferts sol-plante de radionucléides. *PhD thesis, INPL, Vandoeuvre-les-Nancy, France. Abstract in English. 222 p.*
12. Ashworth et D.J., G. Shaw, L. Ciciani and A.P. Butler (2002) Transport of iodine and technetium radionuclides in soils and plant uptake from near-surface contaminated groundwater. *In proceedings of International Workshop Andra-IUR "Mobility in Biosphere of Iodine, Technetium, Selenium and Uranium. Nancy, France, April 3rd-5th.*
13. Denys S. (2001) Prédiction de la phytodisponibilité de deux radionucléides (^{63}Ni et ^{99}Tc) dans les sols. PhD thesis, INPL, France. 153 pp.
14. Ebbs, S. D., Brady, D. J. and Kochian, L. V. (1998) Role of uranium speciation in the uptake and translocation of uranium in plants. *Journal of Experimental Botany,* **49**, 1183-1190.

RADIOECOLOGICAL FACTORS OF THE GEOLOGICAL ENVIRONMENT WITHIN UKRAINIAN TERRITORY

Ye. YAKOVLEV
Ukranian State Geological Institute, Kyiv
UKRAINE

1. Assessment of the current status of Ukraine's geological environment

The Geological Environment (GE) is one of the most important components of the surroundings. Geological, environment implies the upper lithosphere zone which is largely exposed to the impact of human economic activities. Within this zone, a mineral basis-of the biosphere is being formed. GE includes soils, zones of surface aeration and mineral nutrition of the plants, bottom sediments, rocks and ground waters.

The modern concept of GE and its significance in evaluating natural surroundings are based on V.Vernadsky's doctrine on noosphere which is a multicomponent media of human. life and activities in the upper lithospheric horizons. The influence of these activities is comparable with natural geological processes and often brings about the changes of their character and dynamics.

The mankind is closely associated with GE. Any technogenic change of the status of objects included in GE, or of the previously established connections between them, is accompanied by the disruptions in scales and natural velocities of geological processes taking place in the surroundings.

GE is a major accumulator (repository) of polluting substances (radionuclides, heavy-metals, toxic organic compounds), which appear within wastewater effluents and atmospheric emissions.

Underground and open cast mining, construction of features, creation of powerful technogenic fields of physical and chemical nature resulted in irreversible changes in GE which produced either the disruptions of dynamics of natural processes or the appearance of hazardous technogenic processes. GE in its development interacts with adjacent associated systems of natural surroundings (atmosphere, hydrosphere, biosphere) and actively influences the formation of ecological situations.

These circumstances determine the necessity of studying and mapping of the current status and changes of the GE under the conditions of enhance economic development of Ukraine's territory.

Ukraine is characterized, on the one hand, by the intensive grows of industry and agriculture, and on the other hand, high density of the population. As a result, technogenic impacts on GE and inherent physical, chemical and energy related processes are from 5 to 15 times greater then those in the neighbouring countries.

A steady deterioration of the environmental parameters of GE, and associated natural media is formed under the influence on the atmosphere, soils, surface and groundwater of 10-12 million tons of various substances emitted by

F. Brechignac and G. Desmet (eds.), Equidosimetry, 131–146.
© 2005 *Springer. Printed in the Netherlands.*

industries and power plants, alongside with up to 18 million tons of mineral fertilizers, 190 thousand tons of pesticides and dozens of million tons of industrial liquid effluents.

Within the greater part of major industrial urban agglomerations (IUA) and their outskirts, considerable areas were identified where soils are polluted by toxic heavy metals with concentrations substantially exceeding maximum permissible concentrations (Kyiv, Dnipropetrovsk, Dniprodserzhinsk, Gherkasy, Odesa, Chernivtsi, Donetsk, Gorlivka, Lysychansk, Marioupol, Krasnoperekopsko-Armyansky industrial area etc.).

Ukraine's GE became a recipient of all atmospheric, solid and liquid radionuclide emissions end effluents of the Chernobyl NPP accident. In substantial areas, the pollution of landscapes soils and flora by long-lived, isotopes such as Cs-137, Sr-90, Pu -239, 240 was dozens of times greater than the 1963 after-bomb background. As a result, a new nuclear-radiogeochemical province was formed which spreads beyond Ukraine's territory. The long-term radioecological influence of the Chernobyl NPP accident includes:

- a steady and often increasing (due to the decay and redistribution of emitted substances) pollution of upper zone, primarily soils, and subsequently agricultural producers by the radionuclides;

-a high level of risk of their stable penetration into the bottom sediments and the surface runoff of Dnipro river whose water resources are used to supply water to 35 million people (70% of Ukraine's population).

According to the estimates of the Ukrainian State Committee on Geology and the Mineral Recourses and Ukrainian Academy of Sciences, in a number of Ukraine's areas the aggregate negative influence of atmospheric emissions and water effluents on the quality of water resources is dozens of times greater than the maximum permissible concentrations and standards of the protective properties of soils in aeration zone. This led to the development of hundreds hotbeds of actual or potential pollution of ground waters which are regarded as the basic and the most stable source of drinking and technical water supply (Lysychansk, Dnipropetrovsk, Donetsk and other regions, Crimean republic). More than 10% of Ukraine's speculative ground water resources are polluted to a various extent, whereas surface water sources are irreversibly polluted.

Over the last decade chemical soil cultivation became a powerful factor of regional deterioration of the ecological geochemical situation, giving rise to a long term and steady nitrate and pesticide penetration into groundwater in all Ukraine's regions. Furthermore, the penetration depth increased from dozens to hundreds meters. Substantial GS changes are taking place under the influence of major engineering systems (Kiev, Dnipropetrovsk, Odessa, Kharkiv etc.), and Territorial-Industrial Complexes (TIC) such as Donbas, Kryvbas etc.

A large scale equilibrium disturbance in the upper rock zone under the influence of mining, industrial and civil construction, land reclamation, extensive regulation of the surface runoff (28 thousand artificial water storages and ponds were built on the surface of 13000 km^2), with the accumulation of large water masses, led to stirring up exogenic geological processes (EGP) including hazardous ones: landslides, carst, phreatic rise, subsidence etc. Some EGP's are able to accelerate the regional migration of the radionuclides (phreatic rise or technogenic floodings, erosion etc.)

The steady and growing technogenic changes of GE and subsequent deterioration of geological situation necessitate further studying and predicting negative processes in GE in order to provide in advance relevant geological-ecological information to the branches of national economy and to the environment protection entities so that they could plan and implement adequate measures aimed at rational utilization of natural resources, safe location of nuclear power plants and industrial facilities, residential areas etc.

2. Regional technogenic dynamics of the landscape geochemical situation in Ukraine

Practically fifty years after World War II Ukraine was the more developed part of the Soviet Union (SU). Occupying only 3% of SU territory Ukraine had produced about 21-23% of the Total National product (TNP). This was meaning the huge load on all the elements of the environments: surface water objects, biosphere and soil - upper part of the geological environment.

Large-scale development of the agriculture, metallurgical, chemical, built, energetic branches of industry was the base for active ploughing of the landscapes, backwatering of the rivers, active mining of raw mineral materials.

Practically till 1990 Ukrainian economy annually produced about 25 billion cubic m of the polluted water, two billion tons solid wastes (mainly raw mineral materials) and 10-12 million tons air exhausts.

Besides about 12-14 million tons of fertilizers and 160-180 thousands tons pesticides were used within Ukrainian territory (mainly on the territory Dnieper river basin).

To the end of the 20th century Ukrainian territory as the ecological system had the next regional natural-technogenic ecological parameters:

- total backwatering of the large, middle, and small rivers - about 28.5 thousand ponds and water reservoirs with the average groundwater level backwater from the 2-3 till 6-10 metres in the Dnieper's valley;
- destroying of the natural geochemical landscapes due to ploughing, fertilizing (about 70% of Ukrainian territory), air exhaust accumulation and water irrigation with the contaminated river water;
- extracting of the huge volumes of the raw mineral resources within mining industrial centres (Donbass, Krivoy Rog, Carpathian region etc.-50 billion tons within 20000 km^2);
- creation of the geochemical contaminated zones around industrial - urban agglomerations due to influence of the air exhausts, pouring of the wastewater and accumulation of the solid wastes.

That is why the Chernobyl accident radionuclide exhausts in the subsurface air and surface water were the additional regional source of the landscape (first of all ploughed lands) contamination within about 20% of Ukrainian territory (so-called western and southern tracers). Upper zone of the geological environment including the soils and bottom sediments in the water reservoirs were the main "depot" for the accidental radionuclides.

3. Evaluation of radioactive influence on the Chernobyl accident on geological environment

The Chernobyl accident resulted in a substantial deterioration of ecological parameters of GF due to the blow-out of a wide range of radionuclides including long-lived ones:

caesium-137, strontium-90, plutonium-239 and others. The long-term radioecological influence of the NPP accident provoked a stable and steadily increasing pollution of the upper GE zone, primarily soils. The process accelerates with the physical decomposition of radioactive elements leading to the increased mobility of released products. There is also a high degree of risk of a stable radionuclide migration into bottom sediments and the surface runoff of the Dnieper river whose resources provide water supplies for 35 million people.

The extreme contamination of GE took place in zones of the western and southern plumes radioactive nuclide active airborne migration. Environmentally significant soil pollution (evaluated as 10 times greater than the radioactive background value before the accident, or about 0,5 Ci/km^2 for caesium-137) was registered over the surface of nearly 120000 km^2. As a result, approximately 20% of Ukrainian territories have a long-term base for emergence of low-level radiation.

The heaviest pollution with caesium-137 as to the affected area and concentration levels was registered in soils of Kyiv province (up to 15 Ci/km and more), Zhytomyr province (from 1 to 5 Ci/km), and Rivne province (over 1 Ci/km^2).

The following levels of soil pollution with caesium-l37 in Ukraine have been identified:

Concentration

1 - 5 Ci /km^2	40.5 thousand km^2
5-15 Ci /km^2	5.1 thousand km^2
15-40 Ci /km^2	1.1 thousand km^2
40 Ci /km^2 and more	0.8 thousand km^2

The soils polluted with strontium-90 ranging from 0.05 to 0.5 Ci/km^2 practically coincide with areas affected by caesium-137. The strontium-polluted area is negligibly larger in the eastern direction. The sites with a strontium-90 pollution greater than 0.5 Ci/km^2 have been registered only in Kiev province, in the vicinity of the Chernobyl NPP.

In 18 provinces, isolated sites were detected with abnormally high plutonium-239, 240 concentrations in soils. The maximum number of such sites was registered in Chernigiv region. The concentration of plutonium ranges from 0.002 + 0.08 Ci /km^2. In Southern Ukraine, the radioactive environmental situation is being increasingly influenced by radioactive nuclides migration with bottom sediments in water storages of Dnipro river cascade. Another factor of influence is the penetration of radioactive nuclides into irrigated areas as well as into potable water supply and distribution systems. In Kiev water storage reservoir, cesium-137 reserves amount to 8000 Ci, whereas in Kaniv reservoir they are equal to 2400 Ci.

The main sources of cesium-137 penetration into Kaniv water storage reservoir are suspensions, phytoplankton and waters of Kiev water storage reservoir.

In Kremenchuk water reservoir, caesium-137 accumulation in bottom sediments reached, about 400 in 2000. In deeper areas of Dniprodzerzhinsk and Kahovka water reservoirs radioactive nuclides accumulation also takes place. However, the radiogeochemical regime is currently in its initial phase and caesium-137 reserves are increasing.

Ukraine's geological surroundings became not only an accumulating capacity for radioactive nuclides originating from the Chernobyl NPP released but also a long-term source of their penetration into crops and vegetables, milk and meat products, surface and inadequately protected groundwater sources.

An anomalous radioactive-ecogeochemical province has been formed on Ukraine's territory [1, 2].

Table 1 is showing the high concentration of caesium-137 in the different types of soils within Ukrainian territory as in moved and as in the fixed forms (total storage).

Table 1. Caesium-137 concentrations in different soil types in Ukraine.

Type of soils, contaminated with caesium-137	Forms of caesium-137 in soils, %		
	Exchanged	Moving (not exchanged)	Fixed
Semi-clay, semi-sand	21 – 28	6 – 10	61 – 74
Gray (forest, meadow)	8 – 19	5 – 7	83 – 87
Black (steppe, semi steppe)	14 – 16	3 – 4	82 – 85

Besides the big quantity of the moving and fixed forms are witnessing about long time potential of the radio geochemical contamination within agricultural and natural landscapes.

4. Geological environment as the main source for complex influence of low radioactive dose and soil contamination with heavy metals and chemical products

After Chernobyl accident and regional radionuclides contamination of the landscapes within its territory Ukraine could be considered as the artificial radionuclide province. But in the same time we can observe the origination of the technogenic landscape geochemical province within IUAs, mine regions, water irrigation systems etc. Only mountain structures of the Carpathian (Western region) and Southern Crimea can be considered as natural geochemical landscapes (natural geochemical backgrounds).Of course, we can find natural geochemical landscapes within Ukrainian territory but these fragments are not able to stabilize the biogeochemical, hydrochemical and even hydrogeochemical parameters.

The main conclusion in this situation is the following - Ukraine shows the

136

principally new radiogeochemical parameters within the main landscape zones – forest, forest-steppe, steppe, semi-steppe. This indicates the new quality of the foodstuffs, surface and even groundwater quality.In short, we are living in the new ecological parameters of the food chain for the recent and future generations of Ukrainian ethnos. Very often geo-globalists are saying: for the state to be really independent it is necessary the enough stable population, resources of agriculture lands, own base raw mineral materials and energy sources.

Now we can consider Ukraine as example that ecological resources of natural systems within state whole territory is the key resource for the stable development of the state in the near and far future.

Investigations of specialists of the National Academy of Sciences, and State Geological Survey (V.Shestopalov. E.Sobotovich, D.Grodzinsky, Yu.Kutlakhmedov, I.Trakhtenberg et al.) are witnessing that the level of the population illness has a close correlation with total chemical (Radionuclides and Heavy metals) contamination (>0.7).

New generations of Ukrainian population begin to develop and live under stable influence of low radioactive doses and heavy metals. Disturbances of the biogeochemical balance, synergetic effects, long-lived low radioactive doses are the factors of the regional and long–term burden. Today the geological environment is the main base for the Chernobyl accident aftermaths.

Total level of the GE contamination and its correlation with the illness of population is showed on Figure 1and in the Table 2.

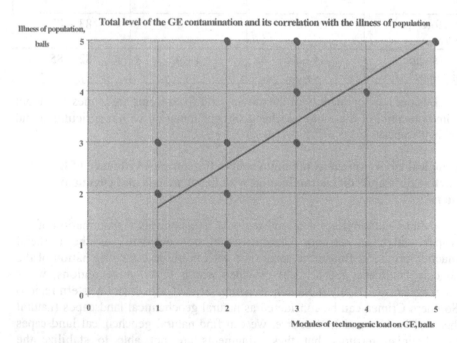

Fig. 1. Level of contamination and its correlation with the illness.

Table 2. Geological environment (GE) contamination and population illness.

NN	Administrative Oblast of Ukraine	Illness of population		Modulus of technogenic load on GE	
		total (on 100000 men)	balls	1000 tons/km^2 year	balls
1	Vinnitska	66550	3	44	3
2	Volynska	66550	3	6	1
3	Dnipropetrovska	72250	4	97	4
4	Donetska	72250	4	99	4
5	Zhytomirska	60850	2	6	1
6	Zakarpatska	55150	1	6	1
7	Zaporizska	72250	4	97	4
8	Iv.-Frankivska	60850	2	15	2
9	Kyivska	77950	5	108	5
10	Kirovogradska	60850	2	10	2
11	Luganska	72250	4	41	3
12	Lvivska	77950	5	25	3
13	Mykolayivska	66550	3	8	1
14	Odeska	66550	3	18	2
15	Poltavska	55150	1	12	2
16	Rivnenska	55150	1	8	1
17	Sumska	55150	1	7	1
18	Ternopilska	55150	1	10	2
19	Kharkivska	77950	5	19	2
20	Khersonska	60850	2	13	2
21	Khmelnitska	55150	1	11	2
22	Cherkaska	66550	3	25	3
23	Chernigivska	60850	2	8	1
24	Chernivetska	60850	2	9	1
25	AR Crimea	77950	5	37	3

This principally new multi-factors ecological situation is creating life-important and new tasks for the geological and radiobiological sciences:

1) special zoning of the Ukrainian territory and agricultural landscapes in accordance with the radiobiogeochemical risks for the human being and biosphere;

2) reconstruction of the State Natural Environment Monitoring System in accordance with the radiobiogeochemical tasks;

3) investigation and substantiation of the long-term permissible level of the low radioactive doses for the different groups of the population in the conditions of regional landscape geochemical contaminations.

References

1. Chernobyl Accident: Reasons and Consequences. The Expert Conclusions, Acad. Nesterenko V., Dr. Sc. Yakovlev Ye. etc. Minsk "Pravo i Economica", 1997

2. Chornobyl Catastrophe. Chief Editor Acad. Baryachtar V.G. Kyiv, Editorial House of Annual Issue "Export of Ukraine", 1997

BEHAVIOUR OF THE RADIONUCLIDES IN PEAT SOILS

G. BROVKA, I. DEDULYA, E. ROVDAN
*Institute for Problems of Natural Resources Use & Ecology, National
Academy of Sciences of Belarus, 10, Staroborisovsky trakt , 220114
Minsk, BELARUS*

1. Introduction

Peatlands occupy 2.9 millions hectares or about 12 % of the territory of the Republic of Belarus. As a result of the Chernobyl Nuclear Power Plant accident most of their areas were subjected to radioactive contamination. From a radiation danger viewpoint long-existing Cs^{137} and Sr^{90} are. at the lead. Prior to the Chernobyl catastrophe the peat soils were intensively used in agriculture and great quantity of fertilisers and pesticides had been introduced . Their combined impact with radionuclides complicates considerably the ecological situation. Ameliorated peat soils are subjected to water and wind erosion promoting the radionuclide transfer.

To minimise the negative influence of the radioactive contamination upon the environment forecasting assessments of radionuclide transfer under the impact of natural and anthropogenic factors are needed, as well as basic means of directed impact upon their transportation processes.

2. Mathematical models

The analysis of literature and our own researches showed that the radionuclides transfer processes in soils in the region of positive temperatures are most completely described by means of the mass exchange equations, regarding the kinetics of their sorption in the system solid phase vs. solution [1, 2]:

$$\frac{\partial C1}{\partial \tau} = K_c \cdot D_0 \frac{\partial^2 C1}{\partial x^2} - V \frac{\partial C1}{\partial \tau} + \frac{\alpha}{W} \cdot \left(\frac{1}{K_d} C2 \quad C1 \right), \tag{1}$$

$$\frac{\partial C2}{\partial \tau} = -\alpha \cdot \left(\frac{1}{K_d} C2 - C1 \right) \tag{2}$$

where C1 and C2 –radionuclide concentration in porous solution and solid phase of the soil respectively; D_0 - the coefficient of the radionuclide molecular diffusion; V– linear velocity of the moisture connective transfer; K_c– liquid phase communication coefficient; K_d - distribution coefficient; W– material moisture content .

In the development of the theory of mass transfer in natural dispersed systems at negative temperatures on the basis of thermodynamics of irreversible processes and hydrodynamics of thin layers of a liquid   with regard to the appearing in these layers of the discharge pressure [3, 4] we have obtained a

F. Brechignac and G. Desmet (eds.), Equidosimetry, 139–146.
© 2005 Springer. Printed in the Netherlands.

differential equation, describing the radionuclides migration in frozen soils:

$$\frac{\partial C_i}{\partial \tau} = D_{io} K_c W_u \cdot \frac{\partial^2}{\partial x^2}\left(\frac{C_i}{K_{di} + W_u}\right) + \frac{K_f}{\rho_{sk}} \frac{\partial}{\partial x}\left(\frac{C_u}{K_{di} + W_u}\right) \times$$

$$\times \left[\frac{\partial P}{\partial x} + \frac{\rho_i L}{T}\frac{\partial T}{\partial x} + \rho_i RT\sum_i v_i \frac{M_w}{M_i}\frac{\partial}{\partial x}\left(\frac{C_i}{K_{di} + W_u}\right)\right] + \qquad (3)$$

$$+ \frac{K_f C_i}{\rho_{sk}(K_{di} + W_u)}\left[\frac{\partial^2 P}{\partial x^2} + \frac{\rho_i L}{T}\frac{\partial^2 T}{\partial x^2} + \rho_i RT\sum_i v_i \frac{M_w}{M_{ci}}\frac{\partial^2}{\partial x^2}\left(\frac{C_i}{K_{di} + W_u}\right)\right]$$

where C_i -concentration of i water soluble compounds is accounting for the mass unit of the rock skeleton; R- specific water gas constant; M_i and M_w - molecular masses of water soluble compound and water respectively; v_i - quantity of ions, on which the molecule of water soluble compound dissociates; ρ_i, ρ_{sk} – the densities of ice and the soil skeleton respectively; W_u - amount of unfrozen water in the soil; K_f – filtration coefficient of the frozen soil; D_{io} – diffusion coefficient of the water soluble compound in porous solution; P–hydrostatic pressure.

According to (3), the variation of the concentration of radionuclides in frozen soils is defined by the radionuclide diffusion processes at the molecular level, by processes of convective transfer along the unfreezing layers under the impact of hydrostatic pressure, the temperature and general concentration of water soluble compounds and divergence of moisture flow due to the non-linearity of the distribution of the moving forces of transfer and phase transfers of water into ice.

In a homogeneous temperature field the velocity of diffusional transfer of radionuclides is determined by the effective diffusion coefficient:

$$D_{ef} = D_o \cdot K_c /(K_d/W + 1), \qquad (4)$$

where W corresponds to the natural dispersed systems moisture content in the region of positive temperatures or to the non-frozen amount of water in that of negative ones.

3. Methods

The models suggested, describing the radionuclides migration (1-3) carrying their phenomenological character, and containing transfer coefficients such as effective diffusion D_{ef}, distribution K_d, diffusions in porous solution D_o, mass exchange α, communication of water ways K_c, filtration of frozen soils K_f should be determined by experiments [5].

The distribution coefficient K_d characterises the radionuclides distribution between the liquid and solid phases of soils and is numerically equal to the ratio of the balanced radionuclide concentrations in solid and liquid phases (5). To determine K_d, the next equation was used:

$$K_d = C_{wm} (1+W)/ C_s - W \qquad (5)$$

where C_{wm} equals the specific activity, accounting the unit of the humid material of the solid residue, C_s - porous solution specific activity of the centrifuged porous solution, W – the moisture content of the solid residue after centrifugation.

The experiments conducted showed that radionuclides distribution coefficients in soils in the sphere of positive temperatures actually do not depend on moisture content and temperature. This makes it possible to consider K_d in soils in negative and positive temperatures spheres, having close values.

The coefficient D_{ps} characterises the radionuclide molecular diffusion in porous solution in soils in thawing state through the unfreezing water layers in negative temperatures spheres:

$$D_{ps} = D_i \cdot \hat{E}_{\tilde{n}} \qquad (6)$$

To determine through tests the effective diffusion coefficients D_{ef} and the diffusion coefficients in porous solution D_{ps}, the integral method has been applied based on the solution of diffusion equation for the two plate-shaped samples, contacting each other at different initial transfer potentials [6] and corresponding to a previously proposed method [7]. For the practical realisation of the integral method of the identification of diffusion coefficients, cassettes have been used made from hydrophobic material, formed as a flat disc of 90 mm diameter and 10 mm height. When studying D_{ef} and D_{ps} in positive temperatures sphere the first cassette was filled with the material containing the salt of the ion studied at a well defined concentration, and the other one with the material of the same moisture content without salt.

While defining experimentally D_{ef} and D_{ps} in the negative temperatures sphere it is mandatory to account for the specific nature of the mass exchange in the frozen soils namely that the convective ingredient of the radionuclide transfer can to a great deal exceed the diffusional one. In order to exclude the convective transfer of radionuclides and to enhance the exactness of D_{ps} determination it is necessary to either conduct the test at a very low radionuclide concentration or additionally definite compound should be inserted in each sample. The compounds then, being in a variety of samples, should be differentiated by chemical or physical ways, and should have close values of the coefficients of molecular diffusion and initial molecular concentrations. The ideal means for that purpose can be compounds including stable and radioactive isotopes.

To obtain the dependence D_{ef} upon the total concentration of the salt of the investigated element in a series of tests the salt concentration of the stable isotope of the element given was varying.

The mass exchange coefficient α characterises the velocity of the fixation of the radionuclides balance in the system solid phase vs. porous solution according to (2). Experimentally α is defined through the kinetics of the change of the specific activity of the porous solution, squeezed by the centrifuge.

The coefficient of the liquid phase communication K_c is regarded as the parameter accounting for all heterogeneities of the porous medium system – windingness and corrugation of pores, availability of blind and closed pores, etc. Numerically, K_c is equal to the ratio of the coefficient of diffusion of unsorbed material versus water-soluble compound in the volumetric solution (6). K_c is mainly the function of moisture content, density and temperature and does not depend on the specific properties of water-soluble compound and its concentration. Having determined K_c of the system by means of already defined compound, the values obtained can be used for other compounds. Preliminary estimation of K_c can be obtained through a more simple conductometrical method.

Considering the labour-intensity and the duration of the tests on the direct determination of effective diffusion coefficients, the main information on the dependence of D_{ef} on the temperature, moisture content and concentration of the radionuclide can be obtained by counting., (4); Experimental data about D_o, K_c, ^{d}K coefficients, and some other limited data, being experimentally obtained, should be used to control the results.

4. Results

The Chernobyl radionuclides behaviour was studied by the use of an ameliorated peat-mire soil (peat deposit "Pogonyanskoye", 21 km off the ChNPP). It is peat of a high decomposition degree (38-38%). Prior to the Chernobyl catastrophe the whole deposit area was drained and used for perennial grass production. The ground water level varies depending on the season from 0 to 1 m.

To control mobility of radionuclides in soils we investigated experimentally all the characteristics of the migration and sorption of radionuclides Cs^{137} and Sr^{90} fen sedge peat, quartz sand, bentonite, kaolin.

The characteristics of the migration and sorption of radionuclides Cs^{137}, Sr in fen peat (2,5 kg/kg), quartz sand (0,2 kg/kg), bentonite (1,25 kg/kg), kaolin (kg/kg), have been experimentally investigated.

It was stated that for Sr^{90} in all systems studied, few values of distribution coefficients are specific and due to this, greater values of the effective diffusion coefficients were found if compared to Cs^{137}. With regard to the sorption capacity of the Cs^{137} ion, the systems studied can be arranged as follows: fen peat > bentonite > kaolin > quartz sand. With regard to the Sr^{90} ion – fen peat > kaolin > bentonite >quartz sand. Thus, effective diffusion coefficients of Cs^{137} are the same in fen peat D_{ef}= $0.9 \cdot 10^{-13}$ m^2/c, quartz sand D_{ef}= $1.8 \cdot 10^{-13}$ m^2/c, bentonite D $2.9 \cdot 10^{-13}$ m^2/c and kaolin D_{ef}= $3.2 \cdot 10^{-14}$ m^2/c. At the same time the effective diffusion coefficients of Sr^{90} are equal in the fen peat D_{ef}= $3.2 \cdot 10^{-12}$ m^2/c, in quartz sand D_{ef}= $1.5 \cdot 10^{-11}$ m^2/c, bentonite D_{ef}= $1.1 \cdot 10^{-11}$ m^2/c, kaolin D_{ef}= $7.0 \cdot 10^{-12}$ m^2/c.

The researches conducted testify for a low Cs^{137} and Sr^{90} diffusion mobility in the sphere of negative temperatures, great mobility of Sr^{90} being preserved. At the temperature of 2^{0}C the effective diffusion coefficient in peat for Cs^{137} is $5.1 \cdot 10^{-15}$ m^2/c, for Sr^{90} being $11.1 \cdot 10^{-15}$ m^2/c. The temperature dependencies of effective diffusion coefficients of radionuclides (sharp drop with decreasing temperature) is determined by the superposition of a two factors-temperature dependency of the connectivity coefficient of water –conducting channels and the amount of non-freezing water in the peat. Following the degree of influence on the Cs^{137} distribution coefficient value in fen peat the electrolytes may be arranged in the row $CsCl > HCl > KCl > CaCl_2$. Of this row the stable Cs influence is strongly marked. Its presence only in porous moisture at an amount of 1.10^{-3} g-eq/ l changes the K_d almost 50 folds and enhances just the diffusion coefficient by 2 orders. This testifies to the difference of the link mechanism with the solid phase of Cs^+ cation from other metals cations. (I do not understand this sentence!!!!!!) The similar row for Sr^{90} looks like follows: $HCl > SrCl_2 > CaCl_2 > KCl > KOH$. Such succession proves that Sr^{90} basic link with the peat soil solid phase appears to be ion-exchangeable.

On the basis of these researches an original way to control radionuclides migration from root soil layer with the aim to decrease plant uptake, has been developed. The method suggests that after the introduction of leaching electrolytes in soils the latter will be isolated from atmospheric precipitation, allowing thence the creation of stationary conditions for radionuclides leaching. The lack of moisture convective flows will permit to prevent the untimely delivery of the electrolytes into the ground waters. Tests of the method of the soil deactivation were tried in field conditions on the Pogonyanskoe peat deposit (Khoinikskiy region, Gomel district, 21 km off the ChNPP). In May 1996 into the top layer of the deposit on the surface of the site (25 m^2) KCl salt has been introduced at an amount of 1 kg/ m^2, which approximately corresponds to its concentration in porous solution of 0,1 g-eq/l. Polyethylene film was used as a moisture isolating material, by which the soil surface was covered for several months. In May and October 1997 the distribution of radionuclides along the peat deposit was determined (Fig.1).

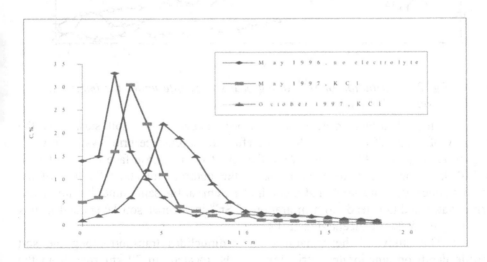

Fig. 1. Effect of electrolytes on Cs^{137} distribution in peat soil.

The analysis of the results showed that the maximum of the radionuclides activity on the salted site shifted into the deep peat deposits and by October 1997 it was at the depth of about 5 cm, whilst on the checked site as well as in the years proceeding, it was found to be at the depth of 2 cm. Though in the test given no practical results were achieved, it reveals the main possibility of controlling the radionuclides mobility with the help of electrolytes solutions. At present, further endeavours are being tested using more effective leaching materials.

The results of the application of subsurface brine water (Gomel region, Belarus Polesye) to the peat soil (peat deposit "Savichi", Gomel district) on the Sr^{90} distribution are given at Fig. 2.

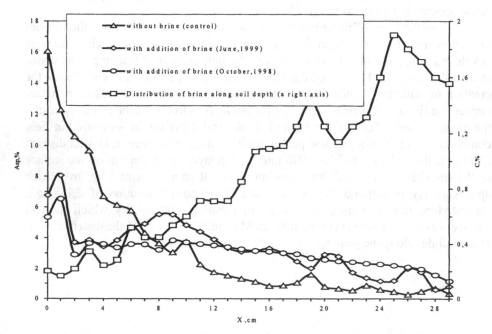

Fig. 2. Distribution of Sr^{90} along peat soil profile under the impact of subsurface brine water.

The subsurface brine water has been added at the soil surface in the quantity of 5 kg m^{-2} in May 1998. The effect of subsurface brine water on Cs^{137} migration is small. At the same time the effect of the subsurface brine water on Sr^{90} distribution is considerable. Thus, on the control plot about 80-90% of the radionuclides are concentrated at a depth of 12 cm; at the same time the subsurface brine water addition leads to a migration of Sr^{90} in deeper soil layers and in this case only 40-50% is concentrated at 12 cm.

The study of the dynamics of radinuclides transport along the soil profile depth on ameliorated agricultural lands, located in 30 km zone from the Chernobyl NPP revealed the vertical migration velocity of radionuclides to be small. On the peat soils, which are not influenced by anthropogenic effects, the depth of migration ^{137}Cs has not exceeded 16 cm, and for ^{90}Sr, 20 cm. The main quantity (about 90 %) ^{137}Cs is in the top 5 cm layer, and for ^{90}Sr in the 9 cm layer. On tillage soils the radionuclides are evenly distributed along the depth of tillage level [8, 9].

The mathematical simulation of ^{137}Cs and ^{90}Sr vertical migration in the ameliorated peat soil (peat deposit "Pogonyanskoye") and comparison with the data of natural observation showed rather a good reproduction by the model used of basic radionuclides transfer processes in a soil. If compared to ^{137}Cs binding with a soil adsorbing complex a rather quick establishment of a sorption balance and a greater migration mobility of ^{90}Sris being stipulated.

The forecast assessment testifies to a slow radionuclide migration in natural conditions: in 20 years in the layer of 0-5 cm about 90% of the summarised reserve in a soil layer of ^{137}Cs and about 53 % of ^{90}Sr will be concentrated.

The moisture increase, the availability of the moisture filtering flow, the distribution coefficient decrease, the mass exchange coefficient increase result in radionuclides migration into more deep soil layers. If as a result of agro-ameliorative exercises (fertilizers application, for instance) ^{137}Cs distribution coefficient decreases 500fold, then it can be expected that in the layer of 0-5 cm in 20 years there will be left 61% of ^{137}Cs . While decreasing ^{90}Sr distribution coefficient to 50fold in the layer of 0-5 cm, there will be 5% of ^{90}Sr left, the distribution maximum will shift into the 15-16 cm layer , and beneath 20 cm there will be 15% of ^{90}Sr.

5. Conclusion

On the basis of theoretical and experimental studies a method of assessment and prognosis of the radionuclides migration soils has been created.

Experimental methods for the determination of transfer characteristics and their assessment, and also the methods for prediction of radionuclide migration in frozen grounds should provide a clear distinction between driving forces and fluxes of the convection and diffusion transfer.

The proposed mathematical models are phenomenological and the transfer coefficients involved in these models are recommended to be determined experimentally.

The methods were developed for investigation of diffusion and convection transfer characteristics of radionuclides in thawing and frozen grounds with the allowance for the sorption by solid phase of the soils. On the basis of the experimental data obtained the information constraints for the mathematical models of the radionuclides migration in soils have been set.

On the basis of researches in the field, lab test and mathematical modelling the impact of physical-chemical factors and the soil component composition changes upon the radionuclides Cs^{137}, Sr^{90} migration and sorption in natural dispersed systems (peat, sand, bentonite, kaolin, sapropel has been investigated.

The impact of electrolytes, including compounds of Cs and Sr stable isotopes upon the radionuclides behaviour in peat soils has been made known, allowing controlling their mobility.

The forecast assessments of radionuclides migration in the peat soil (the peat deposit Pogonyanskoye, the Polessye radiation-ecology reserve) under the natural and anthropogenic factors impact had been carried out.

References

1. Brovka, G.P. & Rovdan, E.N. The influence of the sorption kinetics of radionuclides on processes of diffusion and convective transport them in natural media. Proceedings of the National Academy of Sciences of Belarus. Chemical series , Minsk. 1999, 2 : 63- 67p. (in Russian)
2. Prokhorov V. M. Radioactive pollutants migration in soils. Moskow.: Energoizdat, 1981. 97 p. (in Russian)
3. Deryagin B.V., Churayev N. V., 1980, "Current non-freezing water layers and frost porous bodies goggling". Colloid J. V. 42, 5, 842–852 p. (in Russian)
4. Brovka, G.P. Theoretical analysis of radionuclides migration process in frozen rocks". Colloid J., 1999, V. 61, 6, 752-757 p. (in Russian)
5. Brovka, G.P., Dedyulya, I.V. & Rovdan, E.N.An experimental study of radionuclides migration in frozen grounds. Colloid J., V. 61, 6, 758-763 p. (in Russian)
6. Lykov A. V. The thermal conductivity theory. Moscow: Vyschaya shkola, 1967. (in Russian)
7. Scofield R.K., Graham-Bryce I.J.Diffusion of ions in soil", Nature. V.188, No 4755, 1980 , 1048-1049 p.
8. Rovdan, E.N. "Forecast and control of hydrothermal and radioecological peat soils conditions". Ph.D. Thesis, National Academy of Sciences of Belarus, Institute for Problems of Natural Resources Use and Ecology , Minsk, Belarus, 1997. (in Russian)
9. Lishtvan, I.I., Brovka, G.P., Davidovsky, P.N., Dedyulya, I.V., Rovdan E.N.1997 "Forecasting and control of the migration of radionuclides upper soil layers".One decade after Chernobyl: Summing up the consequences of the accident. Poster presentations-V.2.- Vienna: IAEA, pp. 71-78.

EVALUATION OF THE EFFECT OF DNIPRO RIVER RESERVOIRS ON COASTAL LANDSCAPES

V. STARODUBTSEV, O. FEDORENKO
National Agricultural University, Geroiv Oborony st.15, Kyiv, 03041, UKRAINE

1. Introduction

Construction of large reservoirs and their cascade causes large-scale environmental changes in their vicinity and in the entire basins of rivers with a regulated run-off [1-3]. The achievement of the main objectives of (hydropower utilization, flood control, irrigation, etc) is usually accompanied by the flooding of fertile lands in river valleys, waterlogging and water table elevation, swamping, development of saline and alkaline (sodic) soils and stream-bank erosion. Downstream of the dam some profound changes occur in the characteristics of run-off water and in the landscape of river valleys and deltas, as a result of which the landscapes become steppe-like or desert-like [2-3]. The research work allowed us to identify some general regularities of coastal lands waterlogging and landscape changes which depend on geographic conditions of the region. At the same time, there is a need to perform special investigations of mentioned processes in reservoirs, where some unique practices of hydrotechnical construction were realized to protect the reservoir coastal lands from floods and waterlogging. Such a situation is characteristic for reservoirs on the Dnipro (Dnieper) river in Ukraine. This situation is aggravated by a nuclear-waste pollution of the region [4-6]. The Dnipro river in 50-70s was transformed into a series of reservoirs 855 kilometres long, which crosses the territory of Ukraine from the forests on the north to the dry steppes on the south. A run-off regulation caused a flooding of 0,5 million hectares of lands, changed environmental situation and conditions for farming and forestry on the adjacent territories.

2. The effect of Kyiv reservoir on coastal soils and landscapes

On the territory round the Kyiv reservoir in the forest belt soddy-podzolic soils are spread. These soils (sands and loamy sands) are typical by low organic matter content and acid reaction. They are used primarily in forestry and in farming. Soils along lower banks of the reservoir (left and northern part of right banks) were subject to ground water table elevation on an outstretch of land 0,5-1,0 km wide. A waterlogging was not observed any further along the left bank of the reservoir because of special conservation structures, such as protective dams and drainage channel, which protect the lowland territory between the rivers Dnipro and Desna. There is a strengthening of the processes of gleyzation and sometimes swamp-formation in soils, which became excessively wet under the reservoir impact. This changes the water regime, morphological features and chemical properties of soils. The reaction of formerly acid soddy-podzolic soils

F. Brechignac and G. Desmet (eds.), Equidosimetry, 147–154.

has become neutral under the impact of seepage waters containing calcium hydrocarbonate. Soil organic matter (SOM) content has increased from 0,8-1,0 to 1,2-1,9%. The reduction of hydrolytic acidity caused an increase in percentage base saturation [4]. On excessively wet plots with high water table there takes place the formation of soddy-gley, meadow and meadow-swampy soils gleyed all over the profiles. Organic matter content in their surface horizons is now over 2%.

The research results have enabled us to subdivide the territory associated with the Kyiv reservoir according to changes in the environmental status of soils [4, 6]. Regions I-III were outlined on the left bank. Soils in these regions were wetted by the infiltration water, flowing from the basin of a reservoir to the drainage channel, which was designed along the entire left bank. The degree of soil wetness is increasing from the north (region I) to the south (region III). The landscapes and soils of the region I are characterized by a line of soil profiles (description of the lines of profiles 1-5 was published in [4, 6]). Zonal soils here are suitable for the forestry (coniferous forests). In the lowlands of topography with high level of ground water table (1-2 m below the soil surface) slightly waterlogged gleyed soils with less acid reaction are formed. In the region II an environmental situation is influenced by the wetting of the territory by seepage water, which move towards the drainage channel. According to ground water levels the soils become considerably waterlogged (soddy-gley) on a coastal strip 200-250 m wide. At a distance 300-350 meters from the coast there is a moderate water table elevation and at distance 400-450 meters slightly waterlogged soils are being formed. The coastal territory is used for pastures and grasslands. The western part of the region is forest. In the region III adjoining the dam of a hydropower plant (lines 2 and 2a) a territory between the Dnipro and Desna rivers is protected from floods by a special concrete-faced dam. The soils next to dam are less waterlogged and towards the drainage channel they become moderately and sometimes considerably waterlogged.

On the right bank of the reservoir (region IV, Fig.1, line 3, [4, 6]) soil creep, stream-bank and surface erosion are encountered. On the elevated coasts, 30-40-meters high over the water level of the reservoir, the destruction of soils continues. Only in the southern part of the region near the town of Vyshgorod, stream-bank erosion has ceased and surface erosion becomes weaker due to the practices of erosion control (terracing and afforestation of slopes) and concrete-wall fencing of the banks. In the region V the processes of bank erosion are accompanied by fragmentary waterlogging of soils in the locked depressions of topography (Fig. 2, line 3b, 3v, [4, 6]). In the northern direction the processes of bank erosion gradually disappear, but the waterlogging of soils becomes more intensive. The changes of vegetation correspond to those of the soils: meadow grasses grow on intensively waterlogged soils; deciduous forests with meadow grasses are associated with moderately waterlogged soils; mixed forests - with slightly waterlogged ones. And only beyond any reservoir-associated wetness (Fig.1, line 4b, [4, 6]) there are typical coniferous forests. In the VIth region the processes of stream-bank erosion are practically absent, but soil waterlogging on the territories near the reservoir is intensive (Fig.2, line 6, [4, 6]). Large areas of farmland are waterlogged, primarily grasslands, including those ameliorated by drainage.

Radionuclides migration in soils of the coasts. Nuclear pollution of the territories associated with the Kyiv reservoir is now a serious environmental problem. We investigated the migration of radionuclides in soils of both low (waterlogged) and high coastal plains, the latter having stream-bank and surface forms of erosion (Fig. 1, 2, [4, 6]). It was found that since Chernobyl accident ^{137}Cs radionuclides have penetrated into the soil profiles to the depth 15-25 cm, depending on soil type, organic matter content and texture. The majority of the radionuclides was accumulated in forest litter and sod as well as in the surface 7-10 centimetres layer of the soil rich in organic matter, where pollution was up to 850-1839 Bq/kg. About one-tenth of the initial amount of pollutants has penetrated 15-20 centimetres deep, and even less - to 20-25 centimetres. An exception are the soils on the slopes of the territory near the village of Lyutezh (Fig.2, line 3 and 3a, [4, 6]) as well as the soils subjected to ploughing (line 4a), in which the radionuclides have penetrated deeper [6, 8]. On the slopes where the soils were subject to erosion only one-fourth of the initial pollution remained (269 Bq/kg, line 3a). In general, the downward radionuclide migration into the soddy-podzolic soils under the impact of precipitation still goes on, though slowly. A slight accumulation of nuclear wastes was detected in a capillary fringe of the waterlogged soils on the left bank. Their migration takes place with the seepage of water from the reservoir to the adjacent territories and afterwards with capillary fringe the radionuclides rise to the low horizons of the soils. But on the right bank this process does not occur. Results of our investigation were partly published earlier [4, 6], so we show here last data only (table 1). Along the elevated right bank the newly formed deposits (because of soil erosion) partly overlay the polluted bottom sediments.

Table 1. ^{137}Cs Activity in the Coastal Soils of Kyiv Reservoir

Point of sampling and soils	Depth (cm)	^{137}Cs (Bq/kg)	Point of sampling and soils	Depth (cm)	^{137}Cs (Bq/kg)
Region VI	0-3	1839	Region VI	0-5	425
line 6, profile 1	3-8	216	line 6, profile 3	5-10	98
Soddy Gleyed	8-13	50	Soddy-Podzolic	10-15	44
loamy-sandy	13-18	41	loamy-sandy soil	15-22	51
soil	25-30	36		22-27	44
	62-82	27		64-110	33
	0-6	812	Region V	0-4	212
Region VI	6-10	178	line 3-v, profile 1	4-8	254
line 6, profile 2	10-15	74	Soddy-gleyed	8-12	237
Podzolic-Soddy	15-19	68	Podzolized loamy	12-17	233
Gleyed loamy-	24-30	98	soil	17-22	230
sandy soil	30-50	59		22-27	146

3. The effect of Kaniv reservoir on landscapes and soils

Kaniv reservoir is situated in forest-steppe zone. Parent materials of soils on the high right-side bank are the silt-loam loesses. At the altitude of the reservoir level there are the layers of marl clays, which are helpful in preventing of a stream-bank erosion. On the lower left-side bank there are the alluvial sediments of the sandy terrace of the Dnipro, which further to the east become overlain by the loess-like sediments. There are a series of unique practices of coastal land protection from submerging and waterlogging.

We have identified 8 coastal regions according to soils changes (Fig.3, [4, 6]). 1-st region on the right bank is the polder system with soddy and podzolic-soddy gleyed and meadow-swamp soils that is protected by a dam against the submergence. It could be used for grasslands, pastures and recreation. 2-nd region is a territory occupied by the buildings of Trypilska heat power plant. 3-d region is represented by the high-bank territory covered with chernozems and dark-grey forest soils. The coasts are subject to bank erosion, sheet and gully erosion and landslips. 4-th region is the high-bank coastal territory between the towns of Rzhyschiv and Kaniv covered mainly by dark-grey forest soils. There is fragmentary bank erosion, but high cliff does not form on the bulk of the coasts. However, gully erosion and landslips are encountered here. The region is important for recreation and forestry.

The following regions have been identified on the left bank: 5-th region is a system of polders with mostly soddy, meadow and meadow-swamp waterlogged soils. The pumping stations are regulating their water regime. The lands here are being used as grasslands and pastures, but not effectively. 6-th region is a waterlogged territory containing numerous lakes and coastal bays and covered by meadow and swamp soils. This territory is important for recreation, hunting and fishing. 7-th region is a sandy terrace of the Dnipro. Predominant here is soddy-podzolic soils of sandy and sandy-loam texture. A terrace is being very slightly waterlogged and important for recreation, forestry and a nature protection reserve. 8-th region is a system of polders protected from submergence by a dam with the highway.

Nuclear wastes less pollute landscapes here. The activity of ^{137}Cs in surface horizons (0-5 cm) of Kozyn polder soils reaches 250-300 Bq/kg. Lower in the profile it decreases and only in the capillary fringe reaches 40-50 Bq/kg (Table 2, line 10, [4, 6]). Hydromorphic soils on the left bank are less polluted. Even in the surface horizon the activity of ^{137}Cs does not exceed 100 Bq/kg (line 13, p.3; line 11, p.22). But the radionuclides have penetrated to a greater depth being chelated by the fulvic organic acids. Only near village of Tsybly waterlogged coastal soils are more polluted in upper part of profile. It caused by a capillary raising of polluted ground water and entering of radioactive particles with surface waters in 1986 (line 8, p.1). The soils on the high right-side coasts are very weakly polluted (line 7, 14), except the territory near village of Khodoriv, where the activity of ^{137}Cs in the upper horizons of chernozems reaches 340-370 Bq/kg, (line 15, p.1).

Table 2. Activity of ^{137}Cs in the soils of Kaniv reservoir coasts

Point of sampling	Depth, cm	Cs-137 activity, Bq/kg	Point of sampling	Depth, cm	Cs-137 activity Bq/kg
Right-side bank			Line15, p. 1	0-5	340
Line 10, profile 1 (p.1)	0-2	414		5-10	374
	2-25	82		10-15	354
	25-32	26		20-30	163
	32-38	12		30-50	103
	38-66	16		50-70	98
	66-98	34	Left-side bank		
Line 10, p. 2	0-4	317	Line 5, p. 3	0-4	94
	4-6	32		4-12	138
	16-25	22		12-30	108
	36-58	26	Line 13, p. 3	0-7	93
	58-130	37		7-20	76
Line 7, p. 2	0-15	92		53-63	41
	15-36	111		63-89	31
	36-67	98	Line 6, p. 3	0-20	93
Line 7, p. 4	0-27	184		20-35	52
	70-85	129	Line 11,p. 22	0-5	59
Line 7-a, p.21	0-5	143		15-32	51
	5-27	103		48-60	35
	48-70	86	Line 9, p. 1	0-16	79
Line 14, p. 1	0-5	144		16-31	67
	15-20	139	Line 8, p. 1	0-5	265
	20-25	94		5-10	121
Line 14, p. 2	0-5	144		10-20	153
	5-10	181		20-45	184
	10-15	148		45-55	93

4. The effect of Kremenchuk reservoir on soils

Kremenchuk reservoir was built in 1961 on the south of the forest-steppe zone. The effect of the reservoir on the ecological status of the coastal landscapes is less investigated. So we propose a schematic zoning of the coasts (Fig. 5, [4, 6]). On the right-side bank we identified 4 regions: I – High rough coasts with eroded chernozems, where bank erosion and landslips take place. This territory is suitable for forestry. II – High coasts with typical chernozems where abrasion and surface erosion of soils take place in the thin strip on the shore only. III – Polder system with lakes and swamps. This territory is suitable for fish farming. IV – Lowland territory with complex of waterlogged soils and lakes that has good prospects for fodder grass production.

On the left-side bank we identified 5 regions: V – Lowland territory, divided by numerous riverbeds, covered with a complex of hydromorphic and zonal soils. Waterlogging of soils manifests itself more strongly in northwest part of the region. This territory is fit for a forestry and fodder production. VI – High coastal territory covered with typical chernozems that is not subject for waterlogging. An agricultural land use has not any restrictions. VII – Polder system that can be used mainly for a fishery and hunting. VIII – Territory, which adjoins the polder system on the northwest (Fig.5, 6, line 2). There are salination and alkalisation processes in waterlogged soils. This territory is used as the agricultural lands and grasslands. IX – Lowland territory, adjoining the estuary of river Sula. Swamp, meadow and meadow-chernozemic soils are prevailing here. Processes of a salination are more intensive in waterlogged soils. It is mainly grasslands and pastures. X – Elevated territory covered with typical chernozems, which are a subject for slight waterlogging only (Fig.6, line 1). There are weak processes of salination and alkalisation in deep horizons of waterlogged soils. An agricultural land use has not restrictions so far here.

The most danger processes in the coastal landscapes are salination and alkalisation of the waterlogged soils (Table 3).

Table 3. Soluble salts content in the soils on the Kremenchuk reservoir coasts.

Depth (Cm)	pH	Ions content, %/mg-eq.														Σ, %
		CO₃²⁻		HCO₃⁻		CL⁻		SO₄²⁻		Ca²⁺		Mg²⁺		Na⁺+K⁺		
								Line 2, profile 1-a								
0-2		-		0,417	6,83	0,105	0,96	0,750	15,60	0,020	1,00	0,059	4,90	0,448	19,49	1,779
								Line 2, profile 1								
0-1	8,75	-		0,183	3,00	0,011	0,30	0,250	5,20	0,008	0,38	0,013	1,12	0,161	7,00	0,626
1-7	9,34	-		0,092	1,50	0,008	0,22	0,250	5,20	0,007	0,33	0,003	0,26	0,146	6,33	0,506
7-16	9,92	0,105	3,50	0,390	6,40	0,009	0,24	0,084	1,75	0,045	2,25	0,014		0,195	8,49	0,841
16-23	10,19	0,086	2,88	0,342	5,60	0,010	0,27	0,046	0,95	0,027	1,36	0,021	1,71	0,152	6,63	0,683
23-45	9,52	0,045	1,50	0,220	3,60	0,014	0,38	0,060	1,25	0,021	1,05	0,018	1,48	0,097	4,20	0,474
45-57	10,20	0,024	0,80	0,207	3,40	0,016	0,44	0,050	1,04	0,012	0,60	0,007	0,60	0,103	4,48	0,419
57-76	10,00	0,012	0,40	0,156	2,55	0,013	0,37	0,013	0,26	0,006	0,30	0,019	1,60	0,039	1,68	0,258
76-114	9,30	0,003	0,10	0,107	1,75	0,008	0,22	0,013	0,26	0,014	0,70	0,017	1,40	0,005	0,23	0,167
								Line 2, profile 2								
0-28	6,05	-		0,052	0,85	0,005	0,15	0,013	0,27	0,008	0,40	0,005	0,40	0,011	0,47	0,094
89-123	7,82	-		0,079	1,30	0,007	0,19	0,013	0,27	0,022	1,10	0,005	0,40	0,006	0,26	0,132
200-250	8,10	-		0,092	1,50	0,003	0,07	0,013	0,27	0,006	0,30	0,014	1,20	0,008	0,34	0,136
250-300	8,23	0,001	0,02	0,060	0,99	0,003	0,07	0,012	0,25	0,004	0,19	0,006	0,50	0,015	0,65	0,101
300-380	8,79	0,007	0,24	0,051	0,83	0,002	0,06	0,013	0,27	0,003	0,17	0,008	0,65	0,013	0,58	0,097

5. Conclusion

The processes of land waterlogging, bank erosion and soil salination and alkalisation take place on the coastal landscapes of Kyiv, Kaniv and Kremenchuk reservoirs. We identified some regions on these reservoirs coast according to the intensity and peculiarities of such processes.

An important environmental characteristic of landscapes on Kyiv and Kaniv reservoirs coast is their pollution by nuclear wastes. The radionuclides (^{137}Cs) have penetrated into the soil profile to the depth 15-25 cm, but their predominant part is concentrated in sods, forest litter and the upper 7-10-cm layer of soil. On elevated banks the nuclear wastes are subject to surface erosion while on low banks is a slight accumulation of radionuclides in the capillary fringe of soils waterlogged by seepage waters.

References

1. Reservoirs and Their Impact on the Environment. 1st ed. MAB UNESCO, Moscow : Nauka, 1986. 367 p. In Russian.
2. Starodubtsev, V. M.: The Impact of Reservoirs on Soils. 1st ed. Alma-Ata: Nauka, 1986. 296 p. In Russian.
3. Starodubtsev, V. M.: Changes of Soil-Ameliorative Conditions in River Basins Caused by Water Management Construction. A Precis of Doctor of Biology Thesis. Novosibirsk. 1988. 36 p. In Russian.
4. Starodubtsev, V.M. – Petrenko, L.R. – Kazanina, O.V.: The Effect of Kyiv Reservoir on Environmental Status of Soils. // Journal of Hydrology and Hydromechanics. Bratislava. N 47. 1999. No 5. Pp. 366-377. In English.
5. Starodubtsev, V.M. – Kolodyazhnyy, O.A. - Petrenko, L.R. et al.: Soil Cover and Land Use in Ukraine. Kyiv: Nora-Print, 2000. 98 p. In English.
6. Starodubtsev, V.M. - Petrenko, L.R. - Fedorenko, O.L. et al.: Radionuclides Pollution in Soils of the Kyiv Reservoir Coasts after Chernobyl Accident. / Fifth International Symposium and Exhibition on Environmental Contamination in Central and Eastern Europe, Prague, 2000. P.215. In English.

RADIATION EFFECTS ON THE POPULATIONS OF SOIL INVERTEBRATES IN BELARUS

S. MAKSIMOVA

Institute of Zoology of NASB, Akademicheskaya str., 27, Minsk, 220072, BELARUS

1. Introduction

On April 26, 1986, the Chernobyl Nuclear Power Plant (CNPP), located in the Kiev region of northern Ukraine, 12 km from the Belarusian border, exploded accidently. As a result of the accident 1.95×10^{18} Bq of radioactive material, the largest amount ever reported, was released into the environment [1]. A radionuclide content exceeding 37 kBq m^{-2} resulted in contamination over 46 500 km^2 of surface soil in Belarus. This area amounts to 23 % of the total territory of the Republic. Within this contamination region a 30-km zone around the CNPP was therefore heavily contaminated.

Investigation of the long-term impact of radionuclides on the zoocenosis in the contaminated region is of interest because a precise description of the changes in the faunal complexes will permit a better understanding of the nature and the rate of recovery. To study the effects of ionizing radiation on terrestrial ecosystems it is essential to make observations on soil fauna. They comprise 90 – 95 % of all animal species and 90 – 99 % of total zoomass. Their high population density is stable. Since they also have permanent feeding habits (phytophagous, saprophagous and zoophagous), soil arthropods represent a convenient subject for elucidation of the ways and quantitative patterns of radionuclide migration in biogeocenoses, and may be used as bioindicators of contaminated territories [2].

The following requirements are imposed upon animals used as bioindicators of contamination by radioactive isotopes [3]: (a) high density and metabolism level; (b) great length of life; (c) intensive reproduction; (d) a small home range; (e) sedentary way of life; (f) permanent contact with the anthropogenic factor under study; (g) available collection of mass samples; (h) sensitivity to the factor concerned; (i) fairly large size in order to facilitate dissection.

Investigation of the long-term impact of radionuclides on the zoocenosis in the contaminated region is of interest because a precise description of the changes in the fauna complexes will permit a better understanding of the nature and rate of recovery. As part of a larger study of the impact of the CNPP event on the zoocenosis of the various biogeocenoses , we undertook a study of the soil mezofauna.

The main goal of our study was to investigate the character of effect of a high radioactive background level on soil invertebrates. The objectives of the study were (1) to investigate the main trends in the changes of the soil invertebrate communities subject to radioactive contamination; (2) to reveal the character and degree of the radiocontamination effect on the functional groups; (3) to characterize the state of invertebrate hemolymph under radiological pressure.

F. Brechignac and G. Desmet (eds.), Equidosimetry, 155–161.

2. Study area, material and methods

Material was collected in the Gomel Region, 30 km away from the CNPP in 1986–2000 applying usual pedobiological techniques (soil samples and Barber's pitfall traps) at reference points subjected to radioactive contamination. 105 pitfall traps were used every year. They were emptied 15 times from the middle of April till the end of October, i.e. during the active period in the life of soil-dwelling arthropods. The pitfall traps were plastic cups measuring 72 mm in diameter. They were set at intervals of several meters away from each other. The traps were filled with 4% formalin. The species of the animals collected were identified. At present, the Polessky Radiological Ecological Reserve is located in this site with a contamination level equal to 1500 kBq × m^{-2}. Gamma-radiation at the soil surface was measured with SRP 68–01 and DRG 01T radiometers. As controls, similar biotopes in the Pripaytsky National Park located 150 km to the west of the study area were chosen. In this area, the contamination level was nearly the same as the natural background. The samples of animals, soil and litter were collected in cylindrical plastic containers. The volume of each container varied from 100 to 250 ml. Sometimes, 50-ml containers were used. The containers were placed directly on an ORTEC Ge detector connected to a Canberra 80 series multichannel analyser system. Efficiency calibrations of the detector to all geometries used had previously been performed and checked at intervals with standardised solution of ^{137}Cs.

The ^{137}Cs deposition density at reference point (v. Babchin, pine forest) recalculated as to 1991 was 1110,0 kBq/m^2, ^{90}Sr – 77,7 kBq/m^2. The ^{137}Cs deposition density at reference point (v. Lomachi, oak forest) recalculated as to 1991 was 2331,0 kBq/m^2, ^{90}Sr – 284,0 kBq/m^2.

3. Results and discussion

The studies of zoocenotic characteristics such as population density, trophic structure and species composition of soil invertebrates inhabiting biogeocenoses are aimed at obtaining the definite relations and parameters characterizing the state of soil invertebrate communities - most sensitive to radioactive contamination. Those studies are based on comparing certain parameters characterizing the vital activities of soil fauna in biogeocenoses under contamination with control areas situated far from contamination sources.

The results of out gamma-survey of soil as the invertebrate's habitat show that contamination of soil cover is inhomogeneous in different biogeocenose types. A wide range of gamma-activity and radionuclide content in the soil and years was obtained in biogeocenoses with radioactive contamination. A general trend of maximum radionuclide concentration in a 10–cm soil layer was found for forest biogeocenoses.

3.1. SPECIES RICHNESS AND DOMINANCE IN SOIL MEZOFAUNA COMMUNITIES

The results of ecological studies of invertebrate animals show that radioactive contamination in the 30 km zone has affected the soil fauna, particularly the constant dwellers of the forest litter. The doses were sufficient for providing the death of eggs and the early stages of development of nearly all invertebrates and the adverse effects of the contamination should be associated with disturbances in the process of breeding and regeneration of the population. An initial sharp reduction in animal biodiversity and community structure of soil fauna was observed and was followed by a long-term process of the system returning to the initial parameters.

As it is seen, radioactive contamination exerts the greatest influence on the permanent soil dwellers by decreasing their population level, dynamic density (table 1) especially in deep soil layers, and by abrupt disturbances in the structure of animal communities. The study showed that the species composition is poorer under the increasing level of radioactive contamination.

Table 1. The changes of the dynamic density (ind/100 trap-days) of different groups of soil mezofauna in pine forest (v. Babchin) from 1991 to 1997

Invertebrates	Years of investigations					
	1991	1992	1993	1994	1995	1997
Arachnida	19.98	43.09	51.07	63.16	168.75	29.03
Myriapoda	9.15	9.54	6.97	2.63	2.08	2.15
Insecta (without Hymenoptera)	140.35	196.71	446.72	546.06	2072.9	615.02
	45.57	118.42	75.65	101.32	49.99	107.00
Coleoptera	45.42	118.10	70.05	98.69	35.41	68.96
Carabidae	21.49	78.62	41.49	52.63	20.83	10.75
Staphylinidae	8.52	28.29	19.76	18.42	8.33	8.87
Total:	169.48	249.3	504.9	611.8	2243.74	645.7
Without ants:	*74.85*	*172.5*	*134.9*	*171.0*	*220.82*	*137.6*

After 2 – 3 years marked differences between populations in contaminated and control areas were found, but the species diversity changes in contaminated soils were two times lower.

After 5 – 6 years marked differences between soil invertebrate populations in contaminated and control biogeocenoses were found, but species diversity in contaminated soils was five times lower.

After 9 – 10 years the species diversity of invertebrate community was still reduced by 60 %.

During last years it was observed a slow return of animal diversity and community structure to the initial parameters. Secondary changes and side-effects were registered for phytophagous and saprophagous.

3.2. TROPHIC STRUCTURE

The dynamics of the trophic structure whose changes are indicative of drastic disturbance in the soil invertebrate communities, is an even better indication than density and zoomass. The trophic groups have characterized based on density and zoomass.

It was found that in the soils of studied biogeocenoses (in 1986 – 1989) the density, and, consequently, zoomass decreased in all trophic groups, namely phytophagous, saprophagous (disappear at all) and zoophagous.

We carried out the pedobiological research (in 1994 – 1996) in the forest biogeocenoses with different gamma-radiation background: 1) the weak radioactive contamination (0,15 – 0,19 mr/hr) 2) the strong radioactive contamination (2,28 – 5,58 mr/hr). In the zone of weak radioactive contamination the zoophagous are dominants. In the zone of strong radioactive contamination phytophagous are dominants, but the zoophagous are subdominants.

3.3. RADIONUCLIDE ACCUMULATION

The results of the studies on ^{90}Sr and ^{137}Cs content in the soil, litter and invertebrates in the exclusion zone of the CNPP have been analysed. The dependence of radionuclide accumulation factors on the peculiarities of morphological structure, functional ecology and nutrition type of soil invertebrates was found. The dynamics of ^{137}Cs content in some species of invertebrates was studied (Fig. 1, 2).

The highest contamination level in the invertebrates was found in the year of the accident. It has been found that a year after accident the invertebrates gamma-activity dropped considerably, then its decrease was slower. Similar changes were also observed for gamma radionuclides in the litter and in the soil, but the contents of gamma-radiators in the litter being higher.

For the ^{90}Sr accumulation in the invertebrates, we can arrange in the following decreasing order: Diplopoda (millipedes) – Shelled mollusks – Insecta (insects) – Lumbricidae (earthworms). For ^{137}Cs accumulation in the invertebrates, we can compose the following decreasing order: Insecta – Diplopoda – Shelled mollusks – Lumbricidae.

The radionuclide loading is caused by ecological biotopic and food characteristics and it depends on species of soil animals: saprophagous, which are a final link in the trophic chains, accumulated radionuclides most of all. Then, phytophagous follow, zoophagous being the last.

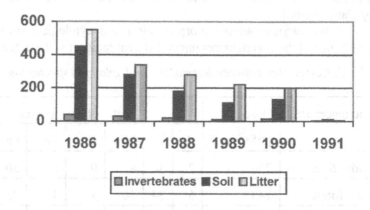

Fig. 1. The changes of activity of ^{137} Cs (kBq/kg) in soil, litter and invertebrates from 1986 to 1991 (pine forest, v. Babchin).

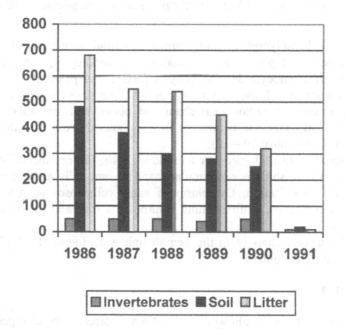

Fig. 2. The changes of activity of ^{137} Cs (kBq/kg) in soil, litter and invertebrates from 1986 to 1991 (oak forest, v. Lomachi).

3.4. CHANGES IN THE HEMOLYMPH

Hemolymph of the invertebrates performs vital functions, metabolic, homeostatic and protective, and its cellular composition is a reliable indicator of the population viability.

We carried out the morphological and cytological investigation of the hemolymph of the invertebrates under different radioactive level (table. 2).

Table 2. Correlation between haemocytes in Diplopoda species under different doses.

Biogeocenoses	Dose	Ratio of haemocytes						
	(mSv/year)	Pr	Ìa	Mi	E	Fp	Fa	D C
Alder forest	200	2	30	8	0	8	30	22
Pine forest	240	5	34	10	1	10	25	15
Oak forest	305	1	25	5	0	4	36	29
Meadow	638	1	18	2	0	2	42	35

Pr – proleukocytes; Ma – macronucleocytes; Mi – micronucleocytes; E – enocytoides; Fa – active phagocytes; Fp – passive phagocytes; DC – dead cells.

In the hemolymph of both normal and irradiation-exposed invertebrates, five haemocyte types were revealed: proleukocytes, macronucleocytes, micronucleocytes, enocytoides; phagocytes (active and passive) (classification of [4]). The proportion of haemocytes varies with the physiological condition. The most typical sings of pathological change of hemolymph are an increase of the amount of dead cells (in control – 2 – 3) and decrease of the amount of proleukocytes (in control 10 – 12).

Morphological changes are as follows: lysis, fission shift of the nucleus to the cell periphery, disturbance of circular shape in some cells, elongation, cytoplasm budding and vacuolisation. Cytoplasm of some cells loses its granular texture, becomes homogeneous, while staining it turns dark blue and the nucleus acquires reddish tinge.

The largest changes in the hemolymph were found at a higher radiation dose absorbed by the objects.

4. Conclusions

Thus, our studies allowed us to estimate correctly the changes induced by radioactive contamination caused by Chernobyl nuclear accident at the level of soil invertebrate communities and to reveal the pattern of their disturbance under constant radiological pressure. All evidence shows that soil mezofauna complexes in different biogeocenoses exposed to irradiation for a long time react clearly by a noticeable suppression. It is concluded from the results that the haematological characteristics can serve as convenient bioindicators of the radioactive contamination in the biogeocenoses. Percentages of all cell types observed and their structural changes may be recommended as criteria for comparison.

5. References

1. M.I. Kuzmenko Radioecological problems in Ukrainian water reservoirs . Gidrobiol. Zh. 34 (6), 1998, pp. 95 – 119 (in Russian).
2. M.S. Ghilarov, D.A. Krivolutsky Radioecological investigations in soil zoology. Zool. Zh. 50 (3), 1971, pp. 329 – 342 (in Russian).
3. D.A. Krivolutsky Radioecology of communities of terrestrial animals. M: Energoatomizdat, 1983, 88 pp. (in Russian).
4. M.I. Sirotina Genesis of blood cells in normal and icteric oak eggar moth larvae and butterflies. Dokl. VASKhNIL, 4, 1965, pp. 22 – 28 (in Russian).

Part 4.

Problems of Estimation of Risks from Different Factors

GENERALIZED ECOSYSTEM INDICES: ECOLOGICAL SCALING AND ECOLOGICAL RISK

V. GEORGIEVSKY

Russian Research Center "Kurchatov Institute", Moscow, RUSSIA

1. Introduction

There are two problems regarding analysis of data for ecological monitoring:

- Reduction of dimensionality data. There it is necessary to convert the contamination data of few ecological chains by few pollutants to system characteristics. It is may be realized by Multidimensional Analysis (Discriminant Analysis, Factor Analysis, Cluster Analysis, etc);
- Interpretation of results of Multidimensional Analysis. This is a key problem for ecological monitoring and is to be considered as a problem of ecological scaling.

Two examples cover the wide spectrum of ecological situations concerning the monitoring of the radioactive contamination territories:

- Analysis of monitoring data in Rovenskaia Area of Ukraine; contamination data of several ecosystem' chains by single radionuclide contamination of soil, milk, potatoes by ^{137}Cs for 236 settlements have been analysed and
- Analysis of monitoring data in Kievskaia Area of Ukraine; contamination data of single ecosystem' chains by several radionuclides (soil contamination by ^{90}Sr, ^{134}Cs, ^{137}Cs, ^{106}Ru. ^{144}Ce for 122 settlements) have been analysed.

2. Methodology

1. The attribute of data from ecological monitoring is the following. Many variables in noise condition are measured but among these variables there is not a key variable. The ecological data are "symptoms" only. The state of ecosystem is determined by effect of all data on the system, not by any particular measured variable. This is in analogy with diagnostics in medicine where it is necessary to indicate disease condition using circumstantial symptoms.

There is apparently a possibility to introduce the Ecological Index as a system characteristics and to develop the ecological scale for ranking the ecological conditions, for the formalization of the concept of ecological risk and for the decision-making [1, 2, 3].

2. The development of ecological scale consists of following steps:
- Reduction of dimensionality of initial data of monitoring and convert it to small system parameters,
- Discrimination and classification,
- Clusterization,
- Development of Multidimensional Reference Ecological Image (MREI),

F. Brechignac and G. Desmet (eds.), Equidosimetry, 165–174.

- Scaling the ecological conditions in terms of MREI.

    In the ecological scale the Scaling Factor is introduced where it acts like a kind of "ecological thermometer".

    An analysis of ecological conditions and ecological decision-making may be realized on the base of this model ("ecological thermometer") by analysis of the ordering and evolution of the ecosystem position on the scale.

    3. The following steps are carried out :.

- Discriminant Analysis [4];
- Classification in space of the canonical roots (in the terminology of Canonical Correlation Analysis [5, 6, 7];
- The procedures of discrimination and classification are iterative. This is a way of uploading the necessary clusters. (The type of countermeasures for Rovenskaia Area in Ukraine was developed with reference to regions in this area. Therefore the accordance of ecological conditions with the administrative affiliation was tested. The administrative affiliation has been changed for reasons of its incorrect classification).
- The pseudovariables 1,2... for uploading clusters are introduced. The centres of these clusters are considered as MREI. The assignment scores 1,2... to the clusters is the expert and iterative procedure (expert knowledge about ecosystem condition as a general rule has a multidimensional appearance);
- In the Multiple linear Regression the pseudovariables 1,2... are dependent variables;
- The mathematical expression for ecological scale is the regression equation where independent variables are the initial ecological data. Linear combination of regression coefficients and ecological data are the Ecological Index, which indicates the position of Ecological condition on the scale and which may be interpreted as Ecological Risk (it is the reflection of the multiple structure of the ecosystem by a single value).

Note: This methodology is some variant of the Multidimensional scaling for ecosystem conditions in terms of Multidimensional Reference Ecological Image (indicated 0, 1, 2. by experts).

    4. This Analysis has been founded on the next hypothesis:

- It is supposed that the ecological conditions depend on distances between the points inside clusters and between the points of different clusters.
- It is supposed that the linear combination of ecological parameters is the approximation to whichever dependency of the ecological conditions on distance.
- The Euclidean distance is used because the canonic roots, which are used for clusterisation, are uncorrelated.

    5. As it is known the Linear Regression is used in the Discriminant Analysis for the classification. Roughly speaking, there the regression function is passed between clusters. In our method the regression functions are used twice:

- on the first stage, it is used for classification (Discrimination Analysis) and there the regression functions pass between clusters, and

- on the second stage, the regression functions pass across clusters. This stage needs further study.

3. Ecological scaling of the radiation monitoring data. Contamination of several ecological chains by single radionuclide - Rovenskaia Area Ukraine, 1992

a) Data source

The procedure will be demonstrated with the example of the analysis of the post-Chernobyl radiation monitoring data for Ukraine that are given in the database [8]. The principal block of this database is the observed data of average contaminations of soil, milk and potatoes by ^{137}Cs for every settlement in all areas of Ukraine.

b) Regions

In our examples selected data regarding Rovenskaia area of Ukraine will be considered. This area consists of 7 Regions: Goschansky, Berezansky, Vladimiretsky, Zarisgnensky, Dubrovitsky, Rokitnevsky, Sarnensky. For our analysis, these data were selected as the list of 236 settlements, as of 1992.

4. The chaotic structure of monitoring data.

The contaminated data into this area have a very chaotic structure. This is illustrated in Figure 1, which shows the concentration of ^{137}Cs in milk (Bq/l) and in potatoes (Bq/kg) versus the ^{137}Cs in soil (Bq/kg).

Contamination ot milk and of potatoes versus contamination of soil.
Rovenskaia Area of the Ukraine, 1992

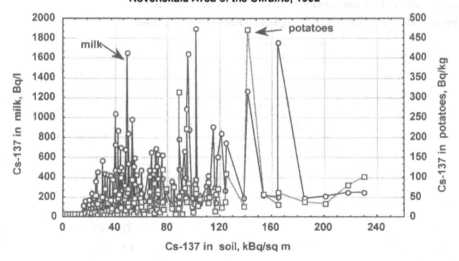

Fig. 1. The chaotic structure of monitoring data for Rovenskaia area of Ukraine.

From Figure 1 it is clear also that levels of contamination of milk and of potatoes may differ on the order of value for same level of contamination of soil or may be the same value, when a contamination differs on the order of value.

Such situation may be associated with greatly different ecological terms inside Rovenskaia Area, for instance, different types of soil.

This example illustrates the typical difficulties of the interpretation of the ecological data monitoring regarding the interpretation, multidimensional ecological data having complex (chaotic) structure and ensemble parameters. The procedures of ecological scaling may remove those difficulties.

5. Discriminant Analysis

5.1 ANALYSIS OF THE ACCORDANCE OF ECOLOGICAL CLASSIFICATION WITH ADMINISTRATIVE CLASSIFICATION

Five variables are used: ^{137}Cs in soil, kBq/m^2; ^{137}Cs in milk, Bq/l; ^{137}Cs in potatoes, Bq/kg; Transfer Factor for milk (TF_{milk}), kBq/l; Transfer Factor for potatoes (TF_{pot}), kBq/kg.

The result of Discriminant Analysis of the contamination data for Rovenskaia Area (Figure1) in space of canonical roots is indicated in Figure 2. (Here the cumulative proportion of variance extracted by root1 and root2 equals 0.93).

From Figure 2 it is clear that for many settlements the ecological conditions do not correspond with the administrative affiliation. Therefore the decision-making will be not correct for many cases (the decision-making regarding countermeasures in contaminated settlements is connected with administrative affiliation).

Discrimination of the settlements in Rovenskaia Area on the base of ecological conditions before change of administrative affiliation

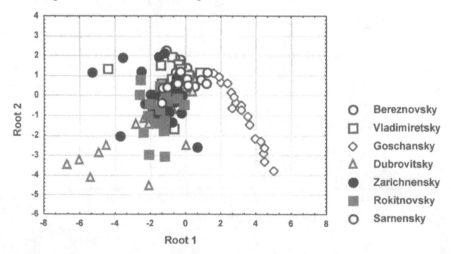

Fig. 2. Analysis of accordance of ecological classification with administrative classification before change of administrative affiliation.

5.2 CHANGE OF THE ADMINISTRATIVE AFFILIATION FOR THE CORRESPONDANCE OF THE ADMINISTRATIVE CLASSIFICATION TO THE ECOLOGICAL CLASSIFICATION

The administrative affiliation for settlement has been changed (the highest classification score has been used in classification functions) and Discriminant Analysis has been repeated. This procedure is iterative. (It is need about 10 iterations)

It appears that only 3 of 236 settlements are incorrectly classified by such procedure. It appears also that Sarnensky Region may be excluded from list of Regions; then the ecological conditions of the settlements including in the Sarnensky Region are classified as Beresnovsky.

The result of such iterative operations of the Discriminant Analysis is indicated in Figure 3. (In legend of Figure 3 there are 6 Regions only).

Discrimination of the settlements in Rovenskaia Area on the base of ecological conditions after change of administrative affiliation

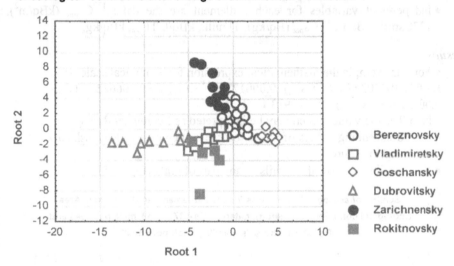

Fig. 3. The correspondence of the ecological classification to the administrative classification.

6. Assignment of the score to MREI

In our example the expert scores 1, 2,...6 for the clusters "Bereznovsky", "Vladimiresky", "Goschansky", "Dubrovitsky", "Zarechensky", "Rokitnovsky" (Figure3) as MREI were assigned. The score "1" was assigned for cluster "Goschansky" which was estimated by experts as the most favourable.

Note:

• The scores 1,...6 are the group centroids in the clusters "Bereznovsky", "Vladimiresky", "Goschansky", "Dubrovitsky", "Zarechensky","Rokitnovsky, respectively.

- The scores for MREI may be changed by experts on the base of test scaling. This procedure is iterative

7. Scaling

a) Method of scaling

The ecological scale is offered developed as Multiple Linear Regression where the score 1,.....6 (MREI for clusters "Bereznovsky", "Vladimiresky", "Goschansky", "Dubrovitsky", "Zarechensky", "Rokitnovsky, respectively) are the pseudovariables and these pseudovariables are dependent variables.

b) Model

The data for Regression Analysis are prepared as following.

- As dependent variables the values 1, 2...6 are assigned for settlements included in clusters "Bereznovsky", "Vladimiresky", "Goschansky", "Dubrovitsky", "Zarechensky","Rokitnovsky, respectively.
- Independent variables for each settlement are the data: $^{137}Cs_{soil}$ (kBq/m^2), 137Csmilk (Bq\l), $^{137}Cs_{pot}$ (Bq/kg), TFmilk, kBq/l, TF$_{pot}$, kBq/kg.

c) Results

- For our example the mathematical expression for ecological scale is:
R= 0.7+0.020377 ∗ ($^{137}Cs_{soil}$)+0.000177 ∗ ($^{137}Cs_{milk}$) + 0.106622∗ (TF$_{milk}$)- 0.001228∗ ($^{137}Cs_{pot}$) + 0.111913 ∗ (TF$_{pot}$),   (1)
where R is the value, which may be interpreted as Ecological Risk (R).
- Sorted ascending values of the R for 236 settlements of Rovenskaia Area are indicated in Figure 4.
- It is possible to show the details of any part of scale (Figure 5).

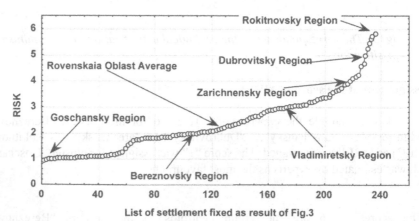

Scaling of ecological conditions for 236 settlements in Rovenskaia Area
Ecological conditions are determined by Cs-137 in soil, milk, potatoes and
Transfer Factors "soil-milk", "soil-potatoes"

Fig. 4. Ecological scale in terms of Risk (Eq.1) for Rovenskaia Area of Ukraine, 1992. (Indication of the mean values of Risk for each Region of Rovenskaia Area).

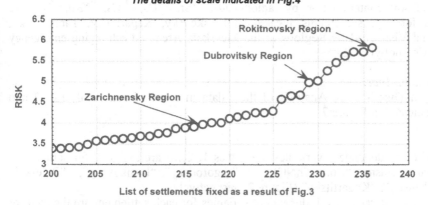

Scaling of ecological conditions for 36 settlements in Rovenskaia Area
Ecological conditions are determinated by Cs-137 in soil, milk, potatoes and
Transfer Factors "soil-milk", "soil-potatoes"
The details of scale indicated in Fig.4

Fig. 5. Details of scale indicated in Figure 4 (Eq.1).

8. Risk in comparison with Effective Doses

As indicated above, R is the reflection of the multiple structure of ecosystem by a single value. In our case R is the linear combination of contamination of several chains in ecosystem.

The Effective Dose due to the ingestion is the reflection of the multiple structures inside Man by a single value. E is the linear combination of contamination of several chains inside Man. The database [8] contains the information about E. There is thus a possibility to compare results of diagnostics of the Rovenskaia Area by both methods R and E (Figure 6).

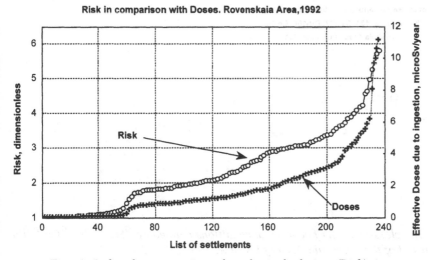

Fig. 6. Index for contaminated ecological chains (Risk) in comparison with Index for contaminated interior of man (Effective Dose).

9. Ecological scaling for territory contaminated by a spectrum radionuclides in the Kievskaia Area of Ukraine, 1987

a) Data source
The data of soil contamination by radionuclides ^{90}Sr, ^{134}Cs, ^{137}Cs, ^{106}Ru, ^{144}Ce are considered for 122 settlements located in Ivankovsky, Kagarlitsky, Tarashansky, Polesky, and Vishgorodsky Regions of the Kievskaia Area and estranging emergency zone near Chernobyl NGS [8].

b) Discriminant Analysis
The result of Discriminant Analysis of these data in space of canonical Root2 and Root3 is indicated in Figure 7.

c) Model
　　1) MREI: As dependent variables the values 1, 2…6 are assigned for settlements included in clusters "Zone ChNGS", "Vishgorodsky", "Ivankovsky", "Polesky", "Tarashansky", "Kagarlitsky" (Figure 7), respectively.
　　2) Independent variables: Independent variables for each settlement are the data for ^{90}Sr, ^{134}Cs, ^{137}Cs, ^{106}Ru, ^{144}Ce, in kBq/m^2.
The Euclidean distance is used because root2 and root3 (Figure 7) are uncorrelated.

d) Results
- The mathematical expression for ecological scale is:
  R= 1.141694 – 0.112609 ∗ (^{90}Sr) –0.094899 ∗ (^{106}Ru) +1.135997 ∗ (^{134}Cs) – 0.175412 ∗ (^{137}Cs) + 0.059925 ∗ (^{144}Ce),　　　　　　　　　(2)
  where R is the value that may be interpreted as Ecological Risk;
- Sorted ascending values of the R for 122 settlements of Kievskaia Area are indicated in Figure 8.

Note: In this example the levels of contamination of part of the Chernobyl zone which are used for analysis, are lower than in the other regions.

Fig. 7. The classification of ecological conditions defined by ^{90}Sr, ^{134}Cs, ^{137}Cs, ^{106}Ru, ^{144}Ce in soil of the Kievskaia Area of Ukraine, 1987 .

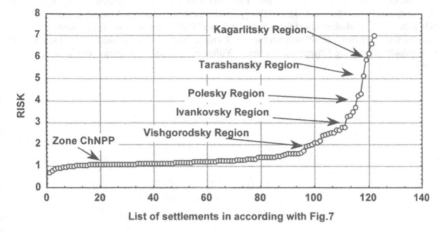

Fig. 8. Ecological scale in terms of Risk (Eq. 2) for Kievskaia Area of Ukraine, 1987. (Indication of the mean values of Risk for each Region of Kievskaia Area).

10. Conclusion

The way of multidimensional ecological scaling is developed. Within the framework of this procedure the ecological monitoring multidimensional data are converted into Ecosystem Index (EI). (This may be considered as a model of "ecological thermometer").

This method is based on the following procedures: reduction of the dimensionality of monitoring data, discrimination, clusterisation, classification and multiple regression.

The scaling factors for ecological scale are introduced in terms of Multidimensional Reference Ecological Images, which are assigned by experts.

The state of the ecosystem is reflected on the scale by value EI which may be interpreted as Ecological Risk.

The evolution of the ecosystem position along scale may be used for decision-making. The method is demonstrated in two examples: contamination of several ecological chains by single radionuclide and contamination of single chain (soil) by several radionuclides.

11. References

1. Georgievsky V.B. (1994). "Ecological and Dose Models for Radiation Accident", "Naukova Dumka" ("Scientific Thought") Publishers, Kiev, 236 pp., 1994 (in Russian);
2. Georgievsky V.B., Kameneva I.P. (1991) "Interactive analysis of data for ecological monitoring", In: Problem of Energy -savings, No 1, p.p.1-10, "Naukova Dumka" ("Scientific Thought") Publishers, Kiev, 1991 (in Russian)
3. Georgievsky V.B., Kameneva I.P., Syrvila A.P. (1992) "Multivariate analysis ecological data monitoring", Institute of Problems of Modeling in Energy of Ukrainian Sciences Academy, 92-45, Kiev, 39 p.p.

174

4. Bartlett M.S (1965). "Multidimensional Statistics" in Theoretical and Mathematical Biology", Edited by Talbot H. Waterman and Harold J. Morowitz, Yale University, Blaisdell Publishing Company New York, Toronto, London
5. Code "Statistica for Windows", StatSoft Inc., Release 4.3, 1993;
6. Maurice G.Kendall, Alan Stuart (1966) "The Advanced Theory of Statistics", V.3, Design and Analysis, and Time – Series, Second Edition, Charles Griffin&Company Limited, London;
7. S.S.Wilks (1967) "Mathematical Statistics", "Science" Publishers, Moscow (in Russian);
8. "Dosimetric certification of settlements of Ukraine which has exposed from radiological contamination resulting Chernobyl Accident", Volumes 1- 5, Kiev, Minzdrav of Ukraine, 1991 - 1995

ECOLOGICAL RISK ASSESSMENTY AS A METHOD FOR INTEGRATING RISKS FROM MULTIPLE STRESSORS AT HAZARDOUS WASTE SITES

R. MORRIS
TREC, Inc., 4276 E 300 N, Rigby, ID 83442,
UNITED STATES OF AMERICA

R. VANHOM
Idaho National Engineering and Environmental Laboratory, Idaho Falls, ID,
83402, UNITED STATES OF AMERICA

1. Introduction

Recent experience with human impacts on the environment has demonstrated a need for methodologies to determine protective levels of radiation dose, toxic chemical concentration, and physical stressors. Although there is a continuing need for research into mechanisms and effects, particularly for non-human components of the environment, the precautionary principle demands development of methodologies to establish sound regulatory standards in the absence of complete knowledge. These methodologies need not necessarily be based on a scientifically elegant or intellectually satisfying understanding of mechanisms, but would instead be pragmatic systems which offer real protection.

Ecological risk assessment is one method developed to analyze potential risks to the environment and allow the setting and application of regulatory standards in the presence of significant uncertainty. Although ecological risk assessment has been successfully applied to radiological, chemical, and physical hazards, methods have not been developed to consider these three types of stressors in a common framework. Analysing the three stressor types in a common framework is desirable because of the synergistic and antagonistic interactions known to exist between them. Separate analyses cannot account for these interactions.

The objective of this paper is to suggest some approaches which might enhance our ability to analyze radiological, chemical, and physical hazards jointly. These approaches are not offered as a final solution to the problem, but as models to stimulate discussion. To this end, it is worthwhile to consider the methods by which each of these stressor types has been analysed separately.

2. Ecological risk assessment

Ecological risk assessment is a science-based process to qualitatively or quantitatively determine potential harm to the environment from human activities. For the purpose of this definition, the environment that can be harmed is defined as populations of non-human biota (plants and animals), individual members of Threatened or Endangered species, or special habitats (e.g., caves, wetlands).

A focus on populations of plants and animals rather than on individual members of those populations is a philosophical stance reflected in most western cultures, reflected in our wildlife and land management laws and regulations. However, it is not shared by all cultures and may be changing in those cultures

F. Brechignac and G. Desmet (eds.), Equidosimetry, 175–185.

where it exists. Thus, while it seems a reasonable initial position, it is important to bear in mind it may not always be tenable to the stakeholders our regulations affect. This subject is discussed in more detail in a recent International Atomic Energy Agency (IAEA) document [1].

The above definition of ecological risk assessment also requires an explanation of which human activities can be evaluated. For this paper, the stressors of interest are grouped into three categories: ionizing radiation, toxic chemicals, and physical disturbance.

Since the inception of radiation protection as a profession, the focus of its activity has been on protecting human health. There are currently no universally agreed criteria or methodologies for explicitly protecting non-human parts of the environment. For many, the International Commission on Radiological Protection (ICRP) provided the basis for protecting the environment with the statement:

The Commission believes that the standard of environmental control needed to protect man to the degree currently thought desirable will ensure that other species are not put at risk [2].

This "belief" is no longer universally accepted [1] because, 1) there is a large degree of variability in radiosensitivity between species, 2) non-human species may be exposed to environmental radioactivity through pathways not normally available to humans, and 3) much of the basis for protection of humans is based on excluding them from contaminated environments. Non-human species cannot always be similarly excluded. Thus it has become necessary for authorities to demonstrate their regulations are protective of non-human organisms and methods are under development in many countries to do so [3, 4, 5]. Many of these methods are based upon an ecological risk assessment framework.

The non-human components of the environment are also adversely affected in a variety of ways by toxic chemicals. These effects range from direct mortality to subtle genetic damage resulting in decreased evolutionary fitness. The ecological risk assessment approach was originally developed to evaluate risks to the environment from toxic chemicals. [6, 7] It has been successfully applied to this problem at hazardous waste sites across the world, and, although the methodology is continually under improvement, it is a well-developed and well-accepted approach.

Physical disturbance of the environment, ranging from direct mortality, to habitat destruction, habitat Fragmentation, or noise disturbance, has not been well analysed in an ecological risk assessment framework. This is not so much because the methods are not applicable, as because of the historical development of ecological risk assessment as an approach for evaluating risks from toxic chemicals.

2.1 GENERAL PRINCIPLES

There are some general principles for ecological risk assessment that are shared in common between the three types of analysis.

2.1.1. Begin from a conceptual model

In every risk assessment, it is important to begin by developing a well-supported conceptual model of how the environment under consideration is organized and how the stressor might adversely affect it. For chemical toxins and radioactivity, this should include consideration of the exposure pathways. The conceptual model allows the assessors to determine and focus on those components of the environment where adverse effects are most likely to be found and, thus, increase the efficiency of their assessment.

2.1.2. Monitor the correct pathways

Most existing environmental monitoring programs for radioactive or chemical stressors are intended to assist in protecting humans from negative impacts of environmental contamination. Thus, they monitor those pathways by which humans are most likely to be exposed. However, non-human organisms are likely to be exposed by transport pathways not usually available to humans. It is important to ensure that the data used in the assessment are appropriate for the organisms of concern. The same general principle applies to assessments of physical hazards. While humans might not be greatly affected by a stream diversion or road construction, the organisms that inhabit the stream or those whose migration corridor is severed by the road might be greatly affected.

2.1.3. Graded approach

Ecological risk assessment professionals have long recognized that a graded approach is the most cost-effective and efficient method of conducting assessments. After an initial data assembly phase, during which appropriate conceptual models are developed, supported, and documented, a screening assessment is conducted. This screening assessment uses general models with conservative parameters, i.e., models which err on the side of showing risk where none exists. If the activity being assessed with these models passes the screen, assessors can be confident that no risk exists. If the activity fails the screen, then a process of iteration with progressively more complex and site-specific models is conducted. This continues until it is either demonstrated the activity poses an acceptable level of risk, or it is concluded unacceptable risks exist which must be mitigated.

2.1.4. Compliance with a standard

In order for risk managers to make practical use of the results of an ecological risk assessment, there must exist some agreed upon standard; a value which regulators and stakeholders agree represents the limit of acceptable risk. In most cases, for ionizing radiation, this has been a dose limit or, more basically, an activity concentration in an environmental medium. For chemical toxins a Lowest Observed Adverse Effect Level (LOAEL) or No Observed Adverse Affect Level (NOAEL) has been applied. There is no generally accepted or universally applied standard for physical disturbance. These general principles are applicable to any ecological risk assessment. It is now useful to consider how they are applied within each of the given hazard types.

3. Assessing ecological risks from ionizing radiation

In this paper, the U.S. Department of Energy's (DOE) Graded Approach [3] is used as the model for assessing risks to the environment from ionizing radiation. This is not because it is the only approach which might be effective, but because it is familiar to the authors and effectively incorporates the general principles defined above.

3.1. U.S. DEPARTMENT OF ENERGY GRADED APPROACH[1]

The Graded Approach was not originally designed as an ecological risk assessment approach, but as a means for DOE facilities to demonstrate compliance with proposed limits for environmental protection. However, it was based on the U.S. Environmental Protection Agency's (EPA) Framework for Ecological Risk Assessment [6] and can be easily integrated into an ecological risk assessment methodology.

The radiation dose standard incorporated into the Approach are derived from an IAEA Technical Report [8] which examined the relevant radioecological literature and found no evidence of population harm when maximally exposed individuals received dose rates of:

- Terrestrial animals: < 1 mGy d^{-1}
- Terrestrial plants: < 10 mGy d^{-1}
- Aquatic Organisms: < 10 mGy d^{-1}

Other standards are possible. Currently the European Union is conducting the Framework for Assessment of Environmental Impact (FASSET) initiative [4] which will likely result in standards different from those on which the Graded Approach is currently based. If other dose standards are found to be more appropriate, the Graded Approach has the flexibility to adjust the default standards.

In the Graded Approach, the dose limits are used to derive Biota Concentration Guides (BCG). These are defined as the soil, sediment, or water concentration of a single radionuclide that would give a limiting dose under the model assumptions. BCGs make it possible to directly use measured concentrations to assess risks to the environment.

Application of the Graded Approach occurs in three major steps which are further subdivided into six minor steps (Fig. 1).

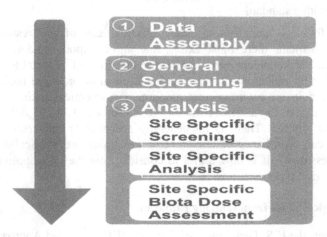

Fig. 1. Schematic of the U.S. Department of Energy's Graded Approach for Assessing Doses to Biota.

3.1.1. Data assembly

This step incorporates all the activities preliminary to the analysis. The conceptual model is developed and evaluated, and the area of the assessment is defined. Radionuclide concentration data are assembled and evaluated. The degree to which the data represent the pathways of concern, as determined by the conceptual model, is of particular interest.

3.1.2. General screening

This is the initial screening step. The highest available substrate (e.g., soil, sediment, or water) contamination found within the area of assessment is divided by the BCG derived using reference organisms and default parameters to create a Hazard Quotient. These BCGs are tabulated in the Graded Approach documentation [3] and are defaults in the software. If the Hazard Quotient is less than 1.0, risk to the populations inhabiting the assessment area is negligible, the results are documented, and the assessment stops. If the Hazard Quotient is equal to or greater than 1.0, the assessment continues with Site Specific Screening.

If more than one radionuclide or more than one environmental substrate is to be assessed, Hazard Quotients are combined using a Sum of Fractions approach. A Hazard Index is calculated as:

$$HI_R = \sum_{ij} \left(\frac{C_i}{BCG_i} \right)_j$$

where:

C_i = Concentration of radionuclide i in medium j

BCG_i = Biota concentration guide for radionuclide i in medium j

The Hazard Index as written above does not account for Relative Biological Effectiveness of different radiations. This is handled in the Graded Approach by using a Quality Factor in calculation of the BCG.

3.1.3. Site-specific screening: step 1

In this step, the same BCG used in the General Screening step divides an "average" substrate contamination in a refined analysis area. As before, if the Hazard Index is less than 1.0, the results are documented and the analysis ends. If the Hazard Index exceeds 1.0, screening continues.

3.1.4. Site-specific screening: step 2

Here, the same average concentration used in the previous step is divided by a new BCG, derived using parameters from organisms actually found on the site and uptake parameters measured on the site. Analysis either ends or continues as before.

3.1.5. Site-specific analysis

This step uses the same concentration and BCG from the previous step, but corrections are made for the relative sizes of the contaminated area and the organism's home range or the residence time of the organism in the contaminated area. Analysis either ends or continues as in previous steps.

3.1.6. Site-specific biota dose assessment

This is the final step in the assessment and is only reached if all the previous screening steps have failed. In this case, it is necessary to complete a comprehensive dose assessment for the organisms of interest from first principles. It will be necessary to assemble an interdisciplinary team to review the requirements and assumptions of the assessment, and either use available tissue concentration data or collect new data. If this step results in a determination that risks are unacceptable, mitigation should be recommended. Of course, at any time during the previous steps, risk managers may choose to opt out of the process and mitigate the risk based on the available results.

4. Assessing ecological risks from toxic chemicals

The methodologies used for assessing risks from chemical toxins to the environment are very similar to those described above, and will not be described in as great a detail. Ecological risk assessment for chemical toxins incorporates the general principles of working from a conceptual model, evaluating the correct pathways, using a graded, iterative, approach, and comparing to a standard. The EPA described the framework for this kind of assessment [6] used throughout the United States of America.

4.1. ECOLOGICAL RISK ASSSMENT AT THE IDAHO NATIONAL ENGINEERING AND ENVIRONMENTAL LABORATORY (INEEL)

The EPA framework (Fig. 2) was used to develop the screening-level ecological risk assessment approach used at the INEEL [9,10,11]. In this approach, the standards for comparison include such things as Lowest Observed Adverse Effect Levels (LOAEL), No Observed Adverse Effect Levels (NOAEL), and water quality criterion drawn from many literature sources. Using these standards, an Ecologically Based Screening Level (EBSL) is defined for each chemical toxin in each medium of interest. The EBSL is defined as the soil, sediment, or water concentration that would give a limiting dose of a specific chemical toxin under the model assumption. Like BCGs, the EBSLs make it possible to screen with measured or modelled substrate concentrations rather than conducting an extensive sampling program.

Similar to the Graded Approach, a Hazard Index derived using the Sum of Fractions is used to combine the risks from different chemicals in different media.

$$HI_C = \sum_{ij} \left(\frac{C_i}{EBSL_i} \right)_j$$

where:
C_i = Concentration of radionuclide i in medium j
$EBSL_i$ = Ecologically based Screening Level for chemical i in medium j.

The use of the Hazard Index is complicated by possible antagonism or synergy between chemicals. This is commonly dealt with by adjusting the acceptable hazard index downward to some value below 1.0. This adjustment has little scientific basis, but it is a pragmatic approach which is intended to ensure protection.

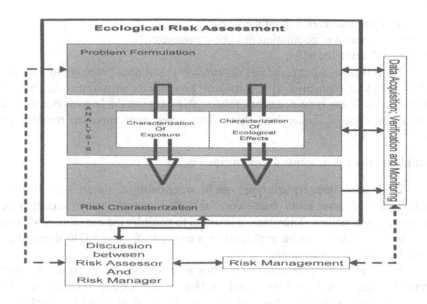

Fig. 2. Schematic of the U.S. Environmental Protection Agency's Framework for Ecological Risk Assessment.

5. Assessing ecological risks from physical hazards

This is a relatively unexplored area in ecological risk assessment. There are no generally accepted measurement endpoints. Possible endpoints might include unacceptable population size changes, levels of habitat destruction, degree of habitat fragmentation, or changes in net primary production, among others. However, although difficult, this is not an insoluble problem. Selection of endpoints can, and should, be negotiated among stakeholders.

In addition, there are no generally accepted standards to which one should compare, i.e., no answer to the question, "How much change is too much change?" This question might be addressed in a number of ways, including negotiating among the stakeholders. However, if one is willing to assume environmental parameters vary about a long-term average, one might establish an acceptable deviation from that mean as, for example, within two standard deviations of the mean. One could then calculate a Hazard Quotient as the actual deviation from the mean, divided by the acceptable deviation from the mean.

Then, one might combine risks from physical hazards using the Sum of Fractions approach to develop a Hazard Index, as for ionizing radiation and chemical toxins.

$$HI_P = \sum_i \left(\frac{M_i - X_i}{M_i - 2SD_i} \right)$$

where:

X_i = Value of population or habitat parameter i

M_i = Long-term mean of population or habitat parameter I

SD_i = Standard deviation of M_i

The primary limitation of this approach is that it does not explicitly account for possible synergy or compensatory mechanisms between the various physical hazards. In addition, because there is no necessary relationship between the hazards one must ask if it is at all meaningful to combine them. This assessment approach requires further study.

6. Combining the assessments across hazard types

While these three hazard types can be successfully analysed individually using an ecological risk assessment framework, this does not allow consideration of the interactions between them. Therefore, it is useful to consider the question of whether the three types of stressors can be evaluated in a common framework to determine their combined risk to the environment.

The number of potential interactions makes this a very large, but, with the appropriate research, not insoluble problem. Hazard indices may provide the common currency necessary to combine the various risks. The difficulty will be in selecting the appropriate method of combination that accounts for the possible synergies and antagonisms between them. This paper makes no pretention of presenting the best and final solution. However, it does offer some possible approaches to the problem. These include multiple regression analysis, principle components analysis, and a Monte Carlo approach.

6.1. MULTIPLE REGRESSION

In this approach, hazard indices for the three hazard types, and all of their possible combinations, are multiplied by coefficients, which represent the relative importance of the individual indices or their combinations, and the results are summed.

$$HI_T = aHI_R + bHI_C + cHI_P + dHI_R HI_C + eHI_R HI_P + fHI_C HI_P + gHI_R HI_C HI_P$$

HI_T = Combined Hazard Index

HI_R = Radiation Hazard Index

HI_C = Chemical Hazard Index

HI_P = Physical Hazard Index

a, b, c, d, e, f, g = coefficients

Of course, the principle difficulty is finding the correct values of the coefficients. It may be, the only possible solution to this is a full factor experiment, subjecting test populations to all possible combinations of stressors and measuring the results. This experiment may obviate the need for the risk assessment. One solution to this problem might be a Monte Carlo approach.

6.2. A MONTE CARLO APPROACH

The application of Monte Carlo procedures may prove useful in using what is known about the coefficients for the regression to determine a distribution for the Combined Hazard Index (Figure3). Also, in this procedure, a simple weighted sum model might be used, where each hazard index is weighted by a coefficient representing their relative contribution, and their results are summed. The hazard indices, as simple linear functions of dose, are non-stochastic parameters in the model. The problem is finding the value of the regression coefficients. In the Monte Carlo approach a set of coefficients is selected randomly from distributions reflecting what is known about them. If one has no knowledge about the potential value of the coefficient, a simple uniform distribution might be selected. On the other hand, if one has fairly complete knowledge about the coefficient, one may be able to specify a more structured distribution for the coefficient.

In any case, these coefficients, randomly selected from assigned distributions, are used to derive a result from the weighted sum model. The result is saved and another set of coefficients is randomly selected. A result from the weighted sum model is again saved and the process is repeated. This goes on until enough results are obtained to sufficiently specify a resulting distribution. This distribution describes the possible values of the combined hazard index, given the uncertainty about the coefficients, and its parameters may be used to determine the presence or absence of risk. If the mean, the upper 95% confidence limit, or some other agreed upon parameter is less than one, there is no evidence of risk.

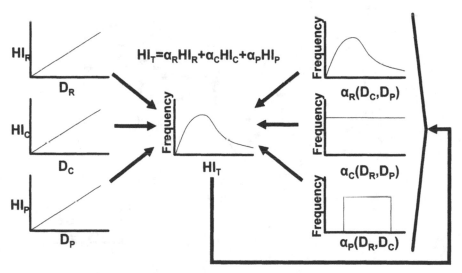

Fig. 3. Schematic of a Monte Carlo approach to determining the distribution of a combined Hazard Index. See the text for details.

184

6.3. PRINCIPAL COMPONENTS ANALYSIS

Principal components analysis is a method to reduce large data sets to a few linear combinations of the data which explain the bulk of the variance. All relevant ecosystem metrics and effects data from a stressed, test area and a similar, non-stressed, reference area may be combined in a single principal components analysis (PCA). The results may then be plotted in the component space (Fig. 4). If the reference area and the test area fall into separate groups, there may be a risk associated with the test area. If not, there is no reason to believe the test area is responding any differently than the reference and no risk is present. As data from various sites are accumulated, it may be possible to rank new sites relative to already assessed sites.

7. Conclusion

It is possible to effectively and efficiently analyze risks to the environment from ionizing radiation, chemical toxins, and physical hazards as individual stressors. However, our ability to effectively manage ecological risk would be greatly enhanced by the ability to analyze these risks in a single framework and it remains to be seen whether this is possible. In this paper, we have attempted to outline some possible approaches to this problem in the hope of stimulating further discussion and research along these lines.

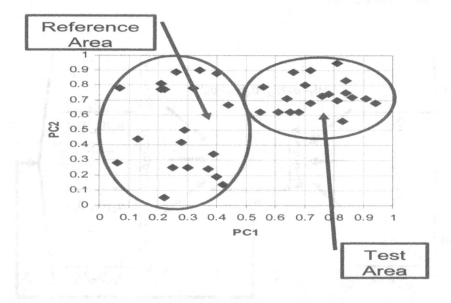

Fig. 4. Hypothetical Principal Components Analysis results to illustrate a result with a difference between the test area and the reference area.

8. References

1. International Atomic Energy Agency, Ethical Considerations in Protecting the Environment from the Effects of Ionizing Radiation: A Report for Discussion, IAEA-TECDOC-1270, IAEA, Vienna (2002).

2. International Commission on Radiological Protection, 1990 Recommendations of the International Commission on Radiological Protection, ICRP Publication 60, Pergamon Press, Oxford and New York (1991).
3. United States Department of Energy, A Graded Approach for Evaluating Radiation Doses to Aquatic and Terrestrial Biota, Interim Technical Standard DOE-STD-XXXX-00, U.S. Department of Energy, Washington, DC (2000).
4. FASSET, http:/www.fasset.org
5. Summary Report of the IAEA Specialists Meeting on Environmental Protection from the Effects of Ionizing Radiation: International Perspectives, 26-29 November 2001.
6. United States Environmental Protection Agency, Framework for Ecological Risk Assessment, EPA/630/R-92/001, U.S. Environmental Protection Agency, Washington, DC (1992)
7. Suter, G.W., Ecological Risk Assessment, Lewis Publishers, Chelsea, MI (1993).
8. International Atomic Energy Agency, Effects of Ionizing Radiation on Plants and Animals at Levels Implied by Current Radiation Protection Standards, Technical Report Series 332, IAEA, Vienna (1992).
9. VanHorn, R.L., N.L. Hampton and R.C. Morris, Guidance manual for conducting screening level ecological risk assessments at the INEL. INEL-95/0190, Lockheed Idaho Technologies Company, Idaho Falls, ID (1995).
10. VanHorn, R.L., N.L. Hampton and R.C. Morris, Methodology for conducting screening-level ecological risk assessments for hazardous waste sites, Part I: Overview. International Journal of Environment and Pollution 9:26-46 (1998).
11. Hampton, N.L., R.C. Morris and R.L. VanHorn, Methodology for conducting screening-level ecological risk assessments for hazardous waste sites, Part II: Grouping ecological components, International Journal of Environment and Pollution 9:47-61 (1998).

COMPARING RISKS FROM EXPOSURE TO RADIONUCLIDES AND OTHER CARCINOGENS AT ARCTIC COAL MINES

G. SHAW, K. VICAT, P. DELARD, S. CLENNEL-JONES
Department of Environmental Science and Technology, Imperial College, Ascot Berkshire, SL5 7PY, UNITED KINGDOM

I. FREARSON
Arctic Research Group, 29 Station Road, Borrowash, Derby, DE72 3LG, UNITED KINGDOM

1. Introduction

The risks posed to humans and ecosystems by environmental pollutants have traditionally been assessed on a substance-by-substance basis. However, it is increasingly recognized that the combined effects of pollutants on organisms and the environment should be evaluated where this is possible. Most locations on the surface of the Earth have received substantial inputs of pollutants and contaminants over recent decades, including artificially enriched heavy metals, organic xenobiotic compounds and radionuclides. The Arctic regions have been the focus of recent concern since they have become contaminated with a complex mixture of pollutants but are still regarded as being relatively unpolluted compared with lower latitude environments [1]. Thus, Arctic ecosystems offer an ideal opportunity for the development of methods for making comparative risk assessments for multiple exposures to contaminants of both human and non-human biota.

Arctic ecosystems receive pollutants from both distant and local sources [2]. Locally, the coal, oil and gas industries can enrich the arctic environment with radionuclides of the uranium and thorium decay series as well as carcinogenic contaminants such as arsenic and polycyclic aromatic hydrocarbons (PAH). On the island of Spitsbergen, in the Svalbard archipelago, coal has been mined since the beginning of the 20th century at several sites, notably Longyearbyen, Ny Ålesund, Sveagruva, Barentsburg and Pyramiden. This has lead to considerable despoilment of the terrestrial environment close to coal mines with coal wastes, comprising coal fragments and the host rock from which the coal was obtained, lying on the ground surface often over large areas [3]. These wastes have even used as in-filling materials for the construction of roads [4]. Any contaminating substances within these wastes are thus introduced into the surface environment and pose a potential hazard to plants, foraging animals and humans. An interesting development is the opening up of the town of Longyearbyen to tourism over the last decade, thereby increasing the potential for human exposure to waste materials from former coal mining activities.

This paper is intended to explore and compare the methodologies by which exposures of humans to both radioactive and non-radioactive contaminants in coal wastes can be assessed. It first attempts to quantify the significance of enhanced *in situ* radiation dose from coal wastes on Spitsbergen. It then attempts to show how the relative exposures and risks from both radioactive and non-radioactive carcinogens can be quantified and compared.

F. Brechignac and G. Desmet (eds.), Equidosimetry, 187–196.

2. Radioactive and non-radioactive contaminants associated with coal mining on Spitsbergen

Table 1 shows results of an analysis of coal sampled from Sveagruva in 1990 and indicates that, as well as having a high potential for soil acidification due to a 4% sulphur content, it also contains relatively high concentrations of potentially toxic elements such as arsenic and chromium. Arsenic is a known carcinogen and exposure to this element *via* environmental pathways is of particular concern.

Table 1. Contaminants in coal sampled in 1990 from Sveagruva, Spitsbergen (units are mg kg^{-1} dry weight).

S	K$_2$O	As	Cr	Pb	Cd
39000	100	19	10	<2.0	<0.2

In addition to potentially toxic elements, coal is known to contain naturally occurring radionuclides of the uranium and thorium decay series, as well as potassium-40. Table 2 indicates that these can become significantly enriched following combustion.

Table 2. Radioactive constituents of coal and in fly ash following coal combustion [5] (units are Bq kg^{-1} dry weight).

	^{40}K	^{238}U	^{226}Ra	^{210}Pb	^{210}Po
Coal	50	20	20	20	20
Fly Ash	265	200	240	930	1700

Monitoring studies over recent years have also shown that arctic areas with active hydrocarbon extraction industries can become significantly contaminated with polycyclic aromatic hydrocarbons (PAH). Measured concentrations of PAH in surface sediments in and around Spitsbergen range from 0.1 to 50 mg kg^{-1} [2] and derive, at least in part, from coal mining activities on the island. These substances have very long environmental half-lives and include certain individual compounds (such as benzo-a-pyrene, anthracene and phenanthrene) which are known to be potent carcinogens.

3. Measured and calculated kerma rates at two coal mines in Spitsbergen

An *in situ* spatial survey of kerma rate (μGy y^{-1}) was carried out in July 1999 at the former coal mine site in Ny Ålesund, situated in north-western Spitsbergen in the Kongsfjorden region. *In situ* measurements were taken at 1m above the ground surface using a Mini Instruments Environmental Radiation Meter Type 6-80 coupled to a Type MC-71 Geiger-Muller tube which has a response range of 0.05 to 75 μGy h^{-1}. Two areas were surveyed. The first was the area immediately adjacent to Ny Ålesund over which the remains of former coal mining activities are clearly visible (including significant quantities of coal wastes). The second survey area was the Brogger peninsula to the north-west of Ny Ålesund which has a similar geology (Carboniferous and Permian) to the mined area in Ny Ålesund but which is unaffected by coal wastes. The Brogger peninsula therefore acted as a control area. The results from these *in situ* surveys are shown as frequency distributions in Figure 1.

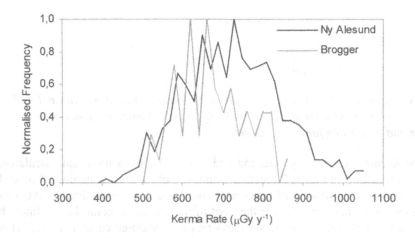

Fig. 1. Frequency distributions of in situ kerma rate at Ny Ålesund and Brogger.

The frequency distributions of *in situ* kerma rate at both the Ny Ålesund and Brogger areas are approximately normally distributed. The arithmetic mean kerma rates were 714 µGy y^{-1} and 663 µGy y^{-1} for Ny Ålesund and Brogger, respectively. A two-tailed t-test showed these means to be significantly different (p<0.001) which indicates that the presence of coal wastes at Ny Ålesund resulted in a generally higher kerma rate than in comparable non-contaminated areas. Though statistically significant, the difference between mean kerma rates in both areas is not dramatic. However, it is perhaps more informative to examine the maximum kerma rates observed in both areas which were 1040 µGy y^{-1} and 860 µGy y^{-1} for Ny Ålesund and Brogger, respectively. The highest kerma rates at Ny Ålesund were associated with areas in which combustion of coal had occurred, leaving behind coal ash deposits in which radionuclides were probably concentrated as indicated in Table 2. The maximum kerma rate recorded at Ny Ålesund could, in theory, lead to absorbed dose rates in excess of 1 mSv y^{-1} and is, therefore, of potential regulatory significance.

In July 1993, samples of soil and surface debris from another coal mining site in Spitsbergen (Sveagruva, in the Van Mijenfjord region) were obtained during an ecological survey of the area. These were subject to analysis by gamma ray spectrometry in the laboratory: gamma-emitting members of the uranium and thorium decay series were quantified and used to calculate *in situ* kerma rate using the dose conversion factors of Kocher and Sjoreen [6]. Kerma rate at a non-contaminated location, some 5 km to the west of Sveagruva on the northern shore of Van Mijenfjord, was calculated in the same way following analysis of soil samples, also collected in 1993, by gamma ray spectrometry.

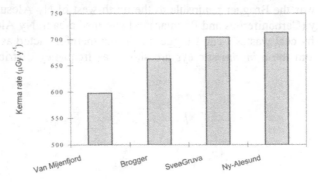

Fig. 2. Comparison of measured and calculated kerma rates at Ny Ålesund and Sveagruva coal mine sites together with their uncontaminated control areas, Brogger and Van Mijenfjord, respectively.

The kerma rates for Ny Ålesund and Sveagruva (measured and calculated, respectively) are shown in Figure 2 together with those of their uncontaminated control areas, Brogger and Van Mijenfjord (measured and calculated, respectively). At both coal mine sites the kerma rate is higher than that of the associated control area although the difference in kerma rate between Sveaguva and its associated control site (18%) is higher than that between Ny Ålesund and its control site (8%). The calculated kerma rate at Sveagruva is very similar to the mean measured kerma rate at Ny Ålesund, suggesting that the method of calculating *in situ* kerma rate from *ex situ* gamma ray spectrometry measurements of samples is reasonable. These results support the conclusion that the presence of coal waste materials has the potential to increase exposure of organisms to gamma radiation at coal mining sites.

Laboratory-based gamma ray spectrometry measurements of soil samples taken from Sveagruva and the Van Mijenfjord control area allow the relative contributions of individual radionuclides to the total kerma rates at each site to be determined. Figure 3 shows the contribution to the total kerma rate of uranium and thorium daughters, ^{137}Cs and ^{40}K.

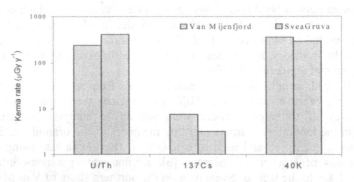

Fig. 3. Contribution to kerma rates at Sveagruva and Van Mijenfjord of uranium and thorium daughters (^{208}Tl, ^{210}Pb, ^{212}Pb, ^{214}Pb, ^{214}Bi, ^{226}Ra and ^{228}Ac summed), ^{137}Cs (derived from atmospheric weapons testing) and ^{40}K.

On average, [40]K made the largest individual contribution (49%) to the total kerma rates, while ^{137}Cs contributed less than 1%. This is of particular interest since it is generally accepted that 'radiocaesium is the most important anthropogenic radionuclide from a human-health perspective' [1]. This suggests that, in general, exposure to naturally occurring radionuclides is likely to be more significant than exposure to anthropogenic radionuclides in the Arctic. Figure 3 shows that uranium and thorium daughters, combined, contributed a total of approximately 50% to the total kerma rates, though this contribution was almost twice as high at Sveagruva, where coal wastes were present, than at the Van Mijenfjord control site.

4. Concentrations of radionuclides and other carcinogens at coal mine sites

As part of the ecological survey carried out at Sveagruva in 1993, soil samples were taken from 113 sampling points in and around an area containing substantial deposits of coal wastes. These samples were subjected to a multi-element analysis by ICP-AES which yielded the frequency distributions for potassium and arsenic concentrations in soil shown in Figures 4 and 5.

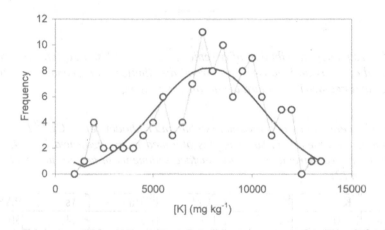

Fig. 4. Frequency distribution of non-radioactive potassium ([39]K) in 113 soil samples at Sveagruva. The solid curve is an idealised normal distribution with arithmetic mean =7584 mg kg[-1] and arithmetic standard deviation =2732 mg kg[-1].

Potassium is a naturally abundant element which displays an approximately normal frequency distribution. Assuming a natural abundance of 1.18×10^{-4} for [40]K, an arithmetic mean and range of [40]K activity concentrations in the Sveagruva soils can be calculated based on the frequency distribution for [39]K, as shown in Table 3. Also shown in Table 3 are the ranges of concentrations for ^{137}Cs, [226]Ra and [75]As (non-radioactive arsenic) determined directly in Sveagruva soil samples. Complete frequency distributions of soil concentrations were not obtained for ^{137}Cs and [226]Ra because of the length of time needed for low-level gamma ray spectrometry analyses. The ranges of concentrations presented for these radionuclides are based on a small number of selected samples. The activity concentrations for ^{137}Cs are equivalent to deposition densities of 0.97-13.9 kBq m^{-2} which can be compared with the range of ^{137}Cs deposition densities of 1 - 5 kBq m^{-2} estimated by AMAP for Spitsbergen [2].

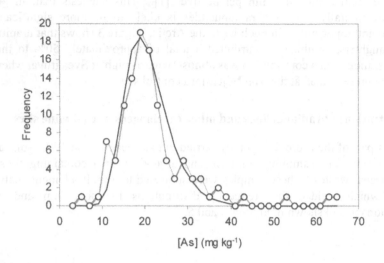

Fig. 5. Frequency distribution of arsenic (^{75}As) in 113 soil samples at Sveagruva. The solid curve is an idealised log-normal distribution with geometric mean = 19.1 mg kg^{-1} and geometric standard deviation = 1.3 mg kg^{-1}.

Table 3. Concentrations of potassium (stable and radioactive), ^{137}Cs, ^{226}Ra and ^{75}As (non-radioactive arsenic) in soil samples obtained from Sveagruva in 1993. PAH concentrations are extreme values for surface sediments in and around Spitsbergen obtained from [2].

	^{39}K	^{40}K	$^{137}Cs*$	^{226}Ra	^{75}As	PAH
Maximum	13100	400.4	31.0	7.3	63.4	50
Mean	7584	232.0	-	-	19.1	-
Minimum	1100	33.6	2.2	2.5	3.8	0.1
	mg kg^{-1}	Bq kg^{-1}	Bq kg^{-1}	Bq kg^{-1}	mg kg^{-1}	mg kg^{-1}

The log-normal distribution for arsenic is typical for many soil contaminants and indicates that high concentrations, potentially representing a high exposure risk, can be found in a relatively small proportion of the area sampled. The same type of distribution is also likely to apply to substances such as PAH. The values in Table 3, however, are the extreme concentrations (maximum and minimum) of PAH in surface sediment in and around Spitsbergen taken from [2]. Direct analysis of samples taken from a site such as Sveagruva is required to verify the applicability of these values to coal mine sites, but the range of PAH values has been inserted in Table 3 for inclusion in the comparative exposure and risk analysis developed in section 5.

5. Comparing environmental exposure risks for radioactive and non-radioactive carcinogens

Assessment of the risks of human exposure to radioactive substances in the environment follows a well-known and accepted procedure, as described by Thorne [7].

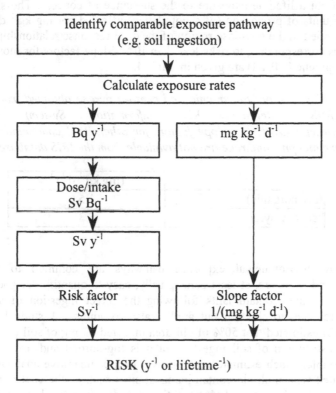

Fig. 6. Flow chart showing comparative calculations of risk posed by radioactive (left) and non-radioactive (right) carcinogens via common exposure pathways.

Exposure pathways are first identified then exposure *via* these pathways quantified, either by measurement, modelling or a combination of these. Estimates of human exposure to radionuclides are usually converted to estimates of radiation dose by using an appropriate dose conversion factor, such as the dose-per-unit-intake conversion coefficients (Sv Bq^{-1}) published by IAEA [8] for ingestion and inhalation pathways. Finally, for low-level radiation exposures a dose-to-risk conversion factor (Sv^{-1}) can be applied to provide an estimate of the fatal cancer risk associated with the exposure scenario being considered. For stochastic (low dose) effects, ICRP [9] has recommended values of 0.04 Sv^{-1} for workers and 0.05 Sv^{-1} for the public. This risk assessment process is represented by the series of linked boxes on the left hand side of Figure 6.

For non-radioactive carcinogens a similar process can be applied, as shown by the linked boxes on the right hand side of Figure 6. In common with the assessment of radionuclide exposure, exposure to a non-radioactive carcinogen must first be calculated following identification of an exposure pathway. Exposures to non-radioactive carcinogens are usually expressed as a daily rate normalized to an individual's body weight (i.e. mg substance kg^{-1} body weight d^{-1}). For prolonged daily exposures throughout an individual's lifetime, risk conversion factors ('slope factors') have been proposed by organizations such as the US Environmental Protection Agency (EPA). These slope factors approximate the 95% confidence limit on the increased cancer risk from a lifetime exposure to the substance of concern. The slope factor is expressed in units of proportion (of a population) affected per mg kg^{-1} d^{-1} and should usually be applied in the low-dose region of the dose-response relationship, specifically for exposures corresponding to risks less than 10^{-2}. Slope factors for inorganic arsenic and benzo-a-pyrene (a PAH) are given in Table 4.

Table 4. Slope factors for arsenic and benzo-a-pyrene obtained from the US EPA IRIS (Integrated Risk Information System) database (http://www.epa.gov/iris/). Slope factors for other PAH compounds such as anthracene and phenanthrene are not available from the IRIS database.

	$(mg\ kg^{-1}\ d^{-1})^{-1}$
As (inorganic)	1.5
benzo-a-pyrene	7.3

Many environmental exposure pathways are common to all types of contaminant and some are, at least conceptually, easy to quantify. A good example is the exposure to soil contaminants following the direct ingestion of soil. This is particularly relevant to children but applies also to adults. A study by Stanek and Calabrese [10] estimated that 50% of children ingested 13 mg of soil per day although the overall distribution of soil ingestion rates is log-normal and 'extreme' ingestion rates are possible. Such estimates can be used in a comparative analysis of exposure risks associated with multiple contaminants at sites such as the coal mines examined here. However, when considering arctic coal mines several caveats should be considered. First, due to the obvious climatic characteristics of arctic sites, which limit outdoor activities and result in snow cover for a large proportion of the year, the opportunities for direct exposure to soil are limited compared with more temperate latitudes. Secondly, the occupancy of the arctic regions is, in general, very low compared to temperate latitudes, although an increase in both permanent occupancy and tourism in Longyearbyen, the principal town of Spitsbergen and a former coal mining site, has occurred over the last decade. Furthermore, due to the very dry nature of the arctic in summer dust migration can be a particular problem and exposure to soil particles via this pathway may be significant.

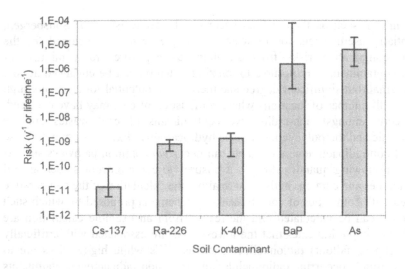

Fig. 7. Comparative cancer risks for exposure to radioactive and non-radioactive carcinogens via soil ingestion at coal mines in Spitsbergen. The columns represent risks calculated using best estimates of contaminant concentrations in soils taken from Table 3 (in the case of ^{137}Cs, ^{226}Ra and benzo-a-pyrene, BaP, this is the mid-point between extreme values) while the bars represent risks calculated using extreme values of soil concentrations.

Following the scheme shown in Figure 6 a comparative risk assessment was performed based on the assumptions that soil ingestion at coal mine sites in Spitsbergen is a significant exposure pathway (13 mg d^{-1}) and that the soils can be contaminated with the substances listed in Table 3 over the full range of concentrations indicated. For arsenic and PAH (benzo-a-pyrene) a constant daily exposure has been assumed while for radionuclides an ingestion period of 365 days has been assumed (in practice this might represent exposure over several years since, as stated above, the soil in arctic regions is only accessible for a small proportion of the year.

Even though the results of this assessment, shown in Figure 7, are based on several assumptions, they indicate that comparative risks associated with radioactive and non-radioactive contaminants in soils can be calculated using concepts and information which currently exist within the disciplines of radioecology and ecotoxicology. While the risks from each substance are generally very low it is the non-radioactive carcinogens which appear to pose a greater comparative risk. Substances such as benzo-a-pyrene pose potentially the highest risk though, due to the lack of site-specific measurements of this and other PAH compounds, the uncertainty on this risk estimate is high and it should be considered to be purely exploratory. The 3 to 4 order of magnitude difference in risk estimates for arsenic and ^{40}K is noteworthy, however, since these comparative estimates are based on actual measurements of each substance at the Sveagruva site. It is also significant that the risk associated with ^{137}Cs, derived from global weapons fallout, is approximately two orders of magnitude lower than the risk associated with ^{40}K, which is naturally occurring.

6. Conclusions

Coal mines, such as those which occur on the arctic island of Spitsbergen, provide potentially useful scenarios in which techniques can be developed for the assessment of comparative risks from simultaneous exposures to a mixture of environmental contaminants. Exposures to ionizing radiation can be enhanced at coal mine sites to a small but significant degree and there is the potential for relatively high exposure at a small number of locations where combustion of coal may have occurred. Of chief concern amongst non-radioactive contaminants of coal wastes are the carcinogens arsenic and the polycyclic aromatic hydrocarbons. Exposures to both these radioactive and non-radioactive contaminants can occur *via* common pathways such as soil ingestion. Following quantification of exposure *via* common routes such as soil ingestion, techniques and data currently exist to allow the calculation of the comparative risks associated with both types of contaminant. A scheme is proposed by which such comparative risks can be calculated and the results from an example calculation are presented. This example indicates that the lowest risks are associated with artificially derived (i.e. weapons fallout) radionuclides such as ^{137}Cs while higher risks are to expected from naturally occurring radionuclides and from non-radioactive contaminants of coal wastes. It would be informative to extend the comparative use of radiation dose assessment methodology and slope factors for non-radioactive contaminants to examine the relative impacts and risks associated with the contamination of the environment in other areas where multiple pollution occurs.

7. Acknowledgements

This study would not have been possible without the kind permission of the Governor of Svalbard and of both the Kongs Bay KulCompanie (Ny Ålesund) and the Store Norske Spitsbergen KulCompanie (Sveagruva). The kind assistance of numerous members of the Arctic Research Group (UK) is also gratefully acknowledged.

8. References

1. AMAP (1997) Arctic Pollution Issues: A State of the Arctic Environment Report. Arctic Monitoring and Assessment Programme, P.O. Box 8100 Dep., N-0032 Oslo, Norway.
2. Crane, K. and J. L. Galasso (1999) Arctic environmental atlas. Office of Naval Research , Naval Research Laboratory, Washington DC, USA.
3. Låg, J. (1988) Soil pollution from coal mines in Svalbard. Saertryk av Jord og Myr, No. 6.
4. Låg, J. (1983) Soil pollution from mining waste material used as filling in Longyearbyen, Svalbard. Trykt i Jord og Myr, 6, 208 – 211.
5. Baxter, M. S. (1993) Environmental radioactivity: a perspective on industrial contributions. IAEA Bulletin, Quarterly Journal of the International Atomic Energy Agency, 35, 33 – 8.
6. Kocher, D. C. & A. L. Sjoreen (1985) Dose rate conversion factors for external exposure to photon emitters in soil. Health Physics, 48, 193 – 205.
7. Thorne, M. C. (2001) Assessing the radiological impact of releases of radionuclides to the environment. Pp. 391 – 446 in 'Radioecology: radioactivity and ecosystems', eds E. Van der Stricht and R. Kirchmann, ISBN 2-9600316-0-1.
8. IAEA (1996) International basic safety standards for protection against ionizing radiation and for the safety of radiation sources. Safety Series No. 115, IAEA, Vienna.
9. ICRP (1991) Recommendations of the International Commission on Radiological Protection. ICRP Publication 60, Annals of the ICRP 21, Nos. 1-3, 1991.
10. Stanek, E. J. & E. J. Calabrese (1995) Daily estimates of soil ingestion in children. Environmental Health Perspectives, 103, 276 – 285.

RISK DUE TO JOINT CHEMICAL AND RADIATION CONTAMINATION OF FOOD

V. GEORGIEVSKY
Russian Research Centre Kurchatov Institute, 1, Kurchatov sq., Moscow, RUSSIA

A. DVORZHAK
Institute of Mathematical Machines and Systems Problems NASU, Kiev, UKRAINE

1. Introduction

There is a well-known problem analysis of joint risk due to chemical and radiation impact to person. Only part of this problem will be considered concerning ingestion of food contaminated by both chemical and radiation pollutants.

The key difficulty in decision of this problem is necessity to estimate threshold and non-threshold functions "dose-effect" in unified terms.

This problem is very important for Ukraine because only threshold "Maximum Permissible Concentration" (MPC) is used as chemical standard there [1].

In this paper it is considered the method of the approximation of the threshold criteria (MPC and Reference Doses [2]) by non-threshold criteria.

This method was used for estimation of chemical risk (due to lead, cadmium, copper, zinc, mercury, DDT in food), radiation risk (due to Cs-137, Sr-90 in food) and combined risk (due to chemical plus radiation contamination of food) for 10 regions (Vinnitska, Volinska, Zhitomirska, Sumska, Ternopilska, Khmelnitska, Kievska, Chernigivska, Cherkaska, Rivnenska) of Ukraine.

2. The basis

1. General principle of developing of the unified Index of Harm is based on the estimation of loss of lifetime for person (for example, ICRP Publication 45, [3]).
2. This principle may be taken as a basis for approximation of the threshold values MPC by non-threshold values. In toxicology the determination of MPC is based on the expert estimation of total harm from the chemical pollutants.

3. Methodology

Hypothesis

1. The values of MPC are standards which were limited both non-carcinogenic and carcinogenic effects;
2. It is possible to evaluate the harm from ingestion of any chemical pollutant in term of risk;
3. The ingestion of any chemical pollutant in quantity of MPC for this pollutant leads to the same risk.
4. Combined risk due to chemical and radiation pollutants in food is the sum of the radiation risk and the chemical risk (which is estimated below).

197

F. Brechignac and G. Desmet (eds.), Equidosimetry, 197–203.

4. Method (chemical risk)

The method is the synthesis of methodology of MPC [1] and methodology of EPA (U.S. Environmental Protection Agency) [2]. It is used the information about slope factor from IRIS [4] for some chemical pollutants and the information about MPC for the same pollutants. The pseudo-slope-factors are introduced in this method. They are calculated on the base of approximation of threshold criteria by non-threshold criteria. The following procedure is used.

Step 1.
The chemical pollutant from Integrated Risk Information System [4] is chosen as Reference Chemical Pollutant. This Reference Chemical Pollutant should have the slope factor and MPC.

Step 2.
Risk for person is estimated as if the person ingests the Reference Chemical Pollutant in quantity of its MPC. (The slope factor for Reference Chemical Pollutant is used for risk estimation).

Step 3.
It is considered any pollutant, which has MPC but does not have slope factor. It is supposed that risk for person from ingestion of this pollutant in quantity of its MPC equals to risk which is estimated in step 2.

Step 4.
It is introduced pseudo-slope-factor for pollutant in step 3 so risk, which is estimated by this pseudo-slope-factor, would be equal to risk in step 2.

Note: This procedure (step1 – step 4) may be repeated for several Reference Chemical Pollutants.

5. Data source

1. For above-mentioned regions of Ukraine the following data were used (1995, 1996):
- Chemical pollutants (mg/kg): lead, cadmium, copper, zinc, mercury, DDT;
- Radiation pollutants (pCi/kg): Cs-137, Sr-90;
- type of food: potatoes, carrot, beet, cucumber, tomato, cabbage;
- the data of milk and meet contamination by both ^{137}Cs and ^{90}Sr for some regions.

2. The data are given for the separate farms.

3. The example of the averaged data (Kievska region) is indicated in the Table 1.

Table 1. Averaged data for Kievska region , 1995

Pollutants	Samples	Mean	Minimum	Maximum	Std.dev
CS-137	85	752.90	166.00	2800.00	542.47
SR-90	85	275.20	13.00	2065.00	336.12
CU	81	.42	0.00	1.50	.21
ZN	81	1.87	.30	15.00	1.92
PB	81	.13	.03	.44	.098
CD	81	.005	.002	.04	.004
HG	81	.009	.002	.048	.009
DDT	81	.028	.011	.051	.009

6. Stages of Analysis

1. Estimation of (a) chemical risk and (b) radiation risk due to food ingestion for every farm in accordance with administrative list (it has been done for 10 regions);
2. Ordering of list of the farms in accordance with ascending of (a) chemical risk, (b) radiation risk and (c) total risk.
3. Estimation of total risk.
4. Analysis of chemical, radiation and total risk structure.

7. Results

Some results are shown by the examples for Kievska region and Volinska region.

1. Risk due to ingestion of chemical and radiation contaminated food in Kievska region is indicated in Figure 1. Radiation risk and chemical risk in 85 districts and farms are given in accordance with the administrative list.

KIEVSKA REGION

Chemical and radiation risk in case contamination of vegetable ration

Chemical pollutants: lead, cadmium, copper, zinc, mercury, DDT

Radiation pollutants: Cs-137, Sr-90

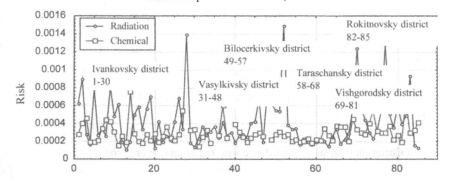

Districts and farms in accordance with administrative list

Fig. 1. Risk due to ingestion food contaminated by radiation and chemical pollutants.

2. Relation of chemical risk and radiation risk is indicated in Figure 2 and Figure 3. In Figure 2 the sequence of districts and farms in Kievska region corresponds to chemical risk ordering, in Figure 3 – to radiation risk ordering.

KIEVSKA REGION

Chemical and radiation risk in case contamination of vegetable ration

Chemical pollutants: lead, cadmium, copper, zinc, mercury, DDT

Radiation pollutants: Cs-137, Sr-90

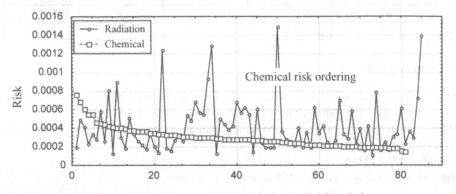

Districts and farms in accordance with chemical risk ordering

Fig. 2. Matching radiation risk with chemical risk. Sequence of districts and farms corresponds to chemical risk ordering.

KIEVSKA REGION

Chemical and radiation risk in case contamination of vegetable ration

Chemical pollutants: lead, cadmium, copper, zinc, mercury, DDT

Radiation pollutants: Cs-137, Sr-90

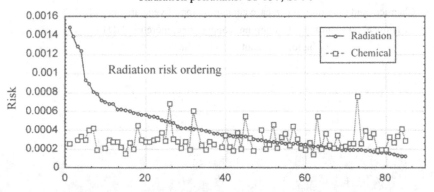

Districts and farms in accordance with radiation risk ordering

Fig. 3. Matching radiation risk with chemical risk. Sequence of districts and farms corresponds to radiation risk ordering.

3. The total (chemical plus radiation) risk in Kievska region is indicated in Figure 4. The consequence of districts and farms corresponds to total risk ordering.

KIEVSKA REGION

Total (chemical plus radiation) risk in case contamination of vegetable ration

Chemical pollutants: lead, cadmium, copper, zinc, mercury, DDT

Radiation pollutants: Cs-137, Sr-90

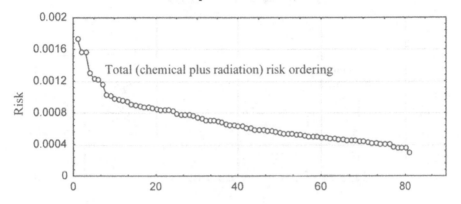

Districts and farms in accordance with total risk ordering

Fig. 4. Total (chemical plus radiation) risk. Sequence of districts and farms corresponds to total risk ordering.

4. The contribution of different pollutants into chemical risk is shown in Figure 5. In this case only vegetable ration contamination is considered the in Kievska region.

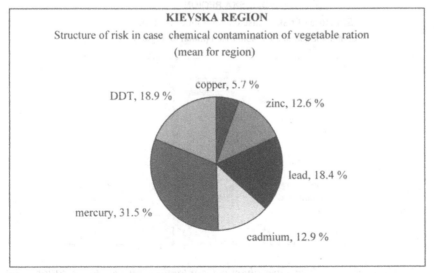

Fig. 5. Contributon different pollutants to chemical risk.

5. Structure of risk is demonstrated in Figure 6 for the case of chemical and radiation food contamination in Kievska region. (The information about chemical milk and meat contamination is absent).

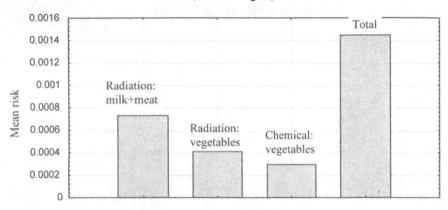

KIEVSKA REGION

Structure of risk in case chemical and radiation contamination of food

(mean for region)

Fig. 6. Structure of risk in case chemical and radiation contamination food. The information about chemical contamination of milk and meat is absent.

6. Chemical risk exceeds radiation risk in situation which is shown in Figure 7. This is the example for Volinska region; only vegetable ration contamination is considered, information about milk and meat contamination is absent.

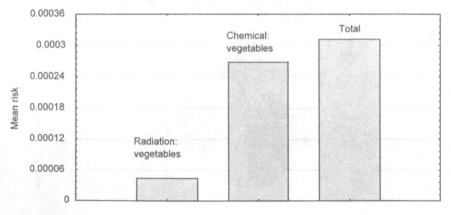

VOLINSKA REGION

Structure of risk in case chemical and radiation

contamination of vegetable ration (mean for region)

Fig. 7. Structure of risk in case chemical contamination vegetable ration. The information about contamination of milk and meat is absent.

8. Conclusion

It is proposed the method of estimation of risk due to combined radiation and chemical food contamination.

This method is the synthesis of "Maximum Permissible Concentration" (MPC) methodology of the Former USSR and methodology of the U.S. Environmental Protection Agency (US EPA). The pseudo-slope-factors are introduced in this method. They are calculated on the base of approximation of threshold criteria (MPC) by non-threshold criteria.

This method was used for analysis of situation in 10 regions of Ukraine. In particular, it is shown that chemical risk may exceed radiation risk. The some results of this analysis are illustrated by examples for Kievska and Volinska regions.

9. References

1. Gabovich R.D., Priputina L.S. Hygienic base of safe food from bad chemical pollutants. Kiev, "Zdorova", 1987
2. Risk Assessment Guidance for Superfund: EPA/540/1-89/002, 92; EPA/630/R-95; Publication 9285.7-01B; Publication 9285.7-01C; Directive 9285.6-03.
3. ICRB Publication 45 (1985), "Quantitative Based for Developing a Unified Index of Harm", International Commission on Radiological protection, Pergamon Press, Oxford, New Yokk, Toronto, Sydney, Frankfurt.
4. The Integrated Risk Information System (IRIS), U.S. Environmental Protection Agency (U.S. EPA), the electronic data base, http://www.epa.gov/iris/intro.htm

Part 5.

Problems of Synergism of Different Pollutants

SYNERGETIC EFFECTS OF DIFFERENT POLLUTANTS AND EQUIDOSIMETRY

V.G. PETIN, G.P. ZHURAKOVSKAYA
Medical Radiological Research Centre, 249036 Obninsk, Kaluga Region,
RUSSIA

Jin Kyu KIM
Korea Atomic Energy Research Institute, 150 Dukjin-dong, Yusong-gu,
Taejon, 305-353
KOREA

1. Introduction

Living organisms and ecosystems are never exposed to merely one harmful agent. Many physical, chemical, biological and social factors may simultaneously exert their deleterious influence to man and the environment. Risk assessment is generally performed with the simplest assumption that the factor under consideration acts largely independently of others. However, the combined exposure to two harmful agents could result in a higher effect than would be expected from the addition of the separate exposures to individual agents [1-3]. Hence, there is a possibility that, at least at high exposures, the combined effect of ionizing radiation with other environmental factors can be resulted in a greater overall risk. The problem is not so clear for low intensity and there is no possibility of testing all conceivable combinations of agents. Moreover, there are contradictions in literatures devoted to synergy problems relative to interaction effectiveness in the dependence of dose of applied agents and their intensity which resulted in various opinions about the importance of the synergistic interaction at low intensity of harmful agents found in biosphere [2-5].

A lot of experimental data were obtained in our laboratories for action of various agents combined with hyperthermia [6-10]. For these cases, it was shown that synergistic enhancement ratio increases, reaches the highest value and then drops with increasing in the ambient temperature. This dependence suggests that the equi-effective synergy may be realized at various temperatures. To gain some more insight into the mode of interaction, we have had to implement the following tasks in this paper. 1) To reveal a common peculiarities of synergistic interaction display. 2) To study post-radiation cell recovery after combined actions. 3) To suggest a unified biophysical conception of synergistic interaction that permits to interpret the revealed regularities and to advance some conclusions to combinations not yet tested. 4) On this basis, to formulate a common mathematical model to describe, optimise and predict the synergistic effects. 5) To prove the condition under which the highest or any equi-effective values of the synergistic enhancement ratio can be achieved. 6) To compare the model predictions with experimental results. 7) To demonstrate that for any fixed intensity of one agent the synergistic effect might be increased, decreased or stayed without change with alteration in the intensity of another agent. The results obtained are discussed from the viewpoint of potential significance of synergistic interaction of deleterious agents delivered at intensities occurred in the human environment.

F. Brechignac and G. Desmet (eds.), Equidosimetry, 207–222.
© 2005 *Springer. Printed in the Netherlands.*

2. Materials and methods

The following yeast strains were used in the experiments: *Zygosaccharomyces bailii* (haploid), *Saccharomyces cerevisiae* (two diploid strains, XS800 and T1), *Endomyces magnusii* (diploid). Cells were incubated before irradiation for 3-5 days at 30°C on a complete nutrient agar layer to a stationary phase. Homogeneous cell populations were treated at stationary phase of growth.

The ^{60}Co γ-ray source was a Gammacell 220 (Atomic Energy of Canada LTD). The γ-ray dose-rate was 10 Gy/min. The electron beam from a 25 MeV pulsed linear accelerator was also used in these experiments. The average dose rates were 5 and 25 Gy/min as determined by ferrous sulphate dosimeter.

For UV exposure, cells were irradiated with germicidal lamps that emitted predominantly UV light of wavelength 254 nm at fluence rates of 0.15 and 1.5 W/m^2. Variation of the intensity was achieved by means of calibrated metal wire nets. The fluence rates were measured using a calibrated General Electrical germicidal meter.

A continuous mode of sonication was accomplished by a 20 kHz ultrasonic unit (Fisher sonic dismembrator). The ultrasonic dose rates were 0.05 and 0.2 W/cm^2 which were measured by the calorimetric method. For sonication, 0.1 ml of cell suspension (10^8 cells/ml) at room temperature was put into 9.9 ml of sterile water which was placed into a metal vessel constructed together with the transducer. To determine the effect induced by ultrasound alone, the absorbed ultrasound heat was completely removed by cooling with water.

Hyperthermia was given in a water bath where a desired temperature ±0.1 °C was maintained. For the simultaneous action of hyperthermia and other agents, the time interval between the introduction of the cells into the preheated water and the beginning of exposure was about 0.1-0.3 min, which was significantly less than the total treatment time. At the end of the treatment, the samples were rapidly cooled to room temperature. Therefore, the duration of physical agent treatment and heat exposure were (was) identical.

After treatment with each agent applied alone or combined simultaneously with heat, a known number of cells were plated so that 150-200 colonies per dish would form by the surviving yeast cells after 5-7 days of incubation at 30°C. All experimental series were repeated 3-5x.

Exposure to radiofrequency radiation (RFR) was performed in a temperature-controlled anechoic chamber. The air temperatures in the chamber were adjusted to 22, 30, and 38 ± 0.2°C. Shinella rabbits were individually exposed to 7-GHz microwave radiation at the following power density: 0, 10, 15, 20, 30 and 100 mW/cm^2. The corresponding specific absorption rate was 0.75, 1.12, 1.50, 2.25, and 7.5 W/kg. A modified medical thermometry unit was used to monitor body temperature during exposure. The synergistic enhancement ratio was determined by the ratio of body temperature increments observed in experiments after simultaneous action of both factors) to that expected for additive summation of thermal effects from microwaves alone and high ambient temperature alone estimated for any identical duration of exposure.

3. Selected experimental results

Fig. 1A provides an example of the basic experimental data used in this investigation. Four types of survival curves were obtained for every condition of thermoradiation action: a heat treatment alone (curve 1), ionizing radiation (or another physical agent or chemical compound) without heating (curve 2), composite simultaneous heat and radiation exposure (curve 4). Curve 3 represents a theoretically expected survival curve that would be obtained if inactivation by composite heat and radiation were completely independent. To estimate quantitatively the sensitisation action of hyperthermia, one can apply the thermal enhancement ratio [11], defined as t_3/t_1 (Fig. 1). This ratio indicates an increase in cell radiosensitivity by high temperature. However, it does not reflect the kind of interaction. To evaluate the synergistic effect we used the synergistic enhancement ratio [6-10], defined as t_2/t_1 (Fig. 1).

As an example, the thermal enhancement ratio (curve 1) and the synergistic enhancement ratio (curve 2) are plotted in Fig. 1B against the irradiation temperature for diploid yeast cells. It is curious that the thermal enhancement ratio increases indefinitely with increasing exposure temperature, while the synergistic enhancement ratio at first increases, then reaches a maximum, which is followed by a decrease. This implies that a synergistic interaction between hyperthermia and ionizing radiation is observed only within a certain temperature range. Noteworthy is the fact that such a dependence of the synergistic effect on temperature under which the exposure occurred was obtained for diploid yeast cells upon the simultaneous combination of hyperthermia with ionizing radiation (Fig. 2A), UV light (Fig. 2B), ultrasound (Fig. 2C), and for mammalian cells inactivated by some chemicals and heat (Fig. 3). Hence, one can conclude that for a given intensity of physical factors or concentration of chemical agents there should be a specific temperature that maximizes the synergistic interaction. Any deviation of temperature from the optimal value results in a decrease of synergism. Another example of similar synergy pattern is shown in Fig. 4. Here again, the dependence of the synergistic enhancement ratio upon the exposure temperature is depicted for relatively thermosensitive (*Endomyces magnusii*, curve 1) and thermoresistance (*Saccharomyces ellipsoideus*, curve 2) diploid yeast strains. One can see that the temperature range strengthening the effect of ionizing radiation has been shifted toward the lower temperatures for temperature-sensitive cell lines. Thus, it can be concluded that the synergistic interaction between hyperthermia and other inactivated agents is realized only within a certain temperature range independently of the object analyzed. For temperatures below this temperature range, the synergistic effect was not observed and cell killing was mainly determined by the damages induced by ionizing radiation. For temperatures above this temperature range, the synergistic effect was also not observed but cell killing was chiefly caused by hyperthermia. It follows that for a given intensity of physical factors or concentration of chemical agents there would be a specific temperature that maximizes the synergistic interaction. Any deviation of the exposure temperature from optimal value will result in a decrease of the synergistic interaction. These results, besides being an important key for searching the synergy, can be considered as an indication of the possibility to optimize and achieve a desirable level of synergy.

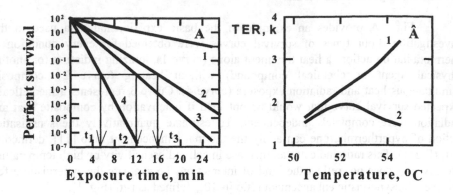

Fig. 1. Example of the basic experimental data. A. Survival curves of haploid yeast cell: curve 1 - heat treatment (45 °C) alone; curve 2 - ionizing radiation (^{60}Co) at about 10 Gy/min and room temperature; curve 3 - calculated curve for independent action of ionizing radiation and heat; curve 4 - experimental curve after simultaneous thermoradiation action. B. The dependence of the thermal enhancement ratio (curve 1) and the synergistic enhancement ratio (curve 2) against the irradiation temperature.

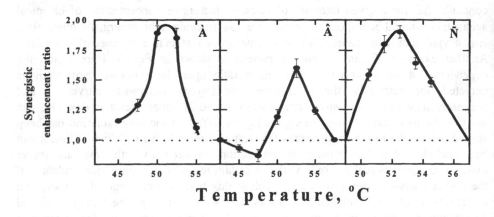

Fig. 2. The dependence of the synergistic enhancement ratio upon the exposure temperature for diploid Saccharomyces cerevisiae yeast cells (strain XS800) simultaneously exposed to heat and ionizing radiation (A), UV light (B) and ultrasound (C). Error bars show inter-experimental standard errors.

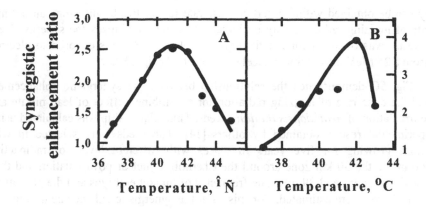

Fig. 3. The dependence of the synergistic enhancement ratio upon the exposure temperature for mammalian cells simultaneously exposed to heat and chemicals: cis-diamminedichloroplatinum (A), tri(1-aziridinyl)phosphine sulphide (B).

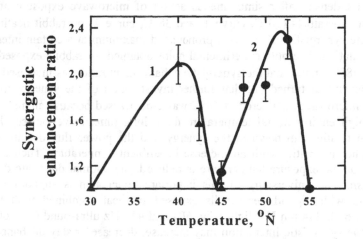

Fig. 4. The dependence of the synergistic enhancement ratio upon the exposure temperature for diploid Endomyces magnusii (curve 1) and diploid Saccharomyces ellipsoideus (curve 2) yeast cells. Error bars show inter-experimental standard errors.

On the basis of the results presented the following conclusion should also be valid: for any constant ambient temperature (or concentration of chemical agent) there should be an optimal intensity of ionizing or UV light radiation resulting in the greatest synergy. Some selected experimental results confirming this prediction are presented in Fig. 5. The data on the simultaneous thermoradiation inactivation of *Bacillus subtilis* spores [12,13] were used to estimate the dependence of the synergistic enhancement ratio on the dose rate for 95°C exposure temperature. The results are depicted in Fig. 5A. One can see that for a constant temperature, at which the irradiation occurs, synergy can be obtained within a certain dose rate range. Inside this range an optimal dose rate of ionizing radiation may be indicated, which maximizes the synergy. Very similar result was obtained for inactivation of diploid yeast cells exposed to electron beam from a 25 MeV pulsed linear accelerator at 51°C (Fig. 5B).

Fig. 5C demonstrates the relationship between the synergistic enhancement ratio and the dose rate of ionizing radiation for a combined effect of lead nitrate and chronic irradiation of *Arabidopsis thaliana* seeds. This relationship was calculated using the experimental results obtained by others [14]. The seeds were selected in wild populations growing for five years in places with different levels of radioactive pollution inside the 30 km zone around the Chernobyl nuclear power station and then treated with lead nitrate (3.39 g/l). The frequency of mutant embryos and the proportion of lethal embryos were estimated. For this case, the synergistic enhancement ratio was defined as the ratio between the increment of the mutant or lethal embryos after the combined action and the sum of these increments after separate action of each agent. As it can be seen from Fig.5C, the synergistic effect has a pronounced maximum at a certain dose rate of ionizing radiation.

Figure 5D shows the relationship between the synergistic enhancement ratio and power flux density after simultaneous action of microwave exposure and high environment temperatures (30°C, curve 1, and 38°C, curve 2) on rabbit heating [15]. Here again, the synergistic effect has a pronounced maximum at a certain intensity of microwave radiation. The only experimental point obtained for rabbits exposed at 100 mW/cm^2 and 38°C, shows that the synergistic enhancement ratio was increased for this power flux density. It turned out that further investigation of the synergy for lower intensities of microwaves delivered at 38°C was embarrassed because animals did not sustain this high environmental temperature for a long time. Nevertheless, this data show that the relationship between the synergy and the power flux density may be shifted to higher intensities with an increase in ambient temperature. Then it can be expected that as the exposure temperature is reduced, the optimal dose rate decreases and *vice versa*. The universality of this important conclusion is supported by our extensive data with diploid yeast cells exposed by heat combined with ionizing radiation (Fig. 6A), 254 nm UV light (Fig. 6B) and 44 kHz ultrasound (Fig. 6C). One can see that the synergistic interaction may increase, decrease or stay unchanged with the decrease in the intensity of any physical factor combined with heat. Nevertheless, the equi-effective values of the synergistic interaction and the whole temperature range, within which the synergy can be occurred, are shifted to lower temperature.

To demonstrate a potential significance of synergistic interaction at low intensity of agents applied, Fig. 7 illustrates the correlation between the intensity of physical factor or the concentration of chemical compound and the exposure.

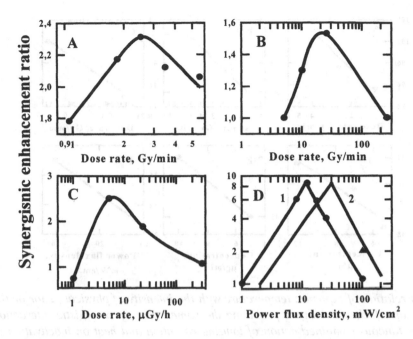

Fig. 5. The dependencies of the synergistic enhancement ratio upon the intensity of physical factors. The role of dose rate for inactivation of bacterial spores Bacillus subtilis (A), diploid Saccharomyces cerevisiae yeast cells (B) exposed to ionizing radiation at 95 and 51℃, respectively. C - the relationship between the synergistic enhancement ratio and the dose rate of ionizing radiation for a combined effect of lead nitrate and chronic irradiation of Arabidopsis thaliana seeds. D - the relationship between the synergistic enhancement ratio and power flux density after simultaneous action of microwave exposure and high environment temperatures (30 ℃, curve 1, and 38℃, curve 2) on rabbit heating.

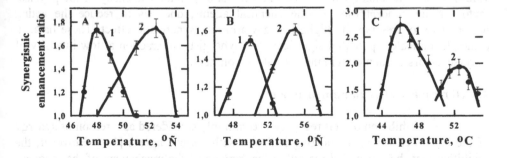

Fig. 6. The dependencies of the synergistic enhancement ratio (k) upon exposure temperature and intensity of physical factors for inactivation of Saccharomyces cerevisiae diploid yeast cells (strain XS800). Heat was applied simultaneously with: A - 25 MeV electron radiation at 5 (curve 1) and 25 Gy/min (curve 2); B - 254 nm UV light at 0.15 (curve 1) and 1.5 W/m² (curve 2); C – 20 kHz ultrasound at 0.05 (curve 1) and 0.2 W/cm² (curve 2). Error bars show inter-experimental standard errors.

214

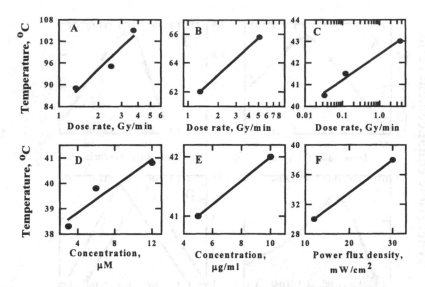

Fig. 7. Correlation of exposure temperature with the intensity of physical factor or the concentration of chemical agent providing the same value of synergistic interaction under simultaneous combined action of ionizing radiation and heat on inactivation of bacterial spores (A), bacteriophage (B) and cultured mammalian cells (C); inactivation of mammalian cells exposed to heat together with cis-diammine-dichloroplatinum (D) and tris(1-aziridinil)-phosphine sulfide (E); rabbit heating by microwave at various power flux density (F).

Temperature, which both provide equi-effective levels of synergistic interaction. To estimate this correlation, the original data were taken from a number of publications for simultaneous thermoradiation action on bacterial spores [12, 13], phage [16], mammalian cells [17,18], and rabbit heating [15]. Original data for mammalian cells exposed to heat combined with cis-DDP or TAPS were taken from the papers [19] and [20], respectively. One can see the linear relationships are found between these values for various biological objects. In all cases, at a smaller intensity of the physical factor or the concentration of the chemical agents, one has to reduce the acting temperature to preserve the highest or any arbitrary synergistic effect. These data, in principle, indicate a potential significance of synergistic interaction at low intensity of adverse factors encountered in the natural environment.

4. Cell recovery after combined actions

The inhibition of cell recovery is commonly considered as a reason of synergy of the combined action of ionizing radiation with various agents [1-4]. However, the inhibition may be proceeded via either the damage of the mechanism of the recovery itself or via the formation of irreversible damage which cannot be repaired at all. Therefore, it would be of interest to estimate quantitatively the probability of recovery per time unit and the fraction of irreversible damage for various combinations.

It is known [21, 22] that the decrease in the effective dose $D_{eff}(t)$ with the recovery time t was fitted to an equation of the form:

$$D_{eff}(t) = D_1[K + (1 - K) e^{-\beta t}], \tag{1}$$

where D_1 is the initial radiation dose; K is an irreversible component of radiation damage; e is the basis of the natural logarithm, and β is the recovery constant characterizing the probability of recovery per time unit. To determine whether the mechanism of synergistic interaction was related to the impairment of the recovery capacity *per se* or to the production of irreversible damages, we estimated the recovery parameters (K and β) after different combined treatments, at which a considerable synergistic interaction occurred. The final results presented in Table 1. One can see that for various biological objects and different conditions of the combined action the irreversible component increased with either the exposure temperature or the concentration of chemicals while the probability of recovery stayed unchanged. It can be concluded on this basis that the recovery process itself is not damaged and the inhibition of recovery is entirely due to the enhanced yield of irreversibly damaged cells.

Table 1. Radiobiological parameters of cell recovery after simultaneous action of UV light and heat, ionizing radiation and heat, as well as after combined treatments with ionizing radiation and pyruvate (Pur), novobiocin (Nov), lactate (Lac) and nalidixic acid (Nal). Original data, used for calculation, have already been published for γ-rays + high temperature [23] and X ray + chemicals [24].

Treatment	Biological Objects	Irreversible component $\hat{E}$	Recovery constant β, hr^{-1}
UV light alone	Yeast cells	0.56	0.071
UV light + 53°C	Yeast cells	0.60	0.071
UV light + 54°C	Yeast cells	0.69	0.071
UV light + 55°C	Yeast cells	0.80	0.071
UV light + 56°C	Yeast cells	0.96	-
UV light + 57°C	Yeast cells	1.00	-
γ-rays alone	Yeast cells	0.41	0.07
γ-rays + 45°C	Yeast cells	0.51	0.063
γ-rays + 50°C	Yeast cells	0.75	0.07
γ-rays + 55°C	Yeast cells	0.90	0.063
X-rays alone	Mammalian cells	0.60	0.15
X-rays + 10 mM Pur	Mammalian cells	0.75	0.16
X-rays + 20 mM Pur	Mammalian cells	0.92	0.16
X-rays + 5 µM Nov	Mammalian cells	0.82	0.14
X-rays + 10 µM Nov	Mammalian cells	0.90	0.14
X-rays + 10 mM Lac	Mammalian cells	0.78	0.14
X-rays + 20 mM Lac	Mammalian cells	0.99	-
X-rays + 5 µM Nal	Mammalian cells	0.74	0.15
X-rays + 10 µM Nal	Mammalian cells	0.82	0.15

5. Mathematical model of synergistic interaction

The results of the previous section, pointing out the negligible role of the recovery inhibition itself in the mechanism of synergistic interaction, strongly invoke the need to elaborate a new theoretical conception of the synergy which, being useful for environmental radiation protection, take into account the regularities revealed. It might be reasonable to assume that some additional lethal lesions produced during combined action are responsible for the synergistic interaction. We suppose that the additional lethal lesions are arisen from the interaction of sublesions induced by both agents and these sublesions are non-lethal when each agent is applied separately. We assume that one sublesion produced for instance by ionizing radiation interacts with one sublesion from another environmental agent (for specificity sake, let it be heat) to produce one additional lethal lesion. It would seem probable to suppose that the number of sublesions was directly proportional to the number of lethal lesions. Let p_1 and p_2 be the number of sublesions that occur for one lethal lesion induced by ionizing radiation and hyperthermia, respectively. Let N_1 and N_2 be the mean numbers of lethal lesions in a cell produced by these agents. A number of additional lesions N_3 arising from the interaction of ionizing radiation and hyperthermia sublesions may be written as

$$N_3 = min\{p_1N_1;\ p_2N_2\}. \tag{2}$$

Here, $min\{p_1N_1;\ p_2N_2\}$ is a minimal value from two variable quantities: p_1N_1 and p_2N_2, which are the mean number of sublesions produced by ionizing radiation and hyperthermia, respectively. Thus, the model describes the mean yield of lethal lesions per cell as a function of ionizing radiation (N_1), hyperthermia (N_2), and interaction $(min\{p_1N_1;\ p_2N_2\})$ lethal lesions. Then the synergistic enhancement ratio k may be expressed as

$$k = (N_1 + N_2 + N_3)/(N_1 + N_2). \tag{3}$$

Taking into account Eqn. 2, the last expression can be rewritten as

$$k = 1 + min\{p_1;\ p_2N_2/N_1\}\ /\ (1 + N_2/N_1). \tag{4}$$

It is evident from here that the highest synergistic interaction will be determined by the least value from the two functions: $f_1 = 1 + p_1\ /\ (1 + N_2/N_1)$ and $f_2 = 1 + (p_2N_2/N_1)\ /\ (1 + N_2/N_1)$. Fig. 8A shows the dependence of both this functions on the ratio of N_2/N_1, calculated for arbitrary chosen p_1 and p_2 $(p_1 = 6,\ p_2 = 4)$. The bold line at this Figure depicts the dependence of the synergistic enhancement ratio on the ratio N_2/N_1, i.e. the ratio of the effects produced by each agent used in combination. Since f_1 decreases while f_2 increases with N_2/N_1, the greatest synergistic effect will be obtained when $f_1 = f_2$, i.e.

$$p_1\ /\ (1 + N_2/N_1) = (p_2\ N_2/N_1)\ /\ (1 + N_2/N_1). \tag{5}$$

From here, the condition of the highest synergistic interaction can be obtained:

$$p_1N_1 = p_2N_2. \tag{6}$$

It means that the highest synergistic interaction occurred when both agents produce equal numbers of sublesions. Taking into account Eqns. 3 and 5, the value of the greatest synergistic enhancement ratio is given by

$$k_{max} = 1 + [p_1 p_2/(p_1 + p_2)]. \tag{7}$$

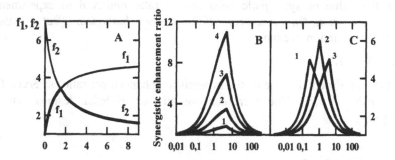

Fig. 8. The calculated dependencies of functions f_1 and f_2 on the N_2/N_1 ratio for the following values of the basic parameter: $p_1 = 6$ and $p_2 = 4$ (A) and theoretically expected dependencies of the synergistic enhancement ratio on the N_2/N_1 ratio for the following values of the basic parameter: B - $p_1 = 5$, $p_2 = 1$ (curve 1), $p_1 = 15$, $p_2 = 3$ (curve 2), $p_1 = 35$, $p_2 = 7$ (curve 3), $p_1 = 60$, $p_2 = 12$ (curve 4); C - $p_1 = 5$, $p_2 = 20$ (curve 1), $p_1 = 10$, $p_2 = 10$ (curve 2), $p_1 = 20$, $p_2 = 5$ (curve 3).

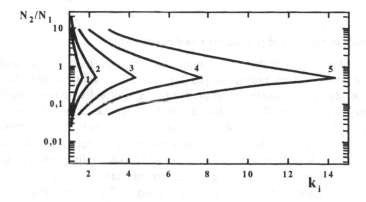

Fig. 9. The prediction of N_2/N_1 ratio, at which equi-effective values of the synergistic enhancement ratio (k_i) can be observed at different values of the basic model parameters: $p_1 = 1$, $p_2 = 2$ (curve 1); $p_1 = 2$, $p_2 = 4$ (curve 2); $p_1 = 5$, $p_2 = 10$ (curve 3); $p_1 = 10$, $p_2 = 20$ (curve 4); $p_1 = 20$, $p_2 = 40$ (curve 5).

Some examples of theoretically predicted dependency of the synergistic enhancement ratio on the N_2/N_1 ratio for various values of the basic model parameters p_1 and p_2 are depicted in Figs. 8B and 8C. If the observed biological effect is mainly induced by heat ($p_1N_1 < p_2N_2$) then taking into account Eqn. 4, the parameter p_1 can be expressed as

$$p_1 = (k_1 - 1)(1 + N_2/N_1), \tag{8}$$

where k_1 is the value of synergistic enhancement ratio observed in experiments *performed in this condition. On the contrary, if the observed biological effect is mainly* induced by ionizing radiation, we have

$$p_2 = (k_2 - 1)(1 + N_1/N_2), \tag{9}$$

where k_2 is the experimental value of the synergistic enhancement ratio observed for the condition $p_2N_2 < p_1N_1$. The corresponding number of lethal lesions can be calculated [25] as

$$N = -lnS, \tag{10}$$

where S is the surviving fraction.

It is easily to demonstrate that the model under consideration can predict two N_2/N_1 ratios, at which equi-effective values of the synergistic enhancement ratio (k_i) can be observed. For the case $p_1N_1 < p_2N_2$, we have

$$N_2/N_1 = (p_1 - k_i + 1) / (k_i - 1), \tag{11}$$

while for the case $p_2N_2 < p_1N_1$

$$N_2/N_1 = (k_i - 1) / (p_2 - k_i + 1). \tag{12}$$

Fig. 9 shows some examples of calculations based on Eqns. 11 and 12.

6. Comparison of the model predictions and experimental data

Several examples of this model application for optimizing and prediction of the synergy have already been published [26-28]. The main value of the mathematical approach presented is the possibility to predict the equi-effective synergy including the highest synergism and the N_2/N_1 ratio at which it can be achieved. We tested the applicability of the model for quantitative description, prediction and optimization of the synergistic interaction observed for various biological objects and test systems. Fig. 10 presents the experimentally obtained (circles) and theoretically predicted (solid lines) relationships between the synergistic enhancement ratio and the N_2/N_1 ratio for simultaneous thermoradiation action on inactivation of T4 bacteriophage (Fig. 10A), *Bacillus subtilis* spores (Fig. 10B), cultured mammalian cells (Fig. 10C). The procedures for calculating of these relationships have been described in detail in the previous section. Initial experimental data used for these calculations were taken from publications associated with bacterial spores [12, 13] bacteriophage [16], and mammalian cells [17, 18]. The similar relationships have been obtained for inactivation of cultured mammalian cells induced by simultaneous exposure to elevated temperature with cis-diammindichloroplatinum (cis-DDP, Fig. 10D) and tri(1-aziridinyl)phosphine sulphide (TAPS, Fig. 10E) as well as for rabbit heating by exposure to microwave radiation at high ambient temperature (Fig. 10E).

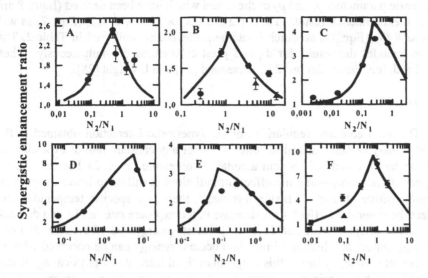

Fig. 10. Experimentally obtained (circles) and theoretically predicted (solid lines) dependencies of the synergistic enhancement ratio k on the N_2/N_1 ratio under simultaneous combined action of ionizing radiation and heat on inactivation of bacterial spores (A), bacteriophage (B) and cultured mammalian cells (C); inactivation of mammalian cells exposed to heat together with cis-diamminedichloroplatinum (D) and tris(1-aziridinil)-phosphine sulfide (E); rabbit heating by microwave at various power flux density and ambient temperatures (F).

Table 2. Parameters of the mathematical model describing the inactivation of various cells and rabbit heating after combined exposure to different environmental factors and hyperthermia (H)

Biological object	Inactivation agents	p_1	p_2	K_{max}	N_2/N_1
Bacillus subtilis spores	$\gamma + H$	2.2	4.5	2.5	0.5
Bacteriophage T4	$\gamma + H$	2.4	1.6	2.0	1.5
Mammalian cells	$\gamma + H$	4.3	30	4.8	0.14
Mammalian cells	Cis-DDP + H	18.0	4.0	4.3	4.5
Mammalian cells	TAPS + H	2.7	13.7	3.2	0.2
Rabbit	Microwave + H	16.0	21.0	10.0	0.76
Yeast cells	$\gamma + H$	2.3	3.0	2.4	0.8
Yeast cells	Ultrasound + H	1.8	25.0	2.7	0.07
Yeast cells	UV light + H	1.5	0.77	1.5	2.0

For these cases, initial experimental data and some previous calculation were taken from earlier publications devoted to microwaves [15] and cytostatic preparations [19, 20]. Predicted values of the synergistic enhancement ratio were estimated by Eqn. 4 using the basic parameters p_1 and p_2 of the model which have been derived (Eqns. 8 and 9) from real experimental results. These parameters together with k_{max} (Eqn. 7) as well as the ratio N_2/N_1 (Eqn. 6), at which it can be achieved, are collected in Table 2. This Table also includes the results for diploid yeast cells exposed simultaneously to heat combined with ionizing radiation [26], ultrasound [27] and UV light [28].

7. Conclusions

The main common regularities of the synergistic interaction obtained in this investigation may be summarized as follows. 1) For any constant rate of exposure, the synergy can be observed only within a certain temperature range. 2) The temperature range synergistically increasing the effects of radiations is shifted to lower temperature for thermosensitive objects. 3) Inside this range, there is a specific temperature that maximizes the synergistic effect. 4) A decrease in the exposure rate results in a decrease of this specific temperature to achieve the greatest synergy and *vice versa*. 5) For a constant temperature at which the irradiation occurs, synergy can be observed within a certain dose rate range. 6) Inside this range an optimal intensity of physical agent may be indicated, which maximizes the synergy. 7) As the exposure temperature reduces, the optimal intensity decreases and *vice versa*. 8) The recovery rate after combined action is decelerated due to an increased number of irreversible damage. 9) The probability of recovery is independent of the exposure temperature for yeast cells irradiated with ionizing or UV radiation. 10) Chemical inhibitors of cell recovery act through the formation of irreversible damage but not *via* damaging the recovery process itself.

The remarkable result of this paper concerns the mathematical model that has been proposed to explain the experimental data of synergistic interaction of hyperthermia with other inactivating agents. The model is based on the supposition that synergism takes place due to the additional lethal lesions arisen from the interaction of non-lethal sublesions induced by both agents. These sublesions are considered noneffective after each agent taken alone. The idea of sublesions is widely used in radiobiology [29-31]. In the model, the synergistic effect is given by $min\{p_1N_1; \ p_2N_2\}$ (Eqn. 2). This means that one sublesion caused by irradiation or chemicals interacts with one sublesion produced by heat. This process is assumed to proceed until the sublesions of the less frequent type is used up. To estimate the basic parameters p_1 and p_2 we have used the experimental values of the synergistic enhancement ratio k_1 and k_2 (Eqns. 8 and 9). It means that the model takes into consideration only the actual interaction determining the synergistic effect. The model predicts the dependence of synergistic interaction on the ratio N_2/N_1 of lethal lesions produced by every agent applied (Eqn. 4), the greatest value of the synergistic effect (Eqn. 7) as well as the conditions under which it can be achieved (Eqn. 6). The model is not concerned with the molecular nature of sublesions, and the mechanism of their interaction remains to be elucidated. In spite of the approximation used in this simplified model, it is evident from the data presented that a good agreement appears to exist between theoretical and experimental results. Moreover, the model discussed here can be used to predict conditions under which the greatest synergy can be observed and its value. The degree of synergistic interaction was found to be dependent on the ratio of lethal damage (N_2/N_1) induced by the two agents applied. The synergistic interaction is not observed at any N_2/N_1 ratios. The effectiveness of synergistic interaction appears to decline with a deviation of the ratio N_2/N_1 from optimal value.

The most interesting results, obtained from the model, is the conclusion that for a lower intensity of physical agents or a lower concentration of chemicals a lower temperature must be used to provide the greatest synergy. Actually, any decrease in the intensity of physical agents would result in an increase of the duration of thermoradiation action to achieve the same absorbed dose. Therefore, the number of thermal sublesion will also be increased resulting in the disruption of the condition at which the highest synergy should be observed (Eqn. 6). Hence, to preserve an optimal N_2/N_1 ratio with any decrease in the dose rate (or the intensity of other agents) the exposure temperature should be decreased. It can be concluded on this basis that for a long duration of interaction, which are important for problems of radiation protection, low intensities of deleterious environmental factors may, in principle, synergistically interact with each other or with environmental heat.

8. References

1. UNSCEAR, *Ionizing Radiation: Sources and Biological Effects*. New York: United Nations Publications, 1982.
2. C. Streffer, W.-U. Müller. Radiation risk from combined exposures to ionizing radiations and chemicals. *Advances in Radiation Biology* 11, 1984, pp. 173-210.
3. UNSCEAR, *Combined Effects of Radiation and Other Agents*. New York: United Nations Publication, 2000.
4. L.A. Dethlefsen, W.C. Dewey (Eds.) *Cancer Therapy by Hyperthermia, Drugs and Radiation*. National Cancer Institute Monograph 61, 1982.
5. A.M. Kuzin (Ed.) *Synergism in Radiobiology*. Pushchino, 1990 [in Russian].
6. V.G. Petin, G.P. Zhurakovskaya. The peculiarities of the interaction of radiation and hyperthermia in *Saccharomyces cerevisiae* irradiated with various dose rates. *Yeast* 11, 1995, pp. 549-554.
7. V.G. Petin, G.P. Zhurakovskaya, L.N. Komarova. Fluence rate as a determinant of synergistic interaction under simultaneous action of UV light and mild heat in *Saccharomyces cerevisiae*. *J. Photochem. Photobiol. B: Biology* 38, 1997, pp. 123-128.
8. V.G. Petin, G.P. Zhurakovskaya and others. Synergism of environmental factors as a function of their intensity. *Russian J. Ecology* 29, 1998, pp. 383-389.
9. J.K. Kim, V.G Petin, G.P. Zhurakovskaya. Exposure rate as a determinant of synergistic interaction of heat combined with ionizing or ultraviolet radiations in cell killing. *J. Radiat. Res.* 42, 2001, pp. 361-365.
10. V.G. Petin, J.K. Kim and others. Some general regularities of synergistic interaction of hyperthermia with various physical and chemical inactivating agents. *Int. J. Hyperthermia* 18, 2002, pp. 40-49.
11. F.A. Stewart, J. Denekamp. Combined X-rays and heating: is there a therapeutic gain? In: *Cancer Therapy by Hyperthermia and Radiation*, (Ed. by C. Streffer) Baltimore-Munich: Urban & Schwarzenberg, 1978, pp. 249-250.
12. M.C. Reynolds, D.M. Garst. Optimizing thermal and radiation effects for bacterial inactivation. *Space Life Sci.* 2, 1970, pp. 394-399.
13. M.C. Reynolds, J.P. Brannen. Thermal enhancement of radiosterilization. In: *Radiation Preservation of Food*. Vienna: International Atomic Energy Agency. 1973, pp. 165-176.

14. Dineva S.B., Abramov V.I., Shevchenko V.A. The genetic effects of treatment of *Arabidopsis thaliana* seeds by the sodium lead of chronic irradiatied population. *Genetics* 29 (1993), pp. 1914-1920 [in Russian].

15. O.I. Kolganova, L.P. Zhavoronkov and others. Thermocompensative rabbit response to the microwave exposure at various environmental temperatures. *Radiation Biology. Radioecology* 41, 2001, pp. 712-717 [in Russian].

16. R. Trujillo, V.L. Dugan. Synergistic inactivation of viruses by heat and ionizing radiations. *Biophys. J.* 12, 1972, pp. 92-113.

17. E. Ben-Hur, M.M. Elkind, B.V. Bronk. Thermally enhanced radioresponse of cultured Chinese hamster cells: inhibition of repair of sublethal damage and enhancement of lethal damage. *Radiat. Res.* 58, 1974, pp. 38-51.

18. E. Ben-Hur. Mechanisms of the synergistic interaction between hyperthermia and radiation in cultured mammalian cells. *J. Radiat. Res,* 17, 1976, pp. 92-98.

19. M. Urano, J. Kahn and others. The cytotoxic effect of cis-diamminedichloroplatinum (II) on culture Chinese hamster ovary cells at elevated temperatures: Arrhenius plot analysis. *Int. J. Hyperthermia* 6, 1990, pp. 581-590.

20. H.A. Johnson, M. Pavelec. Thermal enhancement of thio-TEPA cytotoxicity. *J. Natl. Cancer Inst.* 50, 1973, pp. 903-908.

21. Yu.G. Kapul'tsevich. *Quantitative Regularities of Cell Radiation Damage.* Moscow: Energoatomizdat, 1978 [in Russian].

22. V.I. Korogodin. The study of post-irradiation recovery of yeast: the 'premolecular period'. *Mutation Res.* 289, 1993, pp. 17-26.

23. V.G. Petin, I.P. Berdnikova. Effect of elevated temperatures on the radiation sensitivity of yeast cells of different species. *Radiat. Environm. Biophys.* 16, 1979, pp. 49-61.

24. A. Kumar, J. Kiefer and others. Inhibition of recovery from potentially lethal damage by chemicals in Chinese hamster V79 A cells. *Radiat. Environ. Biophys.*, 24 (1985) 89–98.

25. R.H. Haynes. The interpretation of microbial inactivation and recovery phenomena. *Radiat. Res.* 6 (Suppl.), 1966, pp. 1-29.

26. V.G. Petin, V.P. Komarov. Mathematical description of synergistic interaction of hyperthermia and ionizing radiation. *Mathem. Biosci.* 146, 1997, pp. 115-130.

27. V.G. Petin, G.P. Zhurakovskaya, L.N. Komarova. Mathematical description of combined action of ultrasound and hyperthermia on yeast cells. *Ultrasonics* 37, 1999, 79-83.

28. V.G. Petin, J.K. Kim and others. Mathematical description of synergistic interaction of UV-light and hyperthermia for yeast cells. *J. Photochem. Photobiol. B: Biology* 55, 2000, pp. 74-79.

29. H.P. Leenhouts, K.H. Chadwick. An analysis of synergistic sensitization. *Br. J. Cancer* 37, (Suppl. 3), 1978, pp. 198-201.

30. M.S.S. Murthy, V.V.,Deorukhakar, B.S. Rao. Hyperthermic inactivation of diploid yeast and interaction of damage caused by hyperthermia and ionizing radiation. *Int. J. Radiat. Biol.* 35, 1979, pp. 333-341.

31. M. Zaider, H.H. Rossi. The synergistic effects of different radiations. *Radiat. Res.* 83, 1980, pp. 732-739.

THE BEHAVIOUR OF RADIONUCLIDES AND CHEMICAL CONTAMINANTS IN TERRESTRIAL AND WATER ECOSYSTEMS OF URALS REGION

P. YUSHKOV, A. TRAPEZNIKOV, E. VOROBEICHIK,
YE. KARAVAEVA, I. MOLCHANOVA
Institute of Plant and Animal Ecology, Ural Division of Russian Academy
of Sciences. 8 Marta str., Yekaterinburg, 620144, RUSSIA

1. Introduction

One of the regions of Russia, with extreme ecological conditions is the Ural. On the territory of region some large industrial complexesare located, which structure includes   enterprises of nuclear fuel cycling, non-ferrous and ferrous metallurgy, chemical industry mining, and fuel industry all having an impacton the ecological situation. In this connection ecologists of such a diverse structure studies are developed to investigate the condition of the environment, disclosing any regularities in the evolution of the ecological conditions in the region, with a forecasting goal.

Some results of researches conducted by two scientific collectives of Institute of Ecology of plants and Animals of the Urals division of RAS are represented one of which attends to problems of radioecology, and the other study of the influence on terrestrial ecosystems of chemical contaminants.

The purpose of the work is:

1) to show features of radionuclides contamination of bog-river ecosystems near the NPP and to   evaluate its function as a biogeochemical barrier to migration of radionuclides towards underlying links of a river system;

2) to present a system of   practical and theoretical approaches to be used for the installation of limits of stability of natural ecosystems against the action of pollutants.

2. The content and the behaviour of radionuclides in the bog-river ecosystem in the vicinity of the Nuclear Power Plant (NPP).

The Olkhovka bog is situated 5 km southeast of the Beloyarskaya NPP. It extends from the southwest to the northeast for about 3 km. The south-western part of the bog receives weekly radioactive wastewater from the NPP and municipal sewage from the city of Zarechnyi. The Olkhovka river flows out of the northeastern end of the bog. The eastern and southeastern parts of the bog are occupied by a birch canopy. In addition, there are willow groves (*Salix cinnerea L.* and *S. pentandra L.*) of different size and individual pine trees (*Pinus sylvestris L.*)

The Olkhovka bog-river ecosystem consists of the Olkhovka bog and Olkhovka river, implications from another bog and running into the Pyshma river, which is included into structure of the Ob-Irtysh rivers system. The Olkhovka bog-river ecosystem represents the half-natural formation which has arisen during the work on

F. Brechignac and G. Desmet (eds.), Equidosimetry, 223–230.

the first two reactors of Beloyarskaya NPP. (Beginning of the workson the first reactor was of 1964, second of 1967. These reactors were brought into operation accordingly in 1981 and 1989. In 1980 a third fast neutron reactor is brought into operation . The Olkhovka bog was derived asa result of the dumping on a wood swampy plot of unbalanced and domestic waters from NPP, and also of sewer waters of the city of Zarechny, where the workers of the NPP live.

In general the main radioactive contamination of the bog-river ecosystem has taken place during the operation of the first two reactors (table 1-3). The researches of the last years have shown, that the long-term dumping of unbalanced waters of Beloyarskaya NPP in the Olkhovka bog has also lead to plutonium accumulation . The greatest concentration of ^{239,240}Pu (39-114 Bq/kg) are marked in the base suspension of a beginning of a bog. In region of a source of Olkhovka river in stratums of base adjournment 0-5 ñì and 5-10 ñì the concentration ^{239,240}Pu is equalled accordingly 12,0 and 0,9 Bq/kg. However main contaminants are ^{137}Cs, ^{60}Co and ^{90}Sr. The concentration ^{137}Cs make tens, and ^{60}Co and ^{90}Sr - accordingly units and tenth shares of kBq/kg (tab. 1, 2) [1-5].

Table 1. Concentration of radionuclides in the sediments of bog-river ecosystems for 1978-1986, kBq/kg [1]

Location	Sediments	^{60}Co	^{90}Sr	^{137}Cs
Discharge cannel	Sand	4.0 (1.6-6.0)*	0.07 (0.02-0.10)	16.0 (7.0-25.0)
	Silt	5.0 (3.0-46.0)	0.20 (0.06-0.27)	22.0 (18.0-28.0)
Swamp: beginning	Sand	34.0 (1.0-60.00)	1.80 (0.48-3.80)	58.0 (9.0-112.0)
centre	Silt	3.0 (1.0-7.0)	1.20 (0.07-3.40)	27.0 (7.2-65.0)
Olkhovka river: source	Sand	0.12 (0.10-0.15)	0.03 ((0.02-0.15)	20.0 (10.0-30.0)
	Silt	1.40 (0.77-2.30)	0.19 (0.06-0.87)	30.0 (8.0-51.0)
mouth	Sand	0.30 (0.14-0.50)	0.06 (0.01-0.17)	16.0 (3.0-26.0)
	Silt	0.90 (0.15-2.10)	0.14 (0.01-0.24)	34.0 (7.0-86.0)

* Annual variation

Table 2. Radionuclide content in the soil-vegetative cover of the Olkhovskoye bog region (1) and control lot (2), kBq/m². (Layer of soil 0-10 cm.)

Location	^{90}Sr		^{137}Cs	
	1	2	1	2
Watershed	1.85	1.29	9.09	6.84
Slope	1.20	1.23	4.20	4.74
Bank of the bog	1.85	1.24	47.80	5.31

Table 3. Concentration of radionuclides in the water of the Olkhovskoye bog-river ecosystem, Bq/l [5]

Location	^{60}Co		^{90}Sr		^{137}Cs	
	1976-1986	1989-1991	1976-1986	1989-1991	1976-1986	1989-1991
Discharge cannel	2,6 (2,2-6,4)*	BDL**	0,6 (0,1-1,2)	0,7 (0,3-1,2)	10.0 (2.8-28.3)	9.0 (2.4-15.6)
Bog: beginning	3,3 (2,6-7,6)	BDL	0,8 (0,1-1,6)	1,3 (1,0-1,6)	16.6 (2.3-49.2)	8.4±1.8
centre	2,8 (1,5-10,1)	BDL	0,9 (0,1-1,7)	1,2 (0,7-1,2)	20.7 (4.1-43.4)	17.0 (9.9-24.1)
Olkhovka river: source	2,8 (1,2-7,6)	0,3±0.03	0,8 (0,2-2.5)	1.0 (0.8-1.1)	16.0 (9.7-35.0)	11.6 (4.7-20.2)
mouth	1,3 (1,1-2,6)	0,3±0.02	0,4 (0,2-1.0)	0.5 (0.2-1.0)	10.0 (2.4-16.0)	7.9 (4.1-17.9)

* Annual variation
** Below detection limit

The increased content of radionuclides is established in the soil-vegetative cover of the coastal zone of the bog. Especially high concentrations of ^{137}Cs, exceeding about 3-4 order of magnitudes compared to the background, are detected in wood plants (birch and willows), that grow on the bog and on periodically flooded plots adjacent to land [6,7]. Birch and willow from the territory of the bog cannot be used for economic purposes because of a high ^{137}Cs content.

From 1978 to 1986 onwards., when the Olkhovka bog-river ecosystem was subjected to periodic input from three NPP reactors, the concentration of radionuclides in the water on a section of the discharge channel-mouth of the Olkhovka river was changed a tenfold over the years (table 3). In this period though, on the average, the concentration was below the sanitarian norms for potable water. After the closure from operation of the first two NPP reactors of the inputof ^{60}Co with liquid dumping practically was stopped, the contents in water of the Olkhovka bog-river ecosystem ^{90}Sr was not essentially changed and the ^{137}Cs content has decreased.

The stock of radionuclides (^{60}Co, ^{90}Sr, ^{137}Cs) in the swamp was estimated at more than 100 Ci, with ^{137}Cs a the main component [2].

During the last years dumping of city sewer waters in a bog , whose volume was exceeding the natural charge of the Olkhovka bog, was terminated.This has entailed a modification of the hydrological status of the bog and caused a shrinking of the surface size of this constantly flooded territory. The discontinuing of the l dumping in theunbalanced waters of Beloyarskaya NPP in the Olkhovka bog has resulted in a reduction of the widespread content of radionuclides in this ecosystem . A removal of a small amount of radionuclides occurred with water through Olkhovka river in Pyshma river (table 3) and through radioactive disintegration. A variable velocity of migration ^{90}Sr, ^{137}Cs and Pu in the base suspension and in soil, and also their leaching from the top horizonswill in due course change the character of the vertical distribution of radionuclides in these media.

The analysis of data on the ecological conditions on territory of the Olkhovka bog-river ecosystem has shown, that the levels of contamination of its components are lower than those limits, that are able to cause its deterioration [13]. Therefore, in the given situation the sanitary-hygienic standards are sufficient for safeguarding of the ecosystem.

3. Influence of chemical contaminants on the terrestrial ecosystems

Despite the presence in the Urals region of extensive territories with radionuclide contamination, it is obvious that aquatic and terrestrial ecosystems have been more effected by chemical releases. In a number of places it has lead to the destruction of ecosystems and even to the origination of industrial deserts. It has become clear, that the observance of the sanitary-hygienic standards does not protect a natural medium from a destructive operation by pollutants. It has resulted in the development of an ecological normalization of the smoke-gas emission and liquid dumping of toxicants .

The original methodical approaches for the creation of the ecological specifications were developed on the basis of researches which are being carried out in the vicinity of copper-melting plant , located near the city of Revda (Sverdlovsk District, Middle Urals).

As a methodological basis of the approach to study the influence of chemical contaminants on the terrestrial ecosystems the construction and an analysis of associations between the received pollutants in principal components of the terrestrial ecosystems and the response of natural ecosystems on a concrete toxic level of the load has been chosen. This approach was set up for the installation of threshold levels of the effect (critical loads) of the delivery of heavy metals, in a complex with SO_2 onto the principal components of wood ecosystem: wood and herb-undershrub layers, wood litter, soil biota, epiphytic lichens communities [8-12]. Researches were conducted in a combined operative range with regard to the production draft-quality copper. To study the degree of effect on a natural medium three zones were chosen: impact , buffer and background. The long-term (since 1940) atmospheric contamination has generated a high gradient of toxic load: the content of heavy metals (Cu, Pb, Cd and Zn) in wood litter near the factory exceeds the background level for one to two orders of magnitude (with a maximum of more than 400 times). Actually, it means, that near the emission source the most contrasting technogenic geochemical anomaly is generated. High significant correlation coefficients between the concentrations of separate metals are shown: for the Cu - Pb pair it is equal to 0.95 a,for Zn - Cd equal to 0.94 – and 0.63-0.81 for the in remaining combinations (in all cases $P \ll 0.00001$, $N = 221$). Such close direct correlations allow to correctly present multi-dimensional information about the contamination of the territory as a one-dimensional load index of contamination (average of the concentration, normalized to background levels of elements), which is used in case of the need of analysis of "dose-effect" curves (technique of account is indicated in [8]).

In most cases the response of components on natural ecosystem at a toxic load appears to have a sharply expressed non-linear character (the "dose-effect" curve is S-shaped). It indicates the existence of a trigger effect: there are two metastable conditions (plots with slow modifications), and then a very sharp transition in between(plot of a gradient with fast modifications). Such a response threshold allows to rather objectively installing limits in size of release or dumping of pollutants (critical loads).

The emissionsof SO_2 result in a important acidification of wood litter (Figure 1), which brings about a significant contribution in the expressivity of processes of degradation of wood ecosystems (toxicity of heavy metals, as is known, largely depends on the acidity of the soil solution).

As a convenient integrated parameter of activity of a saprotroph complex of soil biota , which can be effectively used in diagnostics of anthropogenic intrusions of a wood ecosystem, the modification of the thickness of forest litter is used. The increase of the thickness of forest litter - one of the most noticeable manifestations of intrusion of the biological circulation in wood ecosystems, is sensitive to chemical contamination. Emissions from a copper-melting plant (heavy metals and SO_2) in the Middle Urals introduces a 2.7 - 3.9 times increase of the thickness of forest litter. For several variants of biotopes, distinguished by its contours (eluvial, transit and accumulative landscapes), non-linear dose - effect regressive dependencies, connecting thickness of litter to Cu, Pb, Cd and Zn contents, are put up. The critical loads and, accordingly, stability soil biota appear to essentially depend on the connection of the litter thickness to its Cu, Pb, Cd and Zn content in it. The critical loads and, accordingly, stability soil biota essentially moreover depend on the ecotopic conditions: In accumulative landscapes the litter thickness starts to increase at essentially smaller levels of contamination compared to transit and eluvial landscapes.

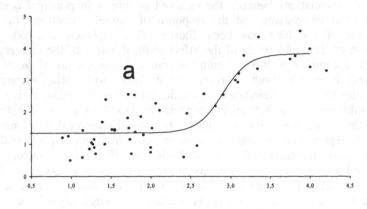

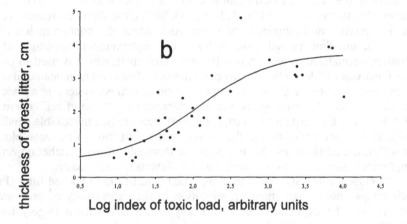

Log index of toxic load, arbitrary units

Fig. 1. Dependence a dose - effects on the thickness of forestry litter in eluvial (a) and accumulative (b) relief elements.

The increased space variation of toxicity of soil and litter in a transitional zone can be one of the key mechanisms causing the observable nonlinearity in the response of ecosystems on contamination. It has caused the necessity of an evaluation of levels of space variation (in scales from ten centimetres up to kilometres) in heavy metal content and acidity of wood litter.

In all zones with a contaminant load the space variation of pollution in the accumulating media both for separate elements and for a resulting parameter, the index of contamination is increased because of the transition from the gross content into exchangeable forms together with a magnification of square of test (coefficient of variation will increase from 10-20 of % till 50-80 of In most cases, the following regularity is observed: the space variability is the lowest in the background zone, is maximum in buffer and the impact zone takes an intermediate position. The space variation both for the gross content, and for the exchangeable forms is higher for the soil in comparison to litter (in soil probably considerably more mechanisms, modifying the contents of elements are active). In the context of the discussed problem the particular importance of a trigger effect is shown by an evaluation of the space range of toxic load in different zones of contamination. The range of concentration of metals in buffer and impact zones is extremely large and reaches to 50-90 % for the complete range of the possible variation.

It also established that the acidity range of of soil and litter both in buffer and impact zones makes 0.7-1.4 units pH and maximum reaches almost two units õĺ, whichsignifies 30-90 % of the complete range of the possible variation. , This condition is probably the key factor that is modifying the space variation of the toxicity of soil and litter for biota.

The suggested approach to standardization is of applied character as there are the following sufficient limits: all reasoning are with assumption that ecosystems in background and transformed condition are obtained for the definite emission structure and after its sufficient changing new standard obtaining is necessary; the research scheme for standard obtaining can be correctly fulfilled only in the region where background environment slightly differs with pre-industrial condition.

Different from the sanitary-hygienic standards, the ecological standards (the extreme allowable ecological loads) should be differentiated on regions (botanical-geographical zones), types of ecosystems, types of productions and nature management regimes.

Acknowledgements

These investigations were prepared with financial support of the Russian Fund Basic Research under Grinds N 01-05-96463, N 01-05-65258.

References

1. Karavaeva, E.N., Kulikov, N.V., and Molchanova, I.V., Radioeological studies on natural ecosystems in the zone liquid waste discharge from the Beloyarskaya Nuclear Power Plant in the Urals. in: The ecology of regions around Nuclear Power Plant, Moscow, 1994, issue 1. pp. 105-143. (in Russian)
2. Karavaeva, E.N. And Molchanova, I.V., Radioeological monitoring in the zone affected by liquid waste discharge from the Beloyarskaya Nuclear Power Plant, Defectoscopia, 1995. ¹ 4. pp. 62-67. (in Russian)
3. Molchanova, I.V., Karavaeva, E.N., and Kulikov, N.V., Some results of the radioecological studies on natural ecosystems in the region of the Beloyarskaya Nuclear Power Plant, Ecologia. 1985. N 5. pp. 30-34. (in Russian)
4. Kulikov, N.V., Molchanova, I.V., and Karavaeva, E.N.,Radioecology of the soil- vegetative cover, Sverdlovsk: Ural. Division of Acad. Nauk SSSR, 1990. p. 169. (in Russian)
5. Molchanova, I.V., and Karavaeva, E.N., Ecological-geochemical aspects of the migration of the radionuclides in the soil- vegetative cover. Yekaterinburg: Urals Division of RAS. 2001. p. 161. (in Russian)
6. Yushkov P.I. Accumulation and distribution of ^{90}Sr and ^{137}Cs in birch in the zone affected by liquid discharge from the Beloyarskaya Nuclear Power Plant. Ecologia. 2000. N 2. pp. 106-112. (in Russian)
7. Yushkov P.I. Accumulation and distribution of ^{90}Sr and ^{137}Cs in willow growing near the Olkhovskoye swamp in the region of the Beloyarskaya Nuclear Power Plant. in: Radiation safety for people and the Environment, Yekaterinburg. 1997. pp. 25-29. (in Russian)
8. Vorobeychik, E.L., Sadykov, O.F., Farafontov M.G., Ecological standardization of terrestrial ecosystems technogenic pollution (local scale). Yekaterinburg: Nauka., 1994. p. 280. (in Russian)

230

9. Vorobeychik, E.L., Modification of a potency wood litter in conditions of chemical contamination, Ecologia. 1995. ¹ 4. pp. 278-284. (in Russian)

10. Vorobeychik, E.L., Population of earthworms (Lumbricidae) of woods of   Middle Urals in conditions of contamination by ejections copper-melting of combines, Ecologia. 1998. ¹ 2. pp. 102-108. (in Russian)

11. Vorobeychik, E.L., Khantemirova Å.V. A response wood phytocenosis on   technogenic contamination: dependence a doze - effect, Ecologia. 1994. ¹ 3. pp. 31-43. (in Russian)

12. Mikhailova I.N., Vorobeychik, E.L., Synusies of epiphytic lichens in conditions of chemical contamination: dependence a doze - effect, Ecologia. 1995. ¹ 6. pp. 455-460. (in Russian)

13. Kutlakhmedov V.I., Korogodin V.I., Kutlakhmedova-Vishnyakova V.Yu., Yaskovec I. The theory and models of a radiocapacity ecosystems in modern radioecology. in: « Modern problems of radiobiology, radioecology and evolution. Proceedings of the International Conference dedicated to Centenary of the Birth of N.W. Òimofeeff-Ressovsky. Dubna, September 6-9, 2000. Dubna JINR, 2001. p. 262-270.

SYNERGISM OF LOW DOSE CHRONIC RADIATION AND BIOTIC STRESS FOR PLANTS

A. DMITRIEV, N. GUSCHA, M. KRIZANOVSKA

Institute of Cell Biology and Genetic Engineering,
148 Zabolotnogo St., Kiev 03143 UKRAINE

1. Introduction

Changes in plant disease resistance and in virulence of plant pathogenic organisms under low dose chronic radiation may pose a threat to agriculture. Since 1986 we carried out studies on plant-pathogen interactions in 30- and 10-km Chernobyl nuclear power plant (ChNPP) zone. Biochemical studies are related mainly to low dose chronic radiation effects on disease resistance of cereals and corn. Test-systems, which could be useful for the estimation of low dose effects, are developed. Our long-term goal is to investigate the population structure of plant pathogenic fungi in the 10-km ChNPP zone. *Puccinia graminis,* the causal agent of stem rust of wheat, rye and oat, is the most damaging disease of these crops. Rust diseases of small grain crops are difficult to control with resistant cultivars, because there are many different pathogenic varieties of the rust fungi. Genes for resistance to wheat and rye rust diseases may be very effective against some varieties, but fail completely against other varieties.

Plants do not have an immune system of the kind known in animals but possess different inducible defences against microbial pathogens. Among the defence responses there are mechanical strengthening of cell walls, formation of pathogenesis-related (PR) proteins, synthesis of inducible antibiotic substances - phytoalexins, increasing of enzyme activities, for example, proteinase inhibitors [1]. Some of these defences can be used as quantitative parameters of plant resistance to biotic and abiotic stress factors of the environment.

2. Decrease of plant disease resistance under low dose chronic irradiation

The results obtained suggest that low dose chronic radiation decreased the plant immunity potential. Analysis of wheat powdery mildew disease of three cultivars (Mironovska-808, Polesska 70, Kiyanka) has shown that the extent of disease was 2-fold higher in plants grown from the seeds, collected in 10-km ChNPP zone, than that in plants grown from control seeds.

The data were confirmed in a set of experiments with artificial inoculation of wheat plants in greenhouse. Seedlings of wheat Kiyanka grown from the seeds, collected in the ChNPP zone, were infected at a second leaf stage by brown rust spores. It turned out that incidence and disease development in the seedlings were 2.6-fold higher than that in seedlings, grown from control seeds. Similar results were obtained for two other cultivars seedlings grown from radionuclide-contaminated seeds.

To elucidate alterations in cereals disease resistance field trials were carried out in 10-km ChNPP zone on three plots with matching soil parameters but differing in dose rates. The absorbed dose during vegetation period was 0.1, 0.8 and 2.7 Gy accordingly. Uncontaminated control wheat seeds were sown on the

F. Brechignac and G. Desmet (eds.), Equidosimetry, 231–236.
© 2005 *Springer. Printed in the Netherlands.*

plots. Leaves were artificially inoculated at the beginning of milk ripeness phase with brown rust spores. Phytopathological analysis carried out 5 and 10 days after inoculation revealed that the incidence and extent of brown rust was more severe on plants grown on heavily contaminated plots. The extent of brown rust disease 5 days after inoculation was 2-fold higher on plot 3 with maximal gamma-background than on plot 1 (Table 1). Ten days after inoculation the extent of the disease increased on all three plots, however it still remained highest (68 %) on plot 3.

Table 1. Incidence and extent of brown rust on wheat plants grown on plots with different level of radionuclide contamination (cv. Kiyanka)

Plot No.	Gamma-background, mR/h	Date of analysis	Incidence of disease, %	Extent of disease, %
1	1-2	15.06	70	23
		20.06	100	38
2	9-11	15.06	99	33
		20.06	100	46
3	35-37	15.06	100	47
		20.06	100	68

It appeared that the differences in wheat brown rust resistance were a result of differences in absorbed dose of ionizing radiation. External radiation dose for plants on plot 3 was 27-fold higher than that for plants on low contaminated plot 1. However, based on the data about specific radioactivity of plants (data not shown), it was found that internal absorbed dose for plants grown on plot 3 was 100-fold higher than that for plants on plot 1.

3. Biochemical mechanisms of decreasing in plant disease resistance

We have also analysed biochemical mechanisms underlying the decrease in plant disease resistance. Three wheat cultivars, rye cv. Saratov, and two lines of corn were grown on the experimental plots. The lines of corn were original (W64A+/+) and high lysine opaque-mutation (W64A o2/o2), which reveals increased sensitivity to stress factors of the environment [2]. The absorbed dose during vegetation period was about 7-8 cGy for cereals and 3 cGy for corn. The activity of plant proteinase inhibitors was determined in the albumine fraction of corn leaves and seeds, and in wheat and rye grains by inhibition of serine proteinases proteolytic activity, estimated by casein method.

The data show that the activity of proteinase inhibitors (trypsin, chemotrypsin, subtilysine) in plants grown on radionuclide-contaminated plots decreased. In wheat and rye grains the activity was decreased up to 15-60 % as compared to the control. The inhibitors could form stable complexes with proteolytic enzymes of pathogens and thus restrict the disease development [3]. It is not clear whether other plant defence responses (phytoalexin synthesis, accumulation of PR-proteins) could be also affected by low dose chronic radiation. But decreasing of proteinase inhibitors activity appear to diminish plant disease resistance.

This assumption was confirmed by experiments with high lysine opaque-mutation of corn. Activity of proteinase inhibitors in the mutant (W64A o2/o2) leaves was decreased (data not shown). Moreover, in the mutant grains the proteinase inhibitors activity was 3-4-fold less than in unirradiated plants and 2-fold less than in irradiated plants of original corn line (Table 2). Therefore, the corn mutation is highly sensitive to ionizing radiation and as well as the *waxy*-mutation of barley [4] it could be a useful tool in understanding low dose effects.

Table 2. Specific activity of proteinase inhibitors in corn grains

N	Sample	Trypsin Inhibitor		Chemotrypsin inhibitor		Subtilisin inhibitor	
		mg/g protein	% of control	mg/g protein	% of control	mg/g protein	% of control
1	W64A+/+ Control	106	-	36	-	118	-
2	W64A+/+ Zone	66	62	23	64	90	76
3	W64Aî2/î2 Control	129	-	20	-	130	-
4	W64Aî2/î2 Zone	34	26	6	30	30	23

Thus results obtained both in greenhouse and field trials demonstrate the decrease in plant disease resistance under low dose chronic radiation.

Simultaneously, changes in virulence and in aggressivity of plant pathogenic fungi could take place in the 10-km zone of ChNPP. We analysed such a possibility using *Puccinia graminis* Pers. (*Uredinales, Basidiomycetes*), a causal agent of stem rust of wheat and other cereals.

4. Changes in population structure of plant pathogenic fungi

Rust fungi are highly variable and well adapted to rapid spread over long distances. New rust varieties may arise by mutations and spread from the 10-km ChNPP zone into other parts of the country and overcome the resistance of cultivars that was effective against the old varieties. Two approaches can improve the durability of rust resistance in wheat and rye: 1) improve our understanding of how rust varieties compete within rust populations in order that we can anticipate virulence shifts in rust populations or manipulate the populations to delay shifts in virulence, and 2) identify new types of rust resistance that are either not variety specific or that do not select for rapid increases in virulence in rust populations.

Determination of areal and nutritional plant species for *Puccinia graminis* was performed by analysis of grain crops on experimental plots. In parallel, we made a search for wild cereals in natural biocenosis' neighbouring to common barberry (*Berberis vulgaris*), the host for the sexual stages of the stem rust fungus, and grain crops. All stages of fungus development – aeciospores, urediniospores and teliospores were analysed. Identification of physiological varieties of *P. graminis tritici* populations was performed on classical Stackman varieties-differentiators with known resistance genes to stem rust [5]. Three types of response reactions were taken into account: resistant (0-2 points), susceptible (3-4 points) and heterogeneous (Õ). Virulence was determined on the basis of monogenic line reactions with resistance genes [6].

Puccinia graminis regularly caused severe epidemics until the mid-1950s. Wheat leaf rust and oat crown rust caused countrywide losses of up to 10% in recent years. These epidemics occurred when new virulent varieties of rust suddenly increased to destructive levels before new cultivars could be developed with resistance to the new varieties.

Table 3. Stem rust lesions of cereal species on experimental plots

	Species	Lesions intensity, % / type of lesions, points			
		"Maneve" region		10-km ChNPP zone	
1	Triticum aestivum Will	50	4	100	4
2	Secale cereale L.	80	4	100	4
3	Hordeum vulgare L.	65	3	80	4
4	Avena sativa L.	90	4	100	4
5	Agrostis alba L.	45	3	100	4
6	Agrostis vulgaris With	53	3	100	3
7	Aspera spica-venti (L.) P.B.	75	3	100	4
8	Calamagrostis epigeios L.	25	4	100	4
9	Dactylis glomerata L.	100	4	100	4
10	Elytrigia repens (L.) P.B.	100	4	100	4
11	Lolium perenne L.	40	3	100	3
12	Poa pratensis L.	70	3	100	3

The advantage of the research was that since 1966 we studied the sexual stage population of P. graminis tritici in "Maneve" region near Kaniv in Ukraine. That allows us to have a distinctive "zero point" to analyze changes in stem rust population structure as a result of ChNPP accident. To characterize the population structure it was necessary to identify varieties with various virulence phenotypes and to determine frequencies of virulent genes. Analysis of wheat, rye, barley, oat, and grasses on experimental plots in 10-km ChNPP zone revealed stem rust development on 12 varieties of cereals. The incidence of disease was about 50-85 % at practically 100 % damaged crop (Table 3).

Three main forms of the fungus were found: 1) wheat (P. graminis tritici), which damages both wheat and barley; 2) rye (P. graminis secale); 3) oat (P. graminis avenae). All the forms were capable to develop on many cereal grass species, which serve as reservoir of infection accumulation duringof the vegetation period. 642 monopustul clones of stem rust were isolated. Among them 8 physiological varieties of the pathogen were revealed: 11, 21, 34, 40, 100, 189, 3ê, but also the variety, whose characteristic was absent in the International register. We named it variety "X".

It was found in 1992 that varieties 3ê (27 %) and 100 (23 %) were dominant in the "Chernobyl " population. Three years after, the most isolates belonged to variety 34 (24 %), but also varieties 11 (18 %), 21 (12 %) and 40 (6 %) were present. Thus only widespread varieties 34, 3ê and the rare variety 189 retained during three years. All varieties demonstrate a high virulence based on their reactions on a set of 12 wheat lines with different genes for rust resistance (Table 4). Analysis of three wheat cultivars (Mironovska-808, Polesska 70 and Kiyanka) inoculated with different varieties of P. graminis revealed a high susceptibility (often "4" points, rarely "3"). Varieties 11, 21 and 34 are worldwide known. Variety 21 was dominant during a few years in ex-USSR countries. Variety 189 was of special interest. It induced very high sensitivity on all varieties-differentiators with known resistance genes to stem rust.

Table 4. Reactions of physiological varieties of stem rust revealed in 10-km ChNPP zone

Variety	Varieties-differentiators					
	Little Club	Marquis	Reliance	Kota	Arnautka	Mindum
11	4	4	3++	3	4	4
21	4	4	0	3	4	4
34	4+	4	4	4	4	4
40	4+	4+	4	4+	4+	4+
100	3	4	3	3	3	3
189	4	4	4	3++	4	4
3ê	4	4	4	4	4	4

Variety	Varieties-differentiators					
	Spelmar	Kubanka	Acme	Eincorn	Vernal	Khapli
11	4	3	3	3	1	1
21	4	4	3	1+	0	1
34	4	4	3	1	1	1
40	4	4=	4	0	4=	1=
100	3	4	1	1	Õ	1
189	4	4	4	4	4	4
3ê	4	4	2	0	0	1

Analysis of genotypes of *P. graminis* on monogenic lines showed that more virulent clones were present with high frequency in the "Chernobyl" population.

5. Conclusions

The results obtained independently both in greenhouse and field trials in 10-km zone of ChNPP demonstrate the decrease in plant disease resistance. Analysis of biochemical mechanisms underlying the decrease in wheat and rye disease resistance revealed reduction in activity of proteinase inhibitors in the plants. An opaque-mutation of corn could be a useful tool in understanding low dose radiation effects. The population structure of *Puccinia graminis*, a causal agent of stem rust and one of the most severe pathogens, has been changed in 10-km ChNPP zone by appearance of a «new» population with high frequency of more virulent clones.

6. References

1. A. Dmitriev. Phytoalexins and their role in plant disease resistance. Kiev: Nauk. dumka, 1999. 208 p.
2. F. Vinnitsenko et al. Influence of a gene opaque-2 on activity of proteinase inhibitors in corn grain // Physiol. Biochem Cult. Plants. 1988. -20. N 5. P. 493-497.
3. L. Metlitskiy et al. Biochemistry of plant immunity, rest and aging. Moscow: Nauka, 1984. 264 p.
4. I. Boubryak, E. Vilensky, V. Naumenko et al. Influence of combined alpha, beta and gamma radionuclide contamination on the frequency of *waxy*-reversions in barley pollen. Sci. Total

Environ., 1992. 112: 29-36.
5. E. Stackman and J. Harrah. Plant pathology. NY: Academic Press.1969.-514 p.
6. L. Semenova L. The methodical recommendations for studies on race structure of stem rust fungi. Moscow: VASCHNIL, 1977. – 144p.

COMPARATIVE STUDY OF THE EFFECTS OF ENDOCRINE DISRUPTOR AND IONIZING RADIATION WITH PLANT BIOASSAY

Jin Kyu KIM
Korea Atomic Energy Research Institute, 150 Dukjin-dong, Yusong-gu, Taejon 305-353, KOREA

Hae Shick SHIN and Jin-Hong LEE
Chungnam National University, Gung-dong, Yusong-gu, Taejon, 305-764, KOREA

Vladislav G. PETIN
Medical Radiological Research Center, 249036 Obninsk, Kaluga Region, RUSSIA

1. Introduction

The use of synthetic chemicals exerts a great influence on the daily life of human beings. The recent reports that some of those synthetic chemicals can act as endocrine disrupting substances in various organisms have drawn special attention in environmental research. Among others, dioxins, DDT, PCBs and bisphenol A (BPA) are classified into endocrine disruptors and are under a strict control in many developed countries. Since a small amount of them can disturb biological systems, the ecological problem can also arise when they exist in excess in the environment. They can give rise to such adverse effects as morphological anomaly, decreased number of sperms, cancer and change of sex [1]. Since the use of the synthetic chemical will steadily increase in modern life, it is necessary to establish assay techniques for monitoring them, and to assess their biological risks.

The reproductive cells are the most sensitive to toxic environmental materials and ionizing radiations. The pollen mother cells, male reproductive cells of *Tradescantia* are highly synchronized in their prophase I and tetrad stages. Chromosomes of this stage are sensitive to physical or chemical mutagens [2]. These induced chromosome aberrations become micronuclei in the synchronized tetrads and they can be easily identified and scored. Based upon these features, the *Tradescantia* micronucleus bioassay was established. The meiotic pollen mother cells have quite long been known to be more sensitive than mitotic chromosomes to radiation [3]. But the fact that the chromosomes of the pollen mother cells were loose in the stage of metaphase I and thus it is very difficult to be scored had prevented the cells to be used for chromosome aberration assay. However, if the treatment is applied to the early prophase I stage and allow the treated meiotic chromosome to go through a proper recovery time [4], the acentric fragments or the sticky chromosome complex would become micronuclei in tetrad stage of meiosis. This was the basis for establishing the *Tradescantia* micronucleus (Trad-MCN) bioassay. This bioassay was first utilized in the study of mutagenic effects of 1,2-dibromoethane [5] and followed by X-rays and known mutagens [6].

Since the micronucleus production was understood as a useful indicator for chromosome damage, the micronucleus assay has been widely applied to plant and animal cells by many workers [7-9].

F. Brechignac and G. Desmet (eds.), Equidosimetry, 237–242.

This research was designed to demonstrate the biological effects of BPA, which is one of the well-known endocrine disruptors, by means of the *Tradescantia* micronucleus (Trad-MCN) assay. The biological effect of gamma rays on the same cell system was also evaluated for comparison with that of BPA to estimate the equidosimetric effectiveness of BPA.

2. Materials and methods

Fresh cuttings of *Tradescantia* clone 4430 (an interspecific hybrid between *T. hirsutiflora* and *T. subacaulis*) were used in this study. Each experimental group consisted of 20 cuttings. The groups for studying radiation effects were irradiated with gamma-rays from ^{60}Co (source strength 150 TBq, Panoramic Irradiator, Atomic Energy Canada Ltd.). During irradiation the stems of the cuttings were immerged in Hoagland's No. 2 (x1/6) solution to prevent them from wilting. Radiation doses are 0, 10, 20, 30, 40 and 50 cGy.

Bisphenol A (4,4'-isopropylidenediphenol, Sigma Co.) was dissolved in absolute ethanol. The solution was diluted with distilled water to the experimental concentrations, which were expressed in mM. Plant cuttings were treated with BPA solution in a beaker for 6 hours, and during this period, the solution was continuously aerated and the plants were illuminated with artificial lights. The negative control DW, solvent control 1% ethanol and tap water were also tested for comparison in these experiments. Right after the treatment, the chemical solution was removed and the stems of the cuttings were rinsed in tap water to remove the remaining BPA solution and then transferred into aerated fresh Hoagland's No. 2 solution (x 1/6) during 24 hours for recovery.

The slide preparation and MCN scoring were carried out according to the standard protocol described in an earlier publication by Ma *et al.* [10].

3. Results

The results of gamma-ray treatment are given in Figure 1. Linear regression analysis of the highest MCN frequencies and radiation doses gave the following equation. Data for the control group will be separately discussed later.

$$F_{MCN} = 1.97D + 4.05, \quad (r^2 = 0.955) \tag{1}$$

where,

F_{MCN} = the maximum MCN frequencies (MCN/100 tetrads),
D = radiation dose in cGy.

The correlation coefficient of the regression was 0.955. No physical damage was observed in the pollen mother cells at all dose levels. According to the above equation, two micronuclei in 100 tetrads can ascribe to 1 cGy of gamma-radiation.

Micronucleus frequencies obtained by scoring slides from the experimental groups are depicted in Figure 2. The MCN frequencies were 8.06±0.70, 12.76±1.06, and 19.67±1.52 in 1, 2, and 4 mM BPA-treated groups, respectively. For the highest MCN frequencies in the BPA-treated groups, linear regression gave the following equation.

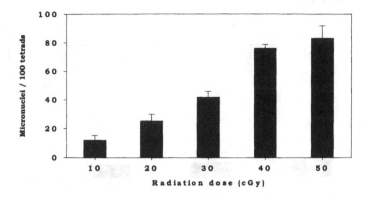

Fig. 1. Radiation dose-response relationship of micronucleus frequencies in pollen mother cells of T. BNL 4430.

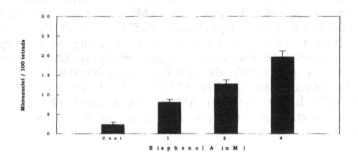

Fig. 2. Bisphenol A concentration-response of micronucleus frequencies in pollen mother cells of T. BNL 4430.

$$F_{MCN} = 4.26C + 3.24, \quad (r^2 = 0.983) \tag{2}$$

where,

   F_{MCN} = the maximum MCN frequencies (MCN/100 tetrads),
   C = concentration of BPA in mM.

   According to the above equation (2), roughly four micronuclei in 100 tetrads can be additively induced by one mM increase in BPA concentration. In the preliminary experiment, the authors found that high concentration (>8 mM) of BPA could affect physically the pollen mother cells themselves, and that such physical damage would be followed by apoptosis of the cells (unpublished data).

For the negative controls, distilled water, solvent (1% Et-OH) and tap water were also tested for their micronucleus induction in the pollen mother cells (Figure 3). The spontaneous MCN frequency in the group without any treatment was 2.75±0.55, and that of 1% EtOH treated group was 2.33±0.62, which were in good accordance with the previously reported background value [11]. In contrast tap water resulted in a wide variation in the induction of micronuclei (2.70±0.72 to 9.00±1.16). However, the unexpectedly high MCN frequency induced by tap water was not reproducible thereafter.

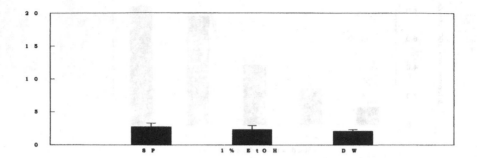

Fig. 3. Micronucleus frequencies induced by 1% ethanol (1% EtOH), distilled water (DW), and spontaneous rate (SP).

Linear regression dose- and concentration-response lines for the gamma-rays and BPA treatments were established by two parallel series of experiments. These two regression lines were compared at the same levels of MCN frequencies induced by these two agents in order to determine the equi-dosimetric effectiveness of BPA as compared with ionizing radiation. Table 1 was constructed by equating the two regression equations and by choosing the dosages of BPA and gamma rays that were needed to induce the same amount of each of the six MCN frequencies. Based on the equations (1) and (2), a unit dose of radiation (cGy) induced 1.97 MCN whereas a unit dose of BPA (mM) did 4.26 MCN. In this context, 1 mM of BPA is equivalent to 2.16 cGy of radiation for the induction of clastogenic damage in the pollen mother cells of *Tradescantia* 4430.

Table 1. Radiation dose and bisphenol A concentration for inducing the same frequencies of micronuclei in Tradescantia pollen mother cells

Bisphenol A (mM)	Micronucleus frequencies(MCN/100 tetrads)	Radiation dose equivalent to BPA concentration (cGy)	Remarks
0.5	5.4	0.68	
1.0	7.5	1.80	
2.0	11.8	3.90	
4.0	20.3	8.20	
8.0	37.3	16.90	
11.8	53.3	25.00	*)

*) External whole-body exposure dose above which clinical symptoms can develop

4. Discussion

Tradescantia is a member of Commelinaceae family. It has two distinct features as a wonderful experimental object in radiobiological and environmental studies. Based on the heterozygosity for the flower colour of *Tradescantia* plants (i.e. clone 02 or 4430), somatic cell mutation assay technique has been developed, which is known as Trad-SH assay [12,13]. The other assay technique based on the chromosome aberrations in the meiotic pollen mother cells is called Trad-MCN assay [3,5]. While another higher plants such as *Allium cepa*, *Vicia faba*, and *Arabidopsis thaliana* can also be used as an botanical assay tool for detecting radiations, mutagens and environmental clastogens, Trad-MCN assay has an advantage over anything else because hundreds of thousands of pollen mother cells can be scored easily and quickly and thus statistically reliable results can be obtained.

The results of the present study confirm that the Trad-MCN bioassay can detect clastogenic effects of endocrine disruptors in form of solution and ionizing radiation, as well. Most of works with BPA have focused on animal studies, only because it gathers attention as an endocrine disruptor. The clastogenic effect of BPA in plant systems is newly reported here. Furthermore, the clastogenic effects of BPA as measured by means of the micronucleus frequencies in the pollen mother cells of *Tradescantia* were compared to those of ionizing radiation in the same cell system for equidosimetry purpose.

The frequencies of micronucleus showed a positive dose-response relationship in the range of 0 to 0.5 Gy, and a clear concentration-response relationship in the experimental range of BPA concentrations. Linear regression dose- and concentration-response lines for the gamma rays and BPA treatments were established by two parallel series of experiments. These two regression lines were compared at the same levels of MCN frequencies induced by these two agents in order to determine the equi-dosimetric effectiveness of BPA as compared with ionizing radiation. Using the regression equations described in the previous section, it is possible to calculate the equivalent doses to each other. A unit dose of radiation (cGy) induces 1.97 MCN whereas a unit dose of BPA (mM) does 4.26 MCN. In this context, 1 mM of BPA is equivalent to 2.16 cGy of radiation in the induction of clastogenic damage in the pollen mother cells of *Tradescantia* 4430.

BPA of 11.8 mM can give rise to 53.3 MCN/100 tetrads, the same frequency that is induced by 25 cGy of gamma rays. It is noteworthy that clinical symptoms can develop after a whole body exposure to ionizing radiation dose higher than 25 cGy in human beings [14].

5. Concluding remarks

This study was designed to demonstrate the biological effects of BPA, one of the well-known endocrine disruptors, by means of the *Tradescantia* micronucleus assay. The biological effect of gamma rays on the same cell system was also evaluated for comparison with that of BPA to estimate the equi-dosimetric effectiveness of BPA. Fresh cuttings of *Tradescantia* BNL 4430 were treated with BPA solutions of 0 to 4 mM for 6 hours for the absorption of the solution through the stem of the plant cuttings. Other groups of the cuttings were irradiated with 0 to 1.0 Gy of gamma rays. The frequencies of micronucleus showed a positive dose-response relationship in the range of 0 to 0.5 Gy, and a

242

clear concentration-response relationship in the experimental range of BPA concentrations. By comparing the two experimental results, it is possible to estimate the BPA concentration and its equivalent radiation dose for a fixed value of MCN frequency. BPA of 11.8 mM can give rise to 53.3 MCN/100 tetrads, the same frequency that is induced by 25 cGy of gamma rays. It is of biological importance that clinical symptoms start to develop after a whole body exposure to radiation higher than 25 cGy. The results indicated that the pollen mother cells were an excellent biological end-point for measuring toxicity of endocrine disrupting chemical bisphenol A. In addition, Trad-MCN assay can be easily applied to measure the biological effects of suspected endocrine disruptors such as octylphenol and nonylphenol.

6. Acknowledgement

This study has been carried out under the National R&D Program and Korea-Russia Scientists Exchange Program by the Ministry of Science and Technology (MOST) of Korea.

7. References

1. D.C. Shin, 1999. What is the endocrine disruptor? *Food Science and Industry*, 33 : 2-18 (1999) (*in Korean*).
2. K. Sax. Chromosome aberration induced by x-rays. *Genetics* 23 : 494-516 (1938).
3. T. H. Ma, G. J. Kontos and V. A. Anderson. Stage sensitivity and dose response of meiotic pollen mother cells of *Tradescantia* to x-rays. *Environ. Expt. Bot.* 20 : 169-174 (1980).
4. J. H. Taylor. The duration of differentiation in excise anthers. *Am. J. Bot.* 37 : 137-140 (1950).
5. T. H. Ma, A. H. Sparrow, L. A. Schairer and A. F. Nauman. Effect of 1,2-dibromoetane (DBE) on meiotic chromosomes of *Tradescantia*. *Mutat. Res.* 58 : 251-258 (1978).
6. T. H. Ma. Micronuclei induced by X-rays and chemical mutagen in meiotic pollen mother cells of *Tradescantia* - A promising mutation test system. *Mutat. Res.* 64 : 307-313 (1979).
7. J. A. Heddle. A rapid *in vivo* test for chromosome damage. *Mutat. Res.* 18 : 187-190 (1973).
8. J. R. K. Savage and D.G. Papworth. An investigation of LET 'finger-prints' in *Tradescantia*. *Mutat. Res.* 422 : 313-322 (1998).
9. W. F. Grant. Higher plant bioassays for the detection of chromosomal aberrations and gene mutations--a brief historical background on their use for screening and monitoring environmental chemicals. *Mutat. Res.* 426 : 107-112 (1999).
10. T. H. Ma, G. L. Cabrera, R. Chen, B. S. Gill, S. S. Sandhu, A. L. Vandenberg and M. F. Salamone. *Tradescantia* micronucleus bioassay, *Mutat. Res.* 310 : 221-230 (1994).
11. J. K. Kim, H. S. Song and S.H. Hyun. Dose-response relationship of micronucleus frequency in pollen mother cells of *Tradescantia*, *J. Kor. Assoc. Radiat. Prot.*, 24 : 187-192 (1999).
12. A. G. Underbrink and A. H. Sparrow. Power Relations as an Expression of Relative Biological Effectiveness (RBE) in *Tradescantia* Stamen Hairs, *Radiat Res.*, 46 : 580-587 (1971).
13. T. H. Ma, G. L. Cabrera, A. Cebulska-Wasilewska, R. Chen, F. Loarca, A. L. Vandenberg and M. F. Salamone. *Tradescantia* Stamen Hair Mutation Bioassay, *Mutat. Res.*, 310 : 211-220 (1994).
14. J. K. Kim and S. H. Hyun. Comparative study on human risk by ionizing radiation and pesticide as biological information about environmental disaster, *J. Kor. Assoc. Radiat. Prot.*, 26 : 385-392 (2001)

Part 6.

Genetic Factors, Environment and Toxicants

CYTOGENETIC EFFECTS IN PLANTS AFTER WEAK AND COMBINED EXPOSURES AND A PROBLEM OF ECOLOGICAL STANDARDIZATION

S.A. GERAS'KIN, V.G. DIKAREV, N.S. DIKAREVA, A.A. OUDALOVA, D.V. VASILIYEV
Russian Institute of Agricultural Radiology and Agroecology, Obninsk, RUSSIA

T.I. EVSEEVA
Biological Institute, Komi Research Center, Urals Department of RAS, Syktyvkar, RUSSIA

1. Difference between sanitary-hygienic and ecological approaches to substantiation of derived levels for ionizing radiation

For the last three decades, a scientific background for radiation protection of biota has been based on principles developed by the ICRP. A substance of the most important among them can be formulated as 'If protection of a human is provided, it will guarantee protection of an environment'. In such an approach, objects of environment are considered as sources of external and internal irradiation for humans, and radionuclides contents are standardized by the sanitary-hygienic approach. At the same time, although standards derived from these principles guarantee the protection of a human, they are not always able to guarantee the equal protection of other objects of living nature. Standardization with this approach assumes in default that doses absorbed in human are close to doses absorbed in other environmental objects and hence if one provides low doses for people it will automatically assure low doses for other alive organisms. This is not necessarily true. So, dose values per unit of radioactive contamination density obtained by plants and animals from radioactive fallouts can exceed values for humans by a factor of 10-100 [1]. With regard to a close radiosensitivity of human and a number of edificator species determining functioning and stability of ecosystems (coniferous trees, many mammalians), such a relationship between doses absorbed in human and environmental objects makes clear that a special attention is to be paid to protection of plants, animals and communities. Thus, taking the sanitary-hygienic approach as the major one in providing radiation security for the population, one should admit the necessity of its complementation with standards of radiation safety for biota.

A form of existence of plants and animals in nature is a population. Therefore, when one changes from a sanitary-hygienic approach to an ecological one not only indexes of standardization norms modify but also the objects to which they should be applied (Table 1). These objects are not individuals as it used to be in frames of the sanitary-hygienic approach but natural or artificial systems of supra-individual levels, that are populations, ecosystems, and biogeocenoses. A distinction of doses delivered to human and biota mainly reveals at severe radiation accidents [1]. But, experience gained from the elimination of consequences of such accidents showed that, even in this case, there was not an overall lethal lesion at the ecosystematic level except at coniferous forests. This regards to very good

F. Brechignac and G. Desmet (eds.), Equidosimetry, 245–255.

possibilities of populations for strength restoration. Thus, up to 85% of young generation in populations of small sparrow birds (brood of a current year) die every year, but the whole population remains quite stable. Therefore, differences between doses, which are able to induce an increase of stochastic effect occurrence in man, and those, which can result in a death of plant and animal populations, are so great that the main principle of the ICRP, -'if a man is protected environment is also saved', - does not overwhelmingly break in current ecological situations.

Table 1. Differences between the sanitary-hygienic and the ecologic approaches at development of derived levels

	Sanitary-hygienic	Ecological
Doses to human and biota	Close	There are radioecological situations when some representatives of biota accumulate doses 10-100 times greater
Object of application	Human	Systems of supra-individual level (populations, ecosystems, biocenosis)
End-points for derived levels development	Deterministic and stochastic effects (cancerogenesis, genetic impairments)	Survival in systems of supra-individual level (populations, ecosystems)

On the other hand, the most important principle that a set of ecological criteria ought to meet is the principle of non-admission of deterioration of the ecological state of the environment. The hierarchic structural-functional organization of living matter supposes a multi-level system of responses to external impact. In this connection, a question emerges: alterations at what level of biological organization really indicate a deterioration of ecology?

Natural and agrarian ecological systems have a diverse variability of responses to irradiation. A range of dose rates at which biota irradiation actually takes place covers five values of magnitude. The derived levels of exposure of man and biota recommended by the international agencies are compared with dose rate values of chronic radiation producing effects at different levels of biological organization in Table 2. From these data, genetic effects and alterations of genetic structure in wild plant and animal populations are observed at dose rates, which are below the level considered by the IAEA as not providing any hazard to biosphere. Nevertheless, it is changes at an underlying, molecular-cellular level of living mature organization that give the earliest and confident information about any negative changes in environment. In fact, according to the principle of attenuation of adverse effects with increasing level of organization of biological systems [4], the earliest changes can be detected at the subcellular and cellular levels of organization of living matter. Genetic tests show direct and specific responses to exogenous stresses. Therefore, they may be of a unique importance for estimating the changes that occur, as a rule, before morphological, physiological, populational, or other deviations from the norm. In addition, the genetic tests reveal the most substantial effects of environmental pollution related to increasing mutagenic pressure on the biosphere. This includes an increasing incidence of cancer and hereditary diseases; an increasing genetic load in humans, animals, and plants; and changing genetic and species structure of biocenoses. Therefore, it is the genetic test-systems that must be used for early diagnosis of the alterations resulting from human economic activity.

Table 2. Comparison of the derived levels of exposure of man and biota recommended by the international agencies with dose rates of chronic radiation producing effects at different levels of biological organization [1-3]

Biological effect	Dose rate, $Gy \cdot year^{-1}$
Dose limit for humans (the ICRP, 1990)	**10^{-3}**
Natural radiation background	$10^{-2}-10^{-3}$
Casual detection of genetic effects	0.05
Steady registration of genetic effects in the most radiosensitive species	0.1
Dose limit for deterministic effects (the ICRP, 1990)	**0.15**
Increase of mean population radioresistance (radioadaptation)	0.2
Doses to biota considered by the IAEA as not providing any hazard	**0.4**
Inhibition of growth and development in radiosensitive species	1-3
Disappearance of sensitive species from a community	4
Radiation damage to ecosystems:	
Coniferous forests	10
Deciduous forests	30
Agricultural crops	50
Herbaceous phytocenoses	70

Let us consider what changes on cytogenetical level can be induced with low doses of ionizing radiation in plants under conditions of single and combined with factors of different nature of action. Important features of biological effect of low-level radiation that do not follow from well-known effects of high and moderate doses are:
- nonlinearity of dose response;
- synergetic and antagonistic effects of combined action of different nature   factors;
- radiation induced replicative instability of genome;
- phenomenon of radioadaptation.

A development of a new concept of radiation protection for human and biota should be based on the clear understanding of these effects and their contribution to response of biological objects to single low-level radiation and combined with other anthropogenic factors influence.

2. Non-linearity of dose response

The analysis of experimentally observed reactions of cells on low level irradiation has shown [5, 6] that regularities of a yield of cytogenetic disturbances in this range are characterized by a intensely expressed non-linearity and have an universal character, differing for miscellaneous objects by values of doses, at which there are changes of the relation character. To check this state an experiment has been carried out on barley seedlings [7]. It is apparent from the results presented in Fig. 1, even without using precise quantitative criteria, that the piecewise linear model based on our concept of biological effect of low-level ionizing radiation [6, 8] describes the experimental data about cytogenetic damage appearance in studied range much better than the linear one. It is important, that the improvement of the quality of approximation is reached not on account of the model complicating but because it is possible to achieve a mutual conformity (functional isomorphism) between a biological phenomenon and its mathematical model on the set of piecewise linear functions. Comparison of approximation quality that can be achieved with models of different complexity by the most common quantitative criteria testifies to it as well (Table 3).

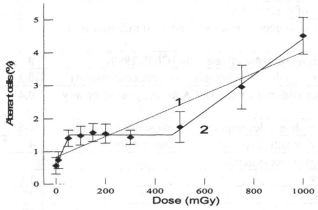

Fig. 1. Frequency of aberrant cells in barley germs exposed to low radiation doses and its approximation with linear (1) and piecewise linear (2) models .

Table 3. Comparison of the approximation quality of the experimental data about the yield of cytogenetic disturbances in irradiated barley seedlings by models of different complexity

Model	SSR	F	R^2, %	T	Hayek
Piecewise	1.35	62.7	88.7	0.34	14.232
Linear	0.03	1829.3	99.7	0.03	
Polynomial of degree 2	0.87	88.5	92.7	0.37	1.951
Polynomial of degree 3	0.49	139.2	95.9	0.33	3.23
Polynomial of degree 4	0.14	435.8	98.9	0.14	6.699
Polynomial of degree 5	4.25	7.2	64.3	6.37	1.653

SSR – sum of residuals squares,

R^2 – multiple correlation coefficient,

T - criteria of structural identification [22],

$$\text{Hayek} = \sqrt{\frac{\left(R_1^2 - R_2^2\right) \cdot \text{d.f.}}{1 - R_1^2}} \,, \quad R_2{}^2 < R_1{}^2$$

An absence of authentic experimental information about biological effects that can be induced by irradiation at doses less than 1 cGy, as well as an existence of alternative approaches to explain the causes of a nontrivial shape of dose dependence at low levels, have stipulated a necessity of new experimental researches. In our studies on cells of stamen filament cells of spiderwort (clone 02) [9] it not only appears possible to confirm existence of the plateau revealed on barley seedlings (for spiderwort, due to the greater sensitivity, this range is shifted towards smaller doses), but also to experimentally support a theoretical prediction of our concept about an existence of dose range (in this case, less 1 cGy) where there is no significant increase of genetic disturbances (Fig.2). Let us remark, that the conclusion about the non-linear character of dose dependence which follows from the presented results is based on experimental data only and does not depend on any hypothesis or extrapolation models.

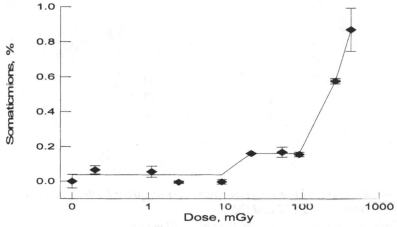

Fig. 2. Frequency of somatic mutations in stamen filament cells of Tradescantia (clone 02) versus γ-irradiation dose.

3. Synergetic and antagonistic effects of combined action of different nature factors

It was shown in course of our experimental studies carried out on spring barley, onion and spiderwort under combined effect of such common agents as γ-radiation, pesticides, heavy natural radionuclides, salts of heavy and alkali metals [9-12] that there is a principal possibility of synergistic and antagonistic effect occurrence. In regard to this consideration, there are two important questions. The first is: are there ranges of doses and concentrations such that nonlinear effects can be observed more often under a combination of exposures from these ranges? It seems there are. It is in the case of combined action of low-level exposures when synergetic effects are induced with an increased probability [9-13]. The second question issues forth from the first: how important is a contribution of nonlinear effects to the response of biological objects? If their contribution is negligible, one can disregard it and estimate a result of a combined influence using available information about single effects of each factor that is more convenient and easier, of course. A study of cytogenetic disturbances induction in intercalar meristem cells of spring barley grown on soil contaminated with low concentrations of Cs-137 and cadmium nitrate [12] has shown that the effect of combined action exceeds the sum of separate effects as much as 70% (Fig.3). On the contrary, the observed effect at soil pollution by radioactive cesium, lead and pesticides averaged only 50% from anticipated one proceeding from the additive model. Therefore, an application of data from a single effect analysis to forecast biological effects of combined effect is unacceptable and causes the essential distortion from picture that can be observed experimentally.

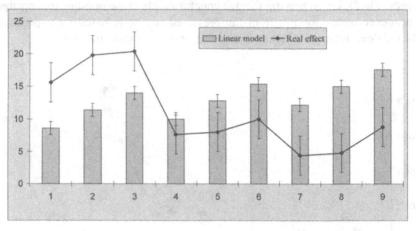

Fig. 3. Cytogenetic damage yield in intercalar meristem of spring barley in conditions of combined soil pollution by Cs-137 and Cd:
  $1 - 1.48\ MBq/m^2 + 2\ mg/kg$; $2 - 1.48\ MBq/m^2 + 10\ mg/kg$;
  $3 - 1.48\ MBq/m^2 + 50\ mg/kg$; $4 - 7.4\ MBq/m^2 + 2\ mg/kg$;
  $5 - 7.4\ MBq/m^2 + 10\ mg/kg$; $6 - 7.4\ MBq/m^2 + 50\ mg/kg$;
  $7 - 14.8\ MBq/m^2 + 2\ mg/kg$; $8 - 14.8\ MBq/m^2 + 10\ mg/kg$;
  $9 - 14.8\ MBq/m^2 + 50\ mg/kg$

4. Radiation induced replicative instability of genome

In 1987-1989 within the ChNPP 10-km zone, we conducted an experiment [14] with one of the purposes being to analyze cytogenetic variability in three successive generations of winter rye and wheat, grown at four plots with different levels of radioactive contamination. A dose on a growing point varied within the limits of 18-717 cGy between plots for the vegetative season 1987-1988, and within the limits of 11-417 cGy for 1988-1989. Fig. 4 shows that in this experiment in autumn of 1989, aberrant cell frequencies in leaf meristem of winter rye and wheat of the second and third generations significantly exceeded these parameters for the first generations; it was significant on experimental plots with the highest levels of radioactive contamination. The distinctions between cytogenetic parameters obtained for the second and third generations are small and statistically not significant,and the observed effect is of a threshold character.

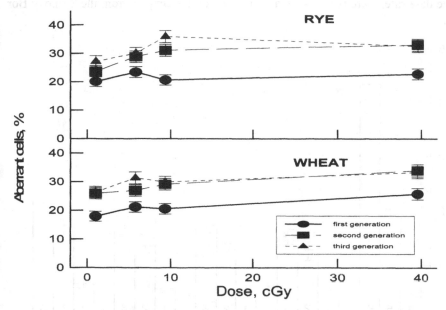

Fig. 4. Modification of aberrant cell frequency in leaf meristem of winter rye and wheat in three generations grown up on contaminated plots.

It is important that it was not root meristem of seedlings but intercalar meristem of plants where the cytogenetic disturbances were analyzed. It means that the most part of radiation-induced alterations accumulated during the previous vegetative season was realized into real mutations long before the samples were fixed for the cytogenetic analysis. As in autumn of 1989 plants of all three generations developed in identical conditions and were exposed to identical doses, the most probable explanation of the detected phenomenon is related to genome destabilization in plants grown from seeds affected by radiation. From these viewpoints, the results observed in this study, indicating a threshold character of genetic instability induction, may be a reflection of the first stage of cytogenetic adaptation, that is, chronic low-dose irradiation appears to be an ecological factor creating a background for alteration of the genetic structure of a population.

5. Phenomenon of radio-adaptation.

A study of cytogenetic damage in vegetative (needles) and reproductive (seeds) organs in the *Pinus sylvestris L.* micropopulations growing at sites with contrast radioactive contamination levels in the 30-km ChNPP zone and also in region of arrangement of the 'Radon' enterprise for processing and storage of radioactive waste (the town of Sosnovy Bor, Leningrad region) is concerned to a question about a role of such alterations on the population structure. From the results of this study presented in Fig. 5, 6, a conclusion follows that there is an expressed genotoxic influence in the investigated regions. And, while the incidence of cytogenetic damage in the samples from the Chernobyl NPP region correlates with the dose rate, there is not such a correlation in the samples from the Sosnovy Bor region.

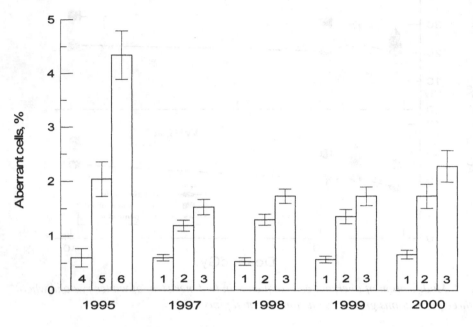

Fig. 5. Aberrant cell frequency within root meristem of Pinus sylvestris L. seedlings in 1995-200. Experimental sites in the Sosnovy Bor region - 1, 2, and 3, in the 30-km zone of the Chernobyl NPP –4, 5, and 6.

The revealed fact obliged us to make a more careful examination of these experimental data. The analysis of the structural mutations spectrum [15] has shown presence in the Sosnovy Bor data and absence in all control and Chernobyl variants of tripolar mitoses, a rather rare type of cytogenetic alterations, which appearance possibly links to spindle damage [16]. It was shown in [17] that an increase in the yield of tripolar mitoses in *Syringa vulgaris L.* and *Armeniaca vulgaris Lam.* was associated with the contamination of the local soils with a heavy metal mixture. From this, together with the dosimetric data, a considerable contribution of genotoxic chemicals to the environmental contamination in the Sosnovy Bor region may be supposed, but not as opposed to the 30-km ChNPP zone.

Now, when an anthropogenous impact to biota has become one of the most significant ecological factors, it is pertinent to consider the adaptive potentials of natural populations. One consequence of chronic irradiation of natural populations is an apparent increase in the mean radioresistance - the so-called "radio-adaptation phenomenon"; this has been observed by studies in the East Urals trail region [18, 19] and can be revealed by exposure of seeds to an additional, acute γ-radiation. It was shown later [19, 20] that the population divergence by radioresistance is connected to the selection for changes in the effectiveness of the reparative system and is not accompanied by visible morphologic alterations.

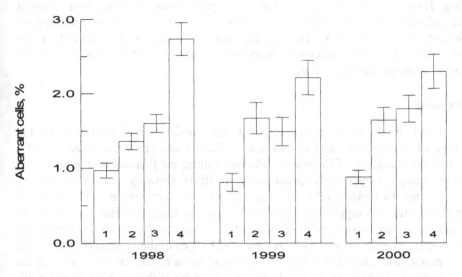

Fig. 6. Aberrant cell frequency in intercalar meristem of needles of Pinus sylvestris L. in 1998-2000. 1, 2, 3, 4 – experimental sites in the Sosnovy Bor region.

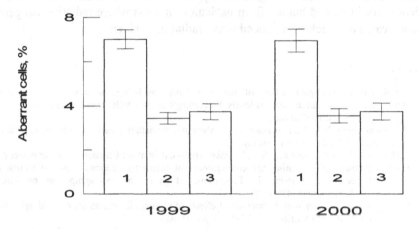

Fig. 7. Aberrant cell frequency in seedling root meristem of Pinus sylvestris L. exposed to acute γ-irradiation of 15 Gy in 1999 and 2000. 1, 2, 3 - experimental sites in the Sosnovy Bor region.

A part of the seeds collected in the Sosnovy Bor region in 1999-2000 has been subjected to an acute γ-ray exposure. The seeds from the Scotch pine populations growing in the town of Sosnovy Bor and at the 'Radon' LWPE site appear to be significantly more resistant to the acute γ-radiation than the controls (Fig. 7). Notice, a number of pine generations replaced during the existence of the 'Radon' LWPE and the Leningrad NPP, unlike the East-Ural radioactive trial, is obviously insufficient for natural selection on repair systems efficiency in its classical meaning. There are different explanations of the phenomenon regarded to a selection against radiosensitive cells [21] as well as an inheritable change of spectrum of functionally active genes [19]. But, in any case, the findings show that there are processes of cytogenetic adaptation in the examined populations that can be revealed by an additive acute γ-radiation of seeds.

6. Conclusion

It follows from the presented results that models, being used now for the estimation of biological effects of low-level radiation and combined effects of low doses and concentrations of factors of different nature, are founded on the linear no-threshold concept. A hypothesis on additivity of effects, allowing even extrapolation , has no strong biological substantiation and is in contradiction with available experimental data. The regularities in the formation of cytogenetic effects of low doses are characterised by vital non-linearity and have a universal character, differing for various objects in dose values at which there are changes of the dependency character. Chronic exposure at doses above a certain value can be an ecological factor altering the genetic structure of a population. The further development of this problem should result in a determination of levels of biological organisation, test-objects and test-systems which can become a basis for ecologically substantiated estimates of biological consequences for biota and human from habitation in areas where radiation-dangerous objects are located or which experienced heavy radiation accidents.

7. References

1. G.N.Romanov, D.A.Spirin. Effect of ionizing radiation on living nature at the levels meeting requirements of the current derived levels. Proceedings of the USSR Academy of Sciences, 1991. V.318. pp. 248-251. (in Russian).
2. V.A.Shevchenko, V.L. Pechkurenkov, V.I. Abramov. Radiation genetics of natural populations. M., Nauka, 1992. 221 pp. (in Russian).
3. R.M.Alexakhin, S.A. Geras'kin, S.V. Fesenko. The accident at the Chernobyl nuclear power plant and the problem of estimating the consequences of radioactive contamination of natural and agricultural ecological systems. in: Proceedings of an international symposium on ionizing radiation. V. 2. Stockholm, 1996. 516-521 p.
4. S.A.Geras'kin, G.V.Koz'min. Estimation of effects of physical factors on natural and agricultural ecological systems. Russ.J.Ecology, 1995. ¹ 6. pp.419-423.
5. S.A.Geras'kin. Critical analysis of up-to-date conceptions and approaches to estimation of biological effect of low-level radiation. Radiat. Biol. Radioekol., 1995. V. 35. ¹ 5. pp. 563-571. (in Russian).
6. S.A.Geras'kin. Regularities of cytogenetic effects of low level ionizing radiation. Thesis. Obninsk, 1998. 50 pp. (in Russian).
7. S.A.Geras'kin, V.G.Dikarev and others. Regularities of cytogenetical disturbances induction by low doses of radiation in barley germ root meristem cells. Radiat. Biol. Radioekol., 1999. V. 39. ¹ 4. pp. 373-383. (in Russian).
8. S.A.Geras'kin. Concept of biological effect of low dose radiation on cells. Radiat. Biology. Radioecology, 1995. V. 35. ¹ 5. pp. 571-580. (in Russian).
9. T.I.Evseeva, S.A.Geras'kin. Separate and combined action of radiation and not radiation nature factors on Tradescantia. Ekaterinburg, Nauka Publishers, 2001. 156 pp. (in Russian).
10. S.A.Geras'kin, V.G.Dikarev and others. The combined effect of ionizing radiation and heavy metals on the frequency of chromosome aberrations in spring barley leaf meristem. Russ.J.Genetics, 1996. V. 32. ¹ 2. pp. 246-254.
11. T.I.Evseeva, S.A.Geras'kin, E.S. Khramova. Cytogenetic effects of separate and combined action of ^{232}Th and Cd nitrates on *Allium cepa* root tip cells. Cytologia, 2001. V. 43. ¹ 8. pp.803-808. (in Russian).

12. S.A.Geras'kin, V.G.Dikarev, N.S.Dikareva. The influence of combined radioactive and chemical (heavy metals, herbicide) contamination on cytogenetical disturbances occurrence in intercalar meristem of spring barley. Radiat. Biol. Radioekol., 2002. V. 42. ¹ 4. (in press) (in Russian).

13. V.G.Petin, G.P.Zhurakovskaya and others. Low doses and problems of synergistic interaction of environmental factors. Radiat. Biol. Radioekol., 1999. V. 39. ¹ 1. pp. 113-126. (in Russian).

14. S.A.Geras'kin, V.G.Dikarev and others. Cytogenetic consequences of chronic irradiation under low doses for agricultural crops. Radiat. Biol. Radioekol., 1998. V. 38. ¹ 3. pp. 367-374. (in Russian).

15. S.A.Geras'kin, L.M.Zimina and others. Bioindication-based comparison of anthropogenic pollution near a radioactive-waste processing facility and in the 30-km control area of the Chernobyl nuclear power plant. Russ. J. Ecology, 2000. ¹ 4. pp. 300-303.

16. I.B.Alieva, I.A.Vorobiev. Cell's behavior and centrioles at multipolar mitosis induced by nocodasolum. Cytology, 1989. V. 31. ¹ 6. pp. 633-641. (in Russian).

17. V.P.Bessonova. The state of pollen as an indicator of the environmental pollution with heavy metals. Russ. J. Ecology, 1992. ¹ 4. pp. 45-50.

18. L.V.Cherezhanova, R.M.Alexakhin. On a biological effect of an increased ionising radiation background and the processes of radioadaptation in populations of herbaceous plants. Russ. J. General Biology, 1975. V. 36. ¹ 2. pp. 303-311. (in Russian).

19. V.A.Shevchenko, V.L.Pechkurenkov, V.I.Abramov. Radiation genetics of natural populations: genetic consequences of the Kyshtym accident. M., Nauka, 1992. 221 pp. (in Russian).

20. S.A.Sergeeva, A.B.Semov and others. Repair of radiation-induced damages to plants growing under conditions of chronic exposure to â-radiation. Radiobiologia, 1985. V. 25. ¹ 6. pp. 774-777. (in Russian).

21. V.A.Kal'chenko, I.S.Fedotov. Genetics effects of acute and chronic ionizing radiation on Pinus sylvestris L. inhabiting the Chernobyl' meltdown area. Russ. J. Genetics, 2001. V. 37. ¹ 4. pp. 427-447.

22. S.A.Geras'kin, B.I.Sarapul'tzev. Automatic classification of biological objects on the level of their radioresistence. Automat. Telemek., 1993. pp. 182-189. (in Russian).

ROLE OF VARIOUS COMPONENTS OF ECOSYSTEMS IN BIOGEOCHEMICAL MIGRATION OF POLLUTANTS OF ANTROPOGENIC ORIGIN IN FORESTS

A. ORLOV, V. KRASNOV
Polesskiy Branch of Ukrainian Scientific Research Institute of Forestry and Forest Amelioration (UkrSRIFA), pr. Mira, 38; Zhitomir, 10004, UKRAINE,
station@zt.ukrpack.net

1. Introduction

At present time the environment is contaminated by a complex of pollutants. After the Chernobyl catastrophe the main attention of researchers, especially in Europe, was paid to radioactive pollution of the environment and therefore such important problems as a contamination of various components of the ecosystem by heavy metals, pesticides etc. had become considered as a minor problem. It should be noted that today the complex evaluation of synergistic influence of pollutants of anthropogenic origin on wild species of biota as well as on man is almost absent. Such evaluation is very pressing now especially in the regions with relatively high levels of contamination of the environment but with concentrations below its permissible values.

One of the main criteria determining the position of certain biogeocenosis in a radioecological classification of forest types, is the ability of the phytocenosis of the given habitat to accumulate the main doseforming technogenic radionuclides (^{137}Cs and ^{90}Sr) with a certain intensity [11]. The analysis of the role of different layers of vegetation in the distribution of the total stock of ^{137}Cs and its biological migration in forests is definite from a geochemical point of view because the vegetative cover can serve as the indicator of autorehabilitation of a forest landscape [3] as well as the index of the recovery of a qualitative and quantitative phytomass structure after radioactive contamination. Biological migration of ^{137}Cs is considered as one of the key processes of redistribution of radionuclide in natural ecosystems [9], determining the possibility of their economic use after radioactive contamination.

In contrast to other types of vegetation, the layers of forest phytocenosis can increase their phytomass a hundred fold during the development of the latter (in particular, tree canopy) and also essentially change the intensity of ^{137}Cs accumulation. In coordination with the development of the tree canopy and especially with closeness of tree crowns begins also the more or less quick development of undergrowth vegetation: undergrowth meaning juvenile tree canopy, grass–dwarf-shrub layer, moss and lichen layers, layer of macromycetes. It should be noticed that there is a close ecological correlation between vegetation and soil in each habitat [4]. The main physical and agrochemical soil parameters not only determine the intensity of radionuclides migration in a system «plant – soil», but also, in turn, essentially depend on the habitability of phytocenosis, which also essentially influences the radionuclides redistribution in the ecosystem [15]. For example, in forest ecosystems the significant part of activity of radionuclides annually returns to the soil with litterfall. Simultaneously radionuclides uptake by tree canopy, dwarf-shrubs and grass by rooted way is

F. Brechignac and G. Desmet (eds.), Equidosimetry. 257–272.

observed [10, 12]. The separate layers of forest vegetation are characterized by a very significant aerial sorption capacity in relation to certain radionuclides.,They do not have a rooted way of radioactive uptake.Examples are sphagnum mosses in forested bogs in ombrotrophic complexes [7], green mosses in pine forests of authomorphic landscapes or epigeous lichens in conditions of extremely poor and dry pine forests. By investigating the regularities of distribution of ^{137}Cs activity in layers of a forest ecosystem we can obtain the basis for prognostic mathematical modeling of migration of ^{137}Cs in forest ecosystems as well as the data for model validation.

2. Objects and methods

The researches were carried out in Central Polessye of Ukraine (Zhitomir region) in summer of 2000 on three experimental plots representing a certain type of forest biogeocenosis of pine woods. The selection of these experimental plots was conducted among 15 sites where main phytocenotic parameters were taken into account i.e. age of tree canopy, affinity of their species composition and cenotic structure, type of forest-growth conditions. Each experimental plot was described according to standard methods [14] and had the area of 1,0 ha.

Two experimental plots represented the extremes of the ecological conditions of forests of the region: first – a pine forest of cenosis *Cladonio-Pinetum Jurassek 1927* in the poorest conditions of high dry sandy dunes. This floristic association belongs to Union *Dicrano-Pinion Libb*. 1933, Ordo *Cladonio-Vaccinietalia K.-Lund 1967*, Class *Vaccinio-Piceetea Br.-Bl. 1939*. Second site represented forested bog complexes of cenosis *Ledo-Sphagnetum magellanici Sukopp 1959 em. Neuhäusl 1969* belonging to Union *Sphagnion magellanici Kästner et Flössner 1933 em Dierss. 1975*, Ordo *Sphagnetalia magellanici (Pawl. 1928) Moore (1964) 1968*, Class *Oxycocco-Sphagnetea Br.-Bl. et R.Tx. 1943*. The third experimental plot represented pine forests of cenosis Molinio-Pinetum Mat. (1973) 1981, belonging to Union *Dicrano-Pinion Libb. 1933*, Ordo *Vaccinio-Piceetalia Br.-Bl. 1939*, Class *Vaccinio-Piceetea Br.-Bl. 1939*. Ecological conditions in the last habitat were close to the optimum for the growth of *Pinus sylvestris L.* in Ukrainian Polessye.

The ecosystem *Cladonio-Pinetum* was situated on the top of a sandy dune with relative altitude of about 9 m. The tree canopy consisted of *Pinus sylvestris* of 40 years old with a mean height of 4,2 m and a mean diameter of 7,3 cm. The layer of juvenile trees was rarefied consisting of *Pinus sylvestris*. The layer of undergrowth also was complex and consisted of *Chamaecytisus ruthenicus (Fisch. ex Wo³.) Klaskova*. Grass–dwarf-shrub layer was complex and had a total projective cover of 10-12%. Its basis was created by *Corynephorus canescens (L.) P.Beauv* (5-7%), *Thymus serpyllum L.* (1-3%), *Calluna vulgaris (L.) Hull* (3-5%). The lichen layer was represented by two sublayers – epigeous and epiphytic. The sublayer of epigeous lichens was dense, with a total projective cover 85-90 % and a phytomass of about 0,2 kg/m^2 of absolute dry weight. Its basis was created by *Cladina arbuscula ssp. mitis (Sandst.) Ruoss* (60-65 %), a smaller participation in the formation of this sublayer was characteristic for *Cladonia uncialis (L.) F.Weber ex F.H.Wigg.* (10-15 %). The sublayer of epiphytic lichens mainly consisted of *Hypogymnia physodes (L.) Nyl*. On the driest and infringed dune sites fragments of moss cover from *Polytrichum piliferum Hedw.* (3-5 %) were observed. The layer of macromycetes was represented mainly by a symbiotic mushroom species - *Lactarius rufus (Scop). Fr., Siullus variegatus (Fr). O.Kuntze, Boletus badius Fr., Paxillus involutus (Batsch.) Fr.*, smaller role in creation of this layer played saprotrophic species – *Cantharellus cibarius Fr.* and *Tricholoma flavovirens (Pers. ex Fr.) Lund et Nannf.*

The ecosystem *Ledo-Sphagnetum magellanici* was situated in a central part of a large ombrotrophic bog "Long Moss" with total squire about 500 ha. Mineral nutrition of mentioned above bog happens mainly from atmospheric precipitation. *Pinus sylvestris* of 40 years old with a mean height of 2,3 m and a mean diameter 7 cm formed a rarefied tree canopy with crown closeness of 0,2-0,3. Layers of undergrowth and juvenile trees were absent. Grass–dwarf-shrub layer had a total projective cover of 30-40 % and mainly consisted of boreal bog species: *Eriophorum vaginatum L.* (20-25 %), *Vaccinium uliginosum L.* (3-5 %), *Ledum palustre L.* (3-5 %), *Oxycoccus palustris (L.) Pers* (3-5 %), *Andromeda polifolia L.* (1-3 %). A moss layer with a total projective cover of 95-98 % mainly created by *Sphagnum fallax (Klinggr.) Klinggr.*, and macromycetes layer – *Suillus variegatus*.

The ecosystem *Molinio-Pinetum* was situated on a plane habitat with soddy-podzolic sandy-loam soils and an average annual depth of ground water level of about 1,3 m. The 45 years old tree canopy of *Pinus sylvestris* had a mean height of 23,5 m, with a mean diameter of 22 cm, and a small participation also of *Betula pubescens Ehrh.* Complex undergrowth with closeness up to 0,2 consisted of *Frangula alnus Mill.* Grass–dwarf-shrub layer was characterized by the total projective cover of 60-75%. Its basis was formed mainly by boreal dwarf-shrubs: *Vaccinium myrtillus L.* (50-60 %), *V. vitis-idaea L.* (5-10 %) and *Calluna vulgaris L.* (1-5 %). As an impurity also species such as *Pteridium aquilinum (L.) Kuhn, Dryopteris carthusiana (Vill.) H.P.Fuchs, Molinia caerulea (L.) Moench.* and *Luzula pilosa L* were met. The moss layer had a total projective cover of about 80-95 % and consisted of green mosses: *Pleurozium schreberi (Brid). Mitt.* (40-50 %) and *Dicranum polysetum Sw.* (30-45 %). The lichen layer was represented by sublayers of epiphytic species, mainly *Hypogymnia physodes* and *Pseudoevernia furfuracea*. The layer of macromycetes was characterized by a rich species composition, including more than 20 species. The most typical among them were *Paxillus involutus, Lactarius rufus* and *Russula paludosa (L.) Kuntze*, fruitbodies of these species also created the basis of total mushrooms fruitbodies biomass in the analyzed forest ecosystem.

For the determination of the distribution of ^{137}Cs activity in ecosystems on experimental plots it was necessary to calculate the weight characteristics of each of its components per unit of the area, and also the specific activity of radionuclide in last ones. For the determination of the main characteristics of tree canopy research was carried out according to standard methods including full account of trees on the area of 1,0 ha [1]. After this procedure the parameters of a mean model tree were calculated for three main stages of trunk thickness. Three model trees were cut down on each experimental plot, each of the model tree represented a certain stage of a trunk thickness. Phytomass of each model tree was divided into separate organs and tissues, which were weighed for fresh weight, and also samples for the further analyses were taken from them. Branches were arranged according to diameter from thin (diameter < 5 mm) to thick (> 5 mm) [6]. The samples from each tree were taken proportionally to their weight from tree crown parts: upper, mean and lower and were further integrated in one sample for each model tree. The trunk was sawn in 1-m sections, from which bark was removed.

Wood and bark were weighed separately. The sample of trunk wood for the analysis was taken from the fixed height of 1,3 m. One integrated sample of needles from branches of a different part of tree crown was taken from each of the model tree, proportional to their weight in crown; samples of annual shoots were also taken similarly. The mass of the thin roots (diameter < 2 mm) and thick roots (> 2 mm) from one model tree was determined according the to literature data [6]. Samples of roots for the analysis of each model tree were collected uniformly on parts of their soil horizon. For research of abovegrowth and undergrowth phytomass of grass–dwarf-shrub layer investigations were conducted in 5 fold on sites with an area of 25 m² each. Total digging of these species were carried out with the consequent division in aboveground and underground phytomass. The roots of vascular plants were washed from soil particles in laboratory. The investigations of epigeous lichens were carried out in 5 fold on sites of 1,0 m², and macromycetes on the area 100 m². Biomass of the macromycetes layer as a whole was calculated as the sum of mass of its fruitbodies sampled on experimental plot and mass of the mushroom mycelium which was accepted as 0,12 kg·m⁻² according to the literature data for similar ecosystem *Molinio-Pinetum* [17]; it was also accepted that the ^{137}Cs specific activity in mushroom fruitbodies and mycelium was identical [18]. Epiphytic lichens were sampled with the divisionper species from model trees before their cutting. The layer of green mosses was studied on experimental plots in automorphic conditions in 5 fold on sites of 500 cm² each and layer of sphagnum mosses in hydromorphic conditions – on an area of 2500 cm². It is important that sphagnum mosses were divided into living parts, dead parts and peat litter, which were weighted and analyzed separately.

The structure of genetic horizons in the three soil profiles was studied on each experimental plot in automorphic landscapes. Separate fractions of forest litter (contemporary, semi-decomposed and decomposed) as well as mineral soil samples were taken from each profile from fixed surface of 500 cm²; the thickness of mineral soil samples was equal to 2 cm and the investigation were carried out up to 30 cm depth. In hydromorphic conditions samples of uninfringed peat were taken after deleting rather thick layers of sphagnum mosses and peat litter, with the help of special turf Giller's core with a diameter of 5 cm; on depth up to 30 cm, peat was divided in 5-cm layers in which the measurement of ^{137}Cs content was carried out separately. In our research the part of soil retaining ^{137}Cs activity was calculated by a computational method and was equaled to a difference of activity of a radionuclide in soil together with fungi mycelium (was determined by results of sampling) and activity deposited in mycelium (was determined by a computational method [17, 18]).

All samples were dried up to air-dry weight during 72 hours at the temperature of 80 ⁰C, for all fractions of phytocenosis the recalculation from fresh into air-dried weight was carried out. The dried samples of vegetation and soil were milled and placed in Marinelli's beaker of 1000 cm³ or smaller cylindrical volumes (75 and 135 cm³). The measurement of ^{137}Cs specific activity in samples was carried out in a spectrometer LP-4900B «ÀFORA» with GeLi-detector. The relative error of measurement of ^{137}Cs specific activity in samples varied in the limits of 10-20% depending on their activity.

In this research we have made the attempt to evaluate the distribution of some pollutants of antropogenic origin proceeding from assumption that ^{137}Cs concentration is the most easily measured one. The first step was the calculation of distribution of weight of forest ecosystem components; the second step was the evaluation of the role of each component in the forest ecosystem in the redistribution of total activity of this radionuclide and selection of key components in this process; the third step was the study of the concentrations of the main heavy metals-pollutants of environment in the selected key components and its comparison with analogues indexes in other ecosystem components. The statistical data processing was made with use of the software package «Statgraphics».

3. Results and discussion

Biogeocenotic research has allowed to describe the distribution of ^{137}Cs specific activity as well as the mass of each of ecosystem component per unit area (1,0 ha). The above mentioned components in each ecosystem were combined in two macroblocks i.e. soil together with forest litter and mycelium of mushrooms and in remaining components. The part of ^{137}Cs retained by separate component at first was calculated within the limits of mentioned above macroblocks, and was recalculated then on the ecosystem as a whole (tables 1, 2, 3).

The data in the above mentioned tables 1, 2, 3 illustrate some important features: significant interspecific differences of ^{137}Cs accumulation by species of the same layer of vegetation in each investigated ecosystem; an exponential decrease of ^{137}Cs specific activity in soil with increasing depth; close dependence of total activity of ^{137}Cs in the specific layer on its biomass and also on intensity of radionuclide accumulation by the root or non-root way.

From generalizing the data of the above mentioned tables it is correct to draw the conclusion that in components of ecosystem the total activity of ^{137}Cs is non-uniformly distributed. In forested bog of ombrotrophic ecosystem Ledo-*Sphagnetum magellanici* the leading geochemical role in migration of ^{137}Cs is played by the layer of sphagnum mosses which retained 84,75 % of total activity of radionuclide of phytocenosis. A significantly smaller role is played by the grass–dwarf-shrub layer (9,61 %) and the tree canopy (4,79 %). The calculated contribution of mushroom mycelium to the total activity in the 30-cm stratum of peat was equal to 33,67%. In the ecosystem Cladonio-Pinetum the main geochemical role in biological migration of ^{137}Cs is played by the tree canopy containing 50,56 % of the total activity of this radionuclide of phytocenosis and also by the layer of epigeous lichens (43,56 %). The contribution of mushrooms mycelium into the total activity of the soil with forest litter was equal to 24,98%.

In the ecosystem Molinio-Pinetum the role of tree canopy and moss layer in radionuclide retaining was comparable – 49,93 % and 46,55 % accordingly. The part of the mycelium of mushrooms in total ^{137}Cs activity of the soil together with forest litter in this cenosis is sharply increased, i.e. up to 58,99 % in comparison with the former ecosystems.

Table 1. Distribution of ^{137}Cs activity in forested bog ecosystem of Ledo-Sphagnetum magellanici, density of ground deposition of ^{137} kBq·m^{-2} (0,41 Ci·km^{-2})

Ecosystem component	Mass, kg·ha^{-1}	^{137}Cs specific activity, Bq·kg^{-1}	^{137}Cs activity, MBq·ha^{-1}	Part of ^{137}Cs total activity, %
Tree canopy	3480	–	5,798	4,79
Wood	1160	1260	1,462	1,21
Bark	480	2450	1,176	0,97
Needles	320	2720	0,870	0,72
Branches	260	2560	0,666	0,55
Roots	560	2900	1,624	1,34
Layer of epiphytic lichens	0,25	7800	0,002	0,001
Grass–dwarf-shrub layer	2050,5	–	11,659	9,614
Oxycoccus palustris (aboveground phytomass)	50	8400	0,420	0,35
(underground phytomass)	0,5	10000	0,005	0,004
Ledum palustre (aboveground phytomass)	100	7800	0,780	0,64
(underground phytomass)	130	9800	1,274	1,05
Eriophorum vaginatum (aboveground phytomass)	900	4400	3,960	3,27
(underground phytomass)	870	6000	5,220	4,30
Layer of Macromycetes (fruitbodies)	0,7	800000	0,560	0,46
Moss layer	9000	–	103,250	84,75
Sphagnum, alive part	3000	16100	48,300	39,83
Sphagnum, dead part	3500	11700	40,950	33,77
Peat litter	2500	5600	14,000	11,55

	PEAT TOGETHER WITH MYCELIUM OF FUNGI			
TOTAL	–	–	121,269	100,00
0-5 cm	20000	4270	85,400	51,34
5-10 cm	50000	527	26,350	15,84
10-15 cm	60000	500	30,000	18,04
15-20 cm	60000	218	13,080	7,86
20-25 cm	70000	100	7,000	4,21
25-30 cm	75000	60	4,500	2,71
including mycelium (in all peat layers)	70	800000	56,000	33,67
TOTAL (peat with fungi mycelium)	335000	–	166,330	100,00

Table 2. Distribution of ^{137}Cs activity in ecosystem of Cladonio-Pinetum, density of ground deposition of ^{137}Cs – 46,4 kBq·m^{-2} (1,25 Ci·km^{-2})

Ecosystem component	Mass, kg·ha^{-1}	^{137}Cs specific activity, Bq·kg^{-1}	^{137}Cs activity, MBq·ha^{-1}	Part of ^{137}Cs total activity, %
Tree canopy	23402	–	9,791	50,56
Wood	8547	126	1,077	5,56
Bark	1155	700	0,809	4,18
Branches	4774	350	1,671	8,63
Annual shoots	150	1210	0,182	0,94
Needles	2079	385	0,800	4,13
Roots thick	6110	730	4,460	23,03
Roots thin	587	1350	0,792	4,09
Juvenile tree layer	5,9	800	0,004	0,026
Undergrowth layer	22	290	0,006	0,031
Grass–dwarf-shrub layer*	*288	*418	0,120	0,62
Lichen layer	1995,8	–	8,197	43,586
Sublayer of epigeous lichens	1995	–	8,192	43,56
Cladina arbuscula ssp. mitis	1400	4180	5,852	30,22
Cladonia uncialis	360	4300	1,548	7,99
Cladonia gracilis	180	4400	0,792	4,09
Cladonia subulata	55	4440	0,244	1,26
Sublayer of epiphytic lichens (Hypogymnia physodes)	0,8	5800	0,005	0,026
Moss layer (Polytrichum piliferum)	70,5	6800	0,479	2,48
Layer of Macromycetes (fruitbodies)	4,76	–	0,524	2,708
Cortinarius sanguineus	0,84	160000	0,134	0.69

Lactarius rufus	0,96	100000	0,096	0,50
Camtharellus cibarius	0,14	11120	0,002	0,008
Siullus variegatus	0,87	79200	0,07	0,36
Siullus bovinus	0,66	58700	0,04	0,20
Boletus badius	0,41	75700	0,03	0,16
Paxillus involutus	0,88	174000	0,153	0,79
TOTAL	–	–	19,367	100,00
SOIL TOGETHER WITH MYCELIUM OF FUNGI				
Forest litter	16400	–	76,310	14,44
Forest litter semi-decomposed	3200	293	0,938	0,18
Forest litter decomposed	13200	5710	75,372	14,26
Mineral soil layers	4935600	–	452,274	85,57
ÍÅ² 0-2 cm	228000	1250	285,000	53,92
ÍÅ² 2-4 cm	290800	330	95,964	18,16
ÍÅ² 4-6 cm	288400	110	31,724	6,00
ÍÅ² 6-8 cm	338000	35	11,830	2,24
ÍÅ² 8-10 cm	289600	20	5,792	1,10
ÍÅ² 10-12 cm	317200	14	4,444	0,84
Ð³ 12-14 cm	332400	6	1,994	0,38
Ð³ 14-16 cm	363200	5	1,816	0,34
Ð³ 16-18 cm	343200	6	2,059	0,39
Ð³ 18-20 cm	335200	6	2,011	0,38
Ð³ 20-22 cm	315200	9	2,837	0,54
Ð³ 22-24 cm	366000	5	1,830	0,35
Ð³ 24-26 cm	392000	6	2,352	0,44
Ð³ 26-28 cm	362800	3	1,088	0,21
Ð³ 28-30 cm	373600	4	1,494	0,28

including mycelium (in all soil layers)	1200	–	132,035	
TOTAL (soil with fungi mycelium)	4012800	–	528,543	100,00

*Note: average-weighted data on whole layer.

Table 3. Distribution of ^{137}Cs activity in ecosystem of Molinio-Pinetum, density of ground deposition of ^{137}Cs – 580,0 kBq·m^{-2} (15,68 Ci·km^{-2})

Ecosystem component	Mass, kg·ha^{-1}	^{137}Cs specific activity, Bq·kg^{-1}	^{137}Cs activity, MBq·ha^{-1}	Part of ^{137}Cs total activity, %
TREE CANOPY (total)	252552	–	1073,8	49,93
^{2} layer (Pinus sylvestris)	248151	–	1057,5	49,17
Wood	185539	2690	499,1	23,21
Bark	9416	13860	130,5	6,07
Branches alive	15213	10090	153,5	7,14
Branches dried up^3	8075	3678	29,7	1,38
Needles	7879	21550	169,8	7,90
Roots	22029	3400	74,9	3,48
^{22} layer (Betula pubescens)	4401	–	16,3	0,76
Wood	1875	1760	3,3	0,15
Bark	230	5645	1,3	0,06
Branches	1020	6470	6,6	0,31
Leaves	235	11060	2,6	0,12

Roots	1041	2400	2,5	0,12
Moss layer (*Pleurozium schreberi*)	20802	48120	1001,0	46,55
Layer of epiphytic lichens (Hypogymnia physodes)	32,8	57920	1,9	0,09
Grass–dwarf-shrub layer	1518	–	66,7	3,10
Vaccinium myrtillus aboveground phytomass	1270	40860	51,9	2,41
underground phytomass	217	44300	9,6	0,45
Dryopteris carthusiana aboveground phytomass	20	162670	3,2	0,15
underground phytomass	11	180000	2,0	0,09
Layer of Macromycetes (fruitbodies)	2,67	–	7,1	0,33
Suillus variegatus	0,4	2258400	0,9	0,04
Paxillus involutus	1,1	3500000	3,9	0,18
Russula paludosa	0,3	1240000	0,4	0,02
Xerocomus badius	0,35	2800000	1,0	0,05
Lactarius rufus	0,3	2900000	0,9	0,04
Cantharellus cibarius	0,22	160000	0,04	0,00
TOTAL	–	–	2150,5	100,00
SOIL TOGETHER WITH MYCELIUM OF FUNGI				
Forest litter	56911	–	2282,1	49,68

Forest litter noncomposed	2544	14036	35,7	0,78
Forest litter semi-decomposed	21745	50500	1098,1	23,90
Forest litter decomposed	32622	35200	1148,3	25,00
Mineral soil layers	3438022	–	2311,8	50,32
ÍÀ 0-2 cm	133138	6660	886,7	19,30
ÍÀ 2-4 cm	172018	3270	562,5	12,24
ÍÀ 4-6 cm	211026	1433	302,4	6,58
ÍÀ 6-8 cm	243158	760	184,8	4,02
ÍÀ 8-10 cm	220323	310	68,3	1,49
ÍÀ 10-12 cm	253241	216	54,7	1,19
À 12-14 cm	272143	140	38,1	0,83
À 14-16 cm	276190	126	34,8	0,76
À 16-18 cm	241667	120	29,0	0,63
À 18-20 cm	234545	110	25,8	0,56
$^2_1$ 20-22 cm	248182	110	27,3	0,59
$^2_1$ 22-24 cm	242500	120	29,1	0,63
$^2_1$ 24-26 cm	240000	100	24,0	0,52
$^2_2$D 26-28 cm	237391	115	27,3	0,59
$^2_2$D 28-30 cm	212500	80	17,0	0,37
including mycelium (in all soil layers)	1200	2258000*	2710,08	58,99
TOTAL (soil with fungi mycelium)	3494933	–	4593,9	100,00

The analysis of distribution of the total stock of ^{137}Cs on the components of ecosystem as a whole (with allowance of radionuclide activity in soil and mycelium of mushrooms) also demonstrates the specific geochemical regularities. In particular, the part of ^{137}Cs activity retained by the main forest layer, the tree canopy, in forest ecosystems of Ukrainian Polessye varies in a wide range , from 1,8 % up to 16 % of a total stock of radionuclide in forest ecosystem as a whole. Due to the growth of phytomass of this vegetation layer increase of part of the radionuclide activity retained by the tree canopy is observed. In ecosystem *Molinio-Pinetum* the tree canopy contained: for the age to 10 years, 3 % of the total stock of ^{137}Cs of ecosystem, for 30 years 7 %, for 55 years - 15 % [19]. In the *Pinus sylvestris* tree canopy in automorphic landscapes the observed increase is mainly due to the biomass of trunk wood and bark (40-60 % of a ^{137}Cs sum activity of tree canopy). In hydromorphic landscapes of ombrotrophic bogs in cenosis *Ledo-Sphagnetum* magellanici the tree canopy of pine retains only 2,03 % of a total stock of ^{137}Cs of ecosystem as a whole, and due to trunk less than 30 %. In 60-year old   forests of cenosis *Serratulo-Pinetum* pine the tree canopy contained 2-3 % of a total stock of ^{137}Cs of ecosystem and in the oak canopies of similar age, 6-9 % is due to the greater intensity of radionuclide accumulation [2, 12, 16, 19].

It is necessary to underline that in deciduous forests the intensity of biological migration of ^{137}Cs also increases due to the intensive annual cycling of significant radionuclide activity with litter-fall and its relatively fast decomposition [12].

The layer of undergrowth (living form – shrubs) of the average crown closeness (0,3-0,5) independently from species composition, in forests of Polessye retained only 0,01-0,05 % of a total stock of ^{137}Cs of forest ecosystem; the close indexes are also characteristic for the layer of juvenile trees (up to 0,01 % of the total ^{137}Cs activity of ecosystem) because their insignificant phytomass per unit area.

Grass–dwarf-shrub layer is characterized by various phytomass in forest cenosis in Ukrainian Polessye – from very rarefied (with the sum projective cover 5-15%) in cenosis *Cladonio-Pinetum* and *Peucedano-Pinetum* up to 60 % of projective cover in forested bogs of cenosis *Ledo-Sphagnetum* magellanici and 95 % in cenosis *Molinio-Pinetum*. Accordingly a part of the sum of ^{137}Cs activity retained by this layer of vegetation in forests of automorphic landscapes varies from 0,01-0,02 % in cenoses *Cladonio-Pinetum* and *Peucedano-Pinetum* up to 1,0 % in cenosis *Molinio-Pinetum* sharply increasing up to 2,5-4,0 % in the forests of hydromorphic landscapes as a result of more intensive ^{137}Cs accumulation by all plant species of the layer as well as because essentially smaller relative part of tree canopy in retaining of ^{137}Cs.

Lichen layer consists of two sublayers – epiphytic (on trunks of trees) which part independently from species composition does not exceed 0,01-0,03 % of a sum activity of ^{137}Cs of forest ecosystem, however this sublayer of lichens plays the main role in the interception of ^{137}Cs stemflow. The sublayer of epigeous lichens plays an important role in the formation of phytomass in cenosis *Cladonio-Pinetum*, where lichens are dominants of undergrowth vegetation of forests, the part of this sublayer reaches 1,5-3 % of a total stock of ^{137}Cs contained in the ecosystem as a whole.

The role of moss layer, the majority of phytocenosis of boreal forests, in retaining of ^{137}Cs by biota is extremely important. As it was shown earlier the part of the sum ^{137}Cs activity circulating in forest ecosystem and retained by layer of green mosses in forests of automorphic landscapes in cenosis *Peucedano Pinetum* varied from 1% in 10-year's pine forests up to 15% in 50-year's cenosis [19]. In forests of the analyzed automorphic landscapes the part of green moss layer changed from 0,1% in cenosis *Cladonio-Pinetum* up to 15 % in cenosis *Molinio-Pinetum*. In hydromorphic landscapes in cenosis *Ledo-Sphagnetum magellanici* the role of moss layer (from sphagnum mosses) increased up to 35-40% in relation to the total stock of ^{137}Cs content in the forest ecosystem.

Also it should be noted that the role of moss layer is determining, as a biogeochemical barrier, the path of vertical ^{137}Cs migration in all types of forest landscapes – from forest litter to mineral soil layers in automorphic landscapes and from peat litter to peat in hydromorphic ones.

Special attention was paid to studying the role of the mushroom complex in retaining of ^{137}Cs activity in forest ecosystems. During mushroom fruitbodies sampling period the productivity of mushrooms on our experimental plots was low – 6-8 times smaller in comparison with average long-term data on the region [12; 13], therefore in spite of significant ^{137}Cs specific activity in mushrooms fruitbodies, their contribution to retaining of the total ^{137}Cs activity of ecosystem was not so large – 0,1-0,2%. However the part of sum ^{137}Cs activity in mushrooms mycelium deducted from macroblock of soil with forest litter changed from 20 % in forested bog ecosystem of cenosis *Ledo-Sphagnetum magellanici* up to 40 % in cenosis *Molinio-Pinetum*.

Thus our calculation has allowed to draw the conclusion that soil (in automorphic landscapes – together with forest litter) contained from 28% of a total stock of radionuclide of ecosystem in cenosis *Molinio-Pinetum* up to 38 % in cenosis *Ledo-Sphagnetum magellanici* and 72% in cenosis *Cladonio-Pinetum*. The results obtained visibly demonstrate that in conditions when the biological circulation of radionuclide is slow because of unfavorable ecological conditions, as for example in cenosis *Cladonio-Pinetum* a large part of total stock of radionuclide remains in the soil and is not taken up by vegetation.

It is necessary to underline that the results obtained in this research are in good agreement with data published earlier [5; 19] in which the relative part of soil with forest litter in retaining of ^{137}Cs varied in the limits 60-65 %, including mycelium of mushrooms in this value.

Our research of ^{137}Cs distribution in three forest ecosystems allowed us to draw a conclusion that key components in distribution of this radionuclide in analyzed biogeocenoses were:

- in *Cladonio-Pinetum* – pine tree canopy and epigeous lichen layer;
- in *Ledo-Sphagnetum magellanici* – *Sphagnum* layer;
- in *Molinio-Pinetum* – pine tree canopy and green moss layer.

So according to the above proposed methodology the heavy metal content was studied in salient key components of ecosystems (table 4).

Analysis of above mentioned data shows that not only ^{137}Cs but also heavy metal content were higher in peak key component. So the distribution of total stock of investigated pollutants should be approximately analogue too. But in other cases essential differences of distribution both in concentrations of pollutants and in its total stock can be observed. Nevertheless, in our mind these regularities of distribution of pollutants are specific and approximately constant in each ecosystem.

Table 4. Average content of ^{137}Cs and heavy metals in some components of different forest ecosystems.

Component	^{137}Cs cont., Bq·kg^{-1} d.w.	Heavy metal content, mg·kg^{-1} d.w.			
		Cu	Zn	Pb	Co
Cladonio-Pinetum					
*Pine tree canopy (needles)	385±45	1,17±0,22	9,85±1,12	0,51±0,07	0,03±0,004
*Epigeous lichen layer	4180±50	3,12±0,38	13,54±1,50	1,48±0,17	0,26±0,03
Soil (mineral 0-2 cm)	1250±140	1,54±0,20	3,00±0,44	1,09±0,21	0,028±0,00
Ledo-Sphagnetum magellanici					
*Sphagnum layer (alive part)	16100±1900	3,40±0,42	21,3±2,75	4,67±0,55	0,52±0,06
*Sphagnum layer (dead part)	11700±2000	2,45±0,29	28,00±3,42	13,65±1,54	0,83±0,11
Cranberry berries	8600±900	17,75±2,01	12,30±1,23	0,02±0,003	0,01±0,00
Pine tree canopy (needles)	2720±300	2,08±0,22	12,04±1,24	0,30±0,05	0,03±0,00
Molinio-Pinetum					
*Pine tree canopy (needles)	21550±2300	4,52±0,47	8,04±0,95	0,17±0,02	0,003±0,00
*Green moss layer	48120±5000	30,14±3,21	19,24±2,22	0,94±0,10	0,17±0,02
Bilberry (shoots)	40860±4000	8,50±1,00	13,92±1,44	0,02±0,003	0,013±0,00
Bilberry (berries)	41000±4200	27,08±3,12	28,84±3,24	0,002±0,00	0,012±0,00
Soil (mineral 0-2 cm)	6660±700	2,08±0,31	4,00±0,43	1,09±0,20	0,03±0,003

Note: key components were marked *.

272

4. References

1. N. Anuchin. Forest accounting. M., Lesnaya promyshlennost, 1977, 512 pp. (in Russian).
2. I. Bulavik. [137]Cs content in soil and arboreal vegetation on various density of radionuclide contamination of the territory. in: Proc. of the conf., Gomel, 1991. pp. 89-99 (in Russian).
3. V. Dolin. Criteria of self-cleaning of terrestrial ecosystems from radioactive contamination. in: Reports of NAS of Ukraine, 1, 2000. pp. 187-190 (in Ukrainian).
4. S. Zonn. Interaction and interdependence of forest vegetation with soils. in: Pochvovedenie, 7, 1956. pp. 80-80 (in Russian).
5. V. Krasnov, V. Nurko, A. Orlov et al. Distribution of ^{137}Cs activity in components of forest biogeocenosis of wet subor. in: Collection of scientific works of the Institute of Forestry of NAS of Belarus, 46, Gomel, 1997. pp. 405-407 (in Russian).
6. V. Myakushko. Pine forests of flat part of Ukraine. Kiev, Naukova dumka, 1978. 256 pp. (in Russian).
7. A. Orlov. The role of sphagnous cover in redistribution of fluxes of potassium and ^{137}Cs in ecosystems of mezo-ombrotrophic bogs. in: Ukrainian botanical journal, 57, 6, 2000, pp. 715-724 (in Ukrainian).
8. A. Orlov, S. Irklienko. The main regularities of ^{137}Cs migration and distribution of its total stock in ecosystems of mezo-ombrotrophic bogs. in: Naukovy visnyk NAU, 20, 1999. pp. 60-68 (in Ukrainian).
9. O. Pushkariov,V. Davydchuk, Yu. Sushchik, I. Shramenko. Evaluation of autorehabilitation capacity of environment. in: Reports of NAS of Ukraine, 2, 2000. pp. 208-213 (in Ukrainian).
10. F. Tikhomirov, A. Shcheglov. Consequences of radioactive contamination of forests in the zone of influence of Chernobyl accident. in: Radiation biology. Radioecology, 37, 4, 1997. pp. 664-672 (in Russian).
11. V. Shubin, I. Maradudin, A. Panfilov. Radioecological classification of forest types. in: Observing information, issue 1-2, Moscow, 1999. 48 pp. (in Russian).
12. A. Shcheglov. Biogeochemistry of technogenic radionuclides in forest ecosystems: on data of 10-years research in the zone of influence of Chernobyl NPP, Moscow, Nauka, 1999. 268 pp. (in Russian).
13. A. Shcheglov et al. To the problem of role of higher macromycetes in biogeochemical migration of ^{137}Cs in forest ecosystems. in: Proceedings Chernobyl-94, vol. 1, 1996. pp. 460-472 (in Russian).
14. A. Yunatov. Foundation of ecological profiles and experimental plots. in: Field geobotany, vol. 3, Moscow-Leningrad, Nauka, 1964. pp. 9-35 (in Russian).
15. P. Bossew, H. Lettner, A. Hubmer. Spatial variability of fallout [137]Cs. in: Proc. of the Intern. Symp. on Radioecology «Ten years terrestrial radioecological research following the Chernobyl accident». Eds. M.Gerzabek, G.Desmet, B.J.Howard et al., Vienna, 1996. pp. 179-186.
16. S. Mamikhin, A. Shcheglov. Dynamics of Cs-137 in the forests of the 30-km zone around the Chernobyl nuclear power station. in: Proc. Seminar on the dynamic behaviour of radionuclides in forests, Stockholm, 1992. p. 10.
17. T. Nylén. Uptake, turnover and transport of radiocaesium in boreal forest ecosystem. Dissertation. Uppsala, 1996.
18. R. Olsen, E. Joner and L. Bakken. Soil fungi and the fate of radiocaesium in a soil ecosystem – a discussion of possible mechanisms involved in the radiocaesium accumulation in fungi and the role of fungi as a Cs-sink in the soil. in: Transfer of radionuclides in natural and semi-natural environment. Eds. G.Desmet et al., London-New York, Elsevier Applied Science, 1990. pp. 657-663.
19. A. Orlov, V. Turko, S. Irklienko, V. Krasnov, A. Kalish. The distribution of the total ^{137}Cs activity in biogeocenoses of pine forest of Ukrainian Polessye. in: Proc. of the VI Symposium, vol. 1, Szarvas, 1999. pp. 79-84.

ECOLOGICAL INSPECTION OF MILITARY OBJECTS

A. LYSENKO, I. CHECANOVA
National Research Center of Defense Technologies and Military Security of Ukraine. build 14, 2/32, Aviation Designer Antonov str. Kiev, 03186, UKRAINE

O. MOLOZHANOVA
Ministry of Health of Ukraine. L.I. Medved's Institute of Ecohygiene and Toxicology 6, Heroiv Oborony Street, Kiev, 03022, UKRAINE

1. Abstract

The methods of military object ecological inspection are proposed and the scheme of interaction between state structures, systems and objects in general system of national ecological safety is developed. The necessity of conducting ecological inspections during solving ecological problems including the problem of radioactive security of military objects of Ukraine is emphasized.

The ecological inspection of military objects is essentially the audit of profitability, efficiency and productivity of usage of resources allocated for nature protection measures by bodies of the Ministry of Defense.

The ecological inspection is organized under the initiative of the supervisor of an object or activity and is ecological self testing or ecological self estimation in character. It is carried out irrespective of state ecological expertise and includes ecological audit. The ecological audit is conducted by specialized auditor organizations having the appropriate qualifying certificate (license). The ecological audit of military - administrative activity is directed on accomplishment of independent checks of the bookkeeping (financial) accountability, operating documentation, tax returns and ecological obligations from the point of view of their conformity to the existing ecological legislation in the field of an environmental protection and using of natural resources.

2. The process of ecological inspection

The problems and contents of ecological inspection are defined by requirements of the legislation of Ukraine:

- concerning the protection of natural environment at locating, constructing, commissioning, operation, modernization, renovation, preservation and using of facilities building or other objects;
- concerning safety accomplishing industrial, Defense and other activities at using of atomic energy.

F. Brechignac and G. Desmet (eds.), Equidosimetry, 273–280.

- concerning the protection of the population and territories at emergency situations of natural and technogenous character at accomplishment of listed kinds of activity;
- concerning the introduction of the financial accounting, payment and accounting documentation, tax returns and registration of other fiscal obligations and demands of economic subjects.

In the mandatory order the ecological inspection is conducted during privatization of state firms, at ecological insurance, in case of submission by enterprises of the applications for obtaining the stamp of high ecological standard of production ("a green mark"), by granting to enterprises a financial support from ecological funds and in other cases, connected with the estimation of ecological costs [1,2].

The process of ecological inspection consists of several stages exhibited in fig. 1.

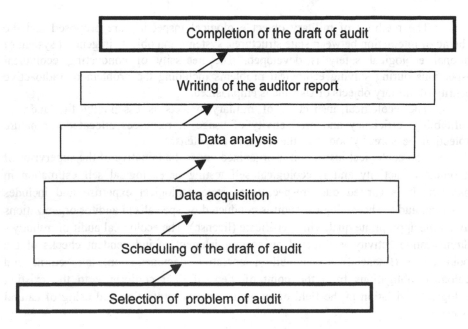

Fig. 1. Process of ecological inspection.

Ecological inspection of military objects is the element of military object environment monitoring system that is based on a systematic approach and includes the documented periodic objective assessment of the management system of Armed Forces of Ukraine and the assessment of compliance of military activities with ecological policy and standard requirements with the purpose of natural environment protection while fulfilling the tasks assigned to Armed Forces of Ukraine.

The subject of ecological inspection of military objects is not so much ecological accountability, as well its actual industrial (nature protection) activity in all aspects, to which refer:

1. 1.Definition of the nature protection purposes and problems.
2. 2.Development of the ecological program and policies.
3. Monitoring, regulation, minimization of volumes of discharges and wastes.
4. Rational usage of natural resources, raw material, materials, and final product.
5. Ensuring personnel safety, including the estimation of risk of emergencies and warning and measures to control them.
6. 6.Connection with bodies of state ecological control and regulation, including licensing of natural resource usage, insurance, certification.
7. 7.Ecological informing, education and training of staff.
8. Reduction of risk for the responsibility of violation of the nature protection legislation, change of payments for natural environment contamination.

For quantitative and qualitative estimation of profitability, efficiency and productivity of military-ecological activity of ecological services of the Ministry of Defense the model "Costs - outcome", presented in a fig. 2 is used.

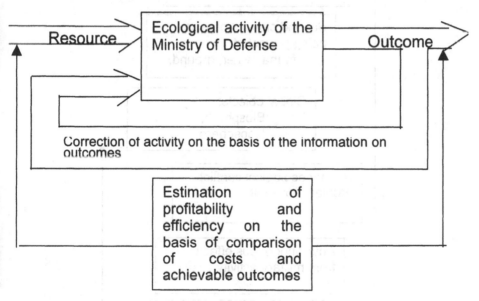

Fig. 2. The model of Ministry of Defense economic activity estimation.

The method of planning of type of military activity ecological audit is presented in a fig. 3.
The method assumes the revealing of a conflict situation between the source of contamination and the ecologically protected object [4]. Ecological inspection is conducted at a level of:

• state structures;
• transnational corporations;
• branches (structural audit - the conformity to a general ecological course of branch is assessed, problem audit - the state of the solution of a particular ecological problem of a branch is valued);
• region;
• facility (object).

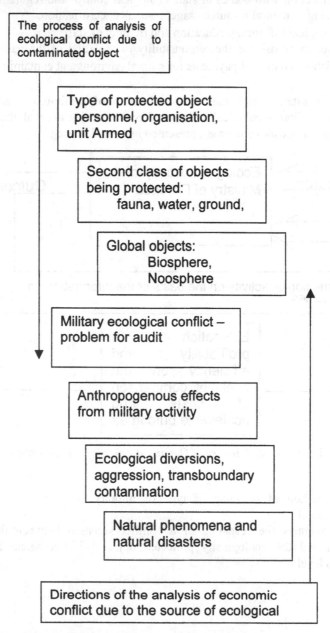

The process of analysis of ecological conflict due to contaminated object

Type of protected object
   personnel, organisation,
unit Armed

Second class of objects being protected:
   fauna, water, ground,

Global objects:
   Biosphere,
   Noosphere

Military ecological conflict – problem for audit

Anthropogenous effects from military activity

Ecological diversions, aggression, transboundary contamination

Natural phenomena and natural disasters

Directions of the analysis of economic conflict due to the source of ecological

Fig. 3. Method of search of military activity ecological conflict situations.

In a course of ecological inspection of military objects the conclusion on the sanitary-hygienic characteristics of the object is confirming the compliance of military activity with the labor protection and professional safety legislation [5, 6].

The analysis of risks of emergency situations, disasters and contamination caused by them is conducted with consideration for operational requirements and consists in determination of risks at the facility, the analyses of possible emergency scenarios and the subsequences for environment and population, the analysis of provided measures and means of warning and minimization of emergency impacts and informing people and local organizations [7].

The analysis of military-activity ecological conflict situation should be carried out with consideration of the hierarchy of the objects of national safety in ecological sphere.

The classification of military objects in the system of national ecological safety is shown in fig. 4.

The ecological inspection of military objects is considered not only in terms of measures of environment (natural objects) impact protection in the course of technogenous activity (ensuring of industrial, radiation - nuclear, fire, technical and technological safety and minimization of threats from the part of the facility being checked – the "primary" source of risks for the natural environment), but also in terms of measures of protection of the population and territories in the case of emergency situations of natural and technogenous character, in particular, of negative impacts of natural objects ("secondary" source of risks the population and territories), ensuring minimization of ecological threat from their part to the vitally important interests of individuals, society and state.

The ecological audit activity in the course of the ecological inspection of military-object is necessary to be conducted in order to assess the effectiveness of ecological policy realization in military sphere. Ecological audit is designed for assisting in search for additional means of studying risks, for ensuring safety of life and for quicker introduction of advanced methods of risk prevention.

Military activity ecological inspection is completed by efficiency and productivity analysis of mechanisms of activity of a multilevel system of ensuring of national ecological safety.

In the field of nature projection activity ecological inspection of military objects improves considerably the state system of ecological monitoring and makes it more effective, flexible and operative. With the help of ecological inspection one may easily reach the results that could not be available by the through the state system of ecological control [3, 4].

The scheme of interaction of different state structures, systems and objects in a general system of national ecological safety is presented in fig. 5.

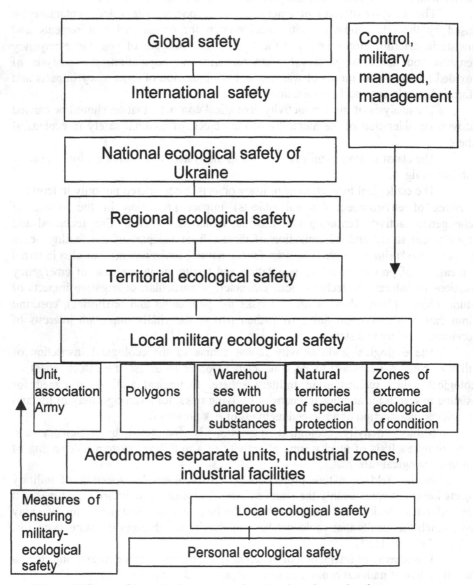

Fig. 4. Military objects in the system of national ecological safety.

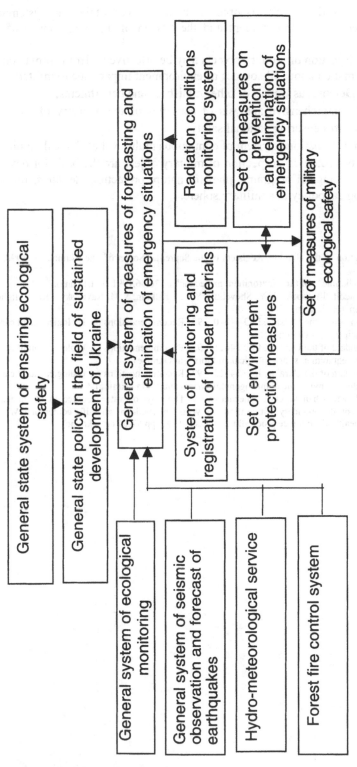

Fig. 5. Mechanism of formation of efficiency, profitability and productivity estimation of military ecological activity of Ministry

3. Conclusions

1. Ecological inspection of military object is an integral part of the state system of stable development control of environment and the system of the civil control of military activities.

2. Ecological inspection of military objects may be effective under the provision of:

objective information about all sources of environment impact due to military activities of Ministry of Defense, as well as of other ministries and departments;

conducting of comprehensive analysis of activities of the Ministry of Defense with other state structures ensuring national safety;

creation of a general national system of continuous multilevel ecological monitoring.

3. The results of ecological inspection of military object are the basis for development of perspective ecological programs, plans of nature protection measures, formation of trends of ecological policy in a military sphere.

4. References

1.   The Constitution of the Ukraine. Registers of the Supreme Soviet of the Ukraine - 1998, № 30 (in Ukraine).
2.   The Law of Ukraine " Of the Environment protection" of 26.06.1991 (in Ukraine).
3.   Economical audit. Textbook . V.Y. Shevchuk, Y.M. Satalkkin, V.M. Navrotsky - Kiev. Higher School 2000 - 344 pp. (in Ukraine).
4.   National Standard of the Ukraine. ISO 14010—97.. Recommendations on ecological audit carrying put. General principles" (in Ukraine).
5.   National Standard of the Ukraine. ISO 14011—97.. Recommendations on ecological audit carrying put. Qualification requirements for ecological auditors" (in Ukraine).
6.   National Standard of the Ukraine. ISO 14012—97.. Recommendations on ecological audit carrying put. Audit procedures. Environment management system audit" (in Ukraine).
7.   Lysenko O.I., Molozhanova O.G., Checanova I.V. The ways of water resource quality management in armed forces on the boundary of two thousand years. International scientific conference. Water quality and human health. Conference proceeding. Odessa –1999. pp 57-61 (in Ukraine).

THE PROCEDURE OF MILITARY SITES REHABILITATION WITH OPTIMAL PLANNING OF TENDER ORDERS

Y. BODRYK, S. CHUMACHENKO, A. NEVOLNICHENKO,
V. SHEVCHENKO
National Research Center of Defense Technologies and Military Security of Ukraine. build 14, 2/32, Aviation Designer Antonov str. Kiev, 03186, UKRAINE

An urgent problem facing the MOD of Ukraine is the necessity of identification, classification and, whenever possible, rehabilitation of the polluted territories occupied by military units and organisations, especially if talking about military sectors.

For effective management of this process it is necessary to develop a procedure that reflects the order of carrying out of standard operations and documenting of the results [1]. The procedure developed on the basis of methods of risk analysis and operation research will allow rationally using available resources and to minimise damage caused to the environment.

The most widespread harmful substances, which are released in the environment due the activity of military units, are hydrocarbons, chemicals, and heavy metals. The sources of hydrocarbons and chemicals in the Armed Forces includes leaks of fuel storage tanks, spills resulting from damaged fuel pipelines, leachate from fire fighter training area. Heavy metals contamination is linked to waste disposal sites, in zones of arms manufacturing and maintenance. Considering the described character of pollution and the necessity of taking in account the site particularities the following procedure of identification, classification and subsequent clean up of the polluted territories can be offered, which defines the whole environment management process in places where military units are allocated or on military sectors.

The process can be broken at some stages, which include site characterisation, identification of major factors influencing the environment, selection of the rehabilitation technology based on the risk analysis, estimation of implied results, implementation of the selected management strategy and confirmation of the results by means of monitoring and testing. At each stage an estimation of contamination and its impact on the environment plays the essential role. The consecutive and complete fulfillment of the stages should ensure that funds are spent in a cost-effective manner.

The selection of a site rehabilitation technology or risk management option will be based on the information obtained during the site characterisation stage. The requirements concerning restoration depend on estimated contamination, its impact on the environment, cost, and the future implications of leaving the contamination in place (higher expenses on restoration are considered at the postponed measures, potential legal cost, risk to human health). During the detailed site investigation the received information is analysed together with the results of risk analysis and the data concerning pollution migration, that allows formulating different variants of the rehabilitation strategy.

F. Brechignac and G. Desmet (eds.), Equidosimetry, 281–284.

In practice the accent should be put on remedial technologies with low cost . For example, depending on site characteristics, it is necessary to examine bio-remedial technologies before other more expensive technologies such as Low Temperature Thermal Desorption are considered. The site characteristics usually determine the successful implementation of various rehabilitation measures; therefore they should be carefully investigated before a choice is made.

Implementation of the project should be based on the effective application of allocated resources. During the tender involving companies, which carry out rehabilitation work, this can be ensured by a technique described below.

First of all, the task of tender planning, which requires special equipment and technologies of "external" in relation to the MOD establishments, arises in case of target or economic inexpediency or impossibility to perform it by military sectors themselves.

Tender distribution of works is carried out from reasons of target efficiency and economic feasibility on the basis of economic parameters estimation of contractors. Optimal planning of the tender orders and performance of works gives till 20-40 % of budget expenses economy and consists in the following.

The initial data for planning are given by the following table:

$$
matr\ C = \begin{array}{c|ccccc} & T_1 & .. & T_j & ... & T_n & \\ \hline R_1 & C_{11} & .. & C_{1j} & ... & C_{1n} & A_1 \\ ... & ... & .. & ... & ... & ... & ... \\ R_i & C_{i1} & .. & C_{ij} & ... & C_{in} & A_i \\ ... & ... & .. & ... & ... & ... & ... \\ R_m & C_{m1} & .. & C_{mj} & ... & C_{mn} & A_m \\ \hline & B_1 & .. & B_j & ... & B_n & \end{array} = \left\| C_{ij} \right\|_{m \times n} , \tag{1}
$$

where: Ri, i = 1 ... m - executors of special works;
A^3, i = 1 ... m - budget of the establishments Ri, i = 1 ... m;
Òj, j = 1 ... n - kinds of special works under the nomenclature;
Bj, j = 1 ... n - required amounts of works of a kind Tj, j = 1 ... n;
Cij - cost of a unit of work of a kind Tj, performed by the executor Ri.

As a plan of the orders and performance of special works is considered the matrix

$$
matr\ X = \left\| x_{ij} \right\|_{m \times n} , \tag{2}
$$

where xij – an amount of works of a kind Tj, that is performed by the contractor Ri. The set of plans of the orders are vectors - columns of the matrix X -

$$Z_j = \left\| z_{ij} \right\|_{m \times n} , \tag{3}$$

the set of plans of performance - vectors - rows of the matrix Õ -

$$P_i = \left\| p_{ij} \right\|_{m \times n} . \tag{4}$$

At the certain plan Õ expenses of the executors of works will make

$$\sum_{j=1}^{n} c_{ij} \times x_{ij} = a_i , \quad i = 1 \ldots m , \tag{5}$$

total amount of the executed works on kinds will make

$$\sum_{i=1}^{m} x_{ij} = b_j , \quad j = 1 \ldots n, \tag{6}$$

and the total cost of the works will make

$$L(x) = \sum_{i=1}^{m} a_i = \sum_{i=1}^{m} \sum_{j=1}^{n} c_{ij} \times x_{ij} . \tag{7}$$

There is the following task of tender planning of orders and performance of the special works: on the set of the plans {Õ}, each of them

$$X = \left\| x_{ij} \right\|_{m \times n} \tag{8}$$

satisfies a system of restrictions

$$\sum_{j=1}^{n} c_{ij} \times x_{ij} \le A_i , \quad i = 1 \ldots m, \tag{9}$$

$$\sum_{i=1}^{m} x_{ij} \ge B_i , \quad j = 1 \ldots n, \tag{10}$$

it is necessary to find such a (optimal) plan

$$X^O = \left\| x_{ij}^O \right\|_{m \times n} \tag{11}$$

that minimizes the total cost of performance of the special works for the customer (MOD)

$$L(X^O) = \sum_{i=1}^{m} \sum_{j=1}^{n} c_{ij} \times x_{ij}^O = \min_{\{X\}} L(X) . \tag{12}$$

It is an integer task of linear programming at matrix argument of the criterion function [2]. Its solution XO permits to detail the plans of the orders (vectors - columns of the matrix 11) and the plans of performance (vectors - rows of the matrix 11) of the special works by contractors.

The task should be included into the structure of special software of the automated environment management system on military sectors.

References

1. Environmental guidelines for the military sector. A Joint Sweden-United States Project Sponsored by the HATO Committee on the Modern Society. March 1996, 73 pp.
2. Zaichenko Y. P. Operation research. Kiev. Vichsha shkola, 1975, 320 pp (in Russian).

MODELING AND SIMULATION OF ^{137}CS MIGRATION IN BOREAL FOREST ECOSYSTEMS

A. KOVALCHUK, V. LEVITSKY
Zhytomyr Institute of Engineering and Technology, Cherniakhovsky str. 103, Zhytomyr, 10005, UKRAINE

A. ORLOV
Polis'kiy Branch of Ukrainian Scientific Research Institute of Forestry and Forest Amelioration, Prospekt Myru, 38, Zhytomyr, 10004, UKRAINE

V. YANCHUK
Zhytomyr Institute of Engineering and Technology, Cherniakhovsky str. 103, Zhytomyr, 10005, UKRAINE

1. Introduction

Modeling of ^{137}Cs in natural ecosystems has a lot of factors that are the basis for mathematical modeling of processes of radionuclides migration in food-chains and for complex analysis of radionuclides migration in natural forest ecosystems [1–3, 6–10, 12, 13, 16 – 18]. Among the variety of models there are some factors for calculation of transfer coefficients for chains soil–plant [1, 10, 18], for assessment of ^{137}Cs content in fruit bodies of fungi or plants [13, 14, 15], for investigation of seasonal variation of ^{137}Cs in game [7, 12], etc. All the types of models mentioned above use different types of analysis and are described by different types of mathematical structures because of analysis of different objects – ecosystem compartments. But all of them use only one or two types of mathematical structure for the process analysis. Such models are useful as we can see they work for a long time giving prediction of processes behavior but they can also be used in another way – the way of multi-step modeling. The main peculiarity of such approach is the multistage process of model forming. Each stage of model creation is connected with a separate process to analyze to using the appropriate existing model.

The approach proposed here includes the following phases:

- Determination of tasks of research of ecological processes, which must be solved in the frame of radionuclides migration study.
- Formal description of compartments of the forest ecosystem. Description must emphasize exchange processes research and anthropogenic factor definition.
- Working towards a conceptual scheme of the ecosystem. The conceptual scheme must take into account tasks selected on the phase1, describe the structure of the ecosystem in terms of "important/negligible", and define relations between important ecosystem compartments.
- Establishing the assumptions concerning balance of radionuclides activity for important ecosystem compartments. The relationships between accumulated

F. Brechignac and G. Desmet (eds.), Equidosimetry, 285–292.
© 2005 *Springer. Printed in the Netherlands.*

activities for compartments are defined on the basis of the preformed conceptual scheme.

• Calculation of the mathematical model parameters using expert knowledge and relationships between accumulated activities of the ecosystem compartments. Determined parameters allow setting up the model in the form of the system of ordinary differential equations, to analyze the model and to interpret the results.

The following example of the mathematical model of radionuclides migration in boreal forest ecosystems shows some details of the above mentioned modeling approach.

2. Site description and description of conceptual scheme of boreal forest ecosystem

The main type of soil is soddy-podzolic sandy-loam, with an average annual depth of ground water level of about 130 cm. This soil is formed from fluvio-glacial sandy-loam deposits. It belongs to the automorphic group and has an average bulk density of 1,15-1,25 $g \cdot cm^{-3}$ per10-cm stratum . The main soil mineral is quartz and its content varies from 60 to 85% (in the 0,05-0,01 mm fraction). The clay content is about 0,5%. About 65% of the soil volume consists of particles exceeding 0.01 mm (physical sand). The soil of the analyzed ecosystem is characterized by low natural fertility and unfavorable hydrophysical properties: high water permeability and low water-holding capacity, this causing rapid deep infiltration of melted snow, while considerable quantities of water are evaporated from the upper layers. But this soddy-podzolic soil is favorable to the growth of boreal species – *Pinus sylvestris, Vaccinium myrtillus,* etc.

Description of soil profile: forest litter thickness reaches 10-15 cm, humus is of a rough moder-type. Forest litter mainly consists of needles of Pinus sylvestris and residuals of mosses (Bryales). The Ah-horizon is grey-black, sandy-loam, its thickness is 8-10 cm , with large amount of roots of dwarf-shrubs, grasses, rhizoids of mosses. E-horizon (eluvium) is light grey, sandy, its thickness varies from 5 to 10 cm, without any roots of plants. Bh-horizon (illuvium) is ferrugineo-brown, loam, dense, with middle diameter roots of *Pinus sylvestris.* Its thickness varies from 5 to 10 cm. Bi-horizon is yellow-brown, sandy-loam or sandy, exceed to the depth of ground water. It contains separate big roots of Pinus sylvestris.

The dominant tree species is pine (*Pinus sylvestris* L.) with sparse birch (*Betula pubescens* Ehrh.). The rising generation includes the same species of trees. The average age of the pine tree is 50 years, birch – 25-30 years. The total amount of pine is 1180 trees, the average height of the pine trees is 22,4 m, diameter – 20 cm. The average density of pine wood biomass is 297,2 metric ton per hectare (in fresh weight).

Vegetation belongs to the floristic association *Molinio-Pinetum* J.Mat. 1981, Union *Dicrano-Pinion* Libb. 1933, Ordo *Vaccinio-Piceetalia* Br.-Bl. 1939 em K.Lund 1967, Class *Vaccinio-Piceetea* Br.-Bl. 1939. This floristic association is one of the most wide spread in Central Polessye of Ukraine (about 40% of total forested square). Understorey vegetation layer is dense, with total projective cover 70-75%, represented mainly by typical boreal dwarf-shrubs – *Vaccinium myrtillus* L. (60-65%), *Vaccinium vitis-idaea* L. (5-10%), *Vaccinium uliginosum* L. (1-3%) as well as the grasses such as *Molinia caerulea* (L.) Mill. (1-3%), *Melampyrum pratense* L. (1%), *Dryopteris carthusiana* (Vill) H.P.Huchs (1%), separate plants also met – *Ledum palustre* L., *Equisetum sylvaticum* L., *Luzula pilosa* L., *Lysimachia vulgaris* L., *Calluna vulgaris* L., *Potentilla erecta* (L.) Raeusch et al. Moss cover was dense (with total projective cover 90-98%), mainly consists of *Pleurozium schreberi* (Brid.) Mitt. (60-65%), *Dicranum polysetum* Sw. (30-33%) and *Polytrichum commune* (1-5%).

The total biomass of understorey vegetation is about 1,3 kg·m^{-2} (13,0 t·ha^{-1}, d.w.), including small trees of the rising generation, and shrubs of *Sorbus aucuparia* and *Frangula alnus* (about 0,1 kg·m^{-2}) d.w. Biomass of mosses is near 1,0 kg·m^{-2}, dwarf-shrubs (mainly *Vaccinium myrtillus* – 0,12-0,14 kg·m^{-2}). The main species of mushrooms are *Boletus badius* Fr., *Cantharellus cibarius* Fr., *Russula paludosa* (L.) Kuntze, *Suillus luteus* Quél, *Boletus edulis* (Bull.) Fr.

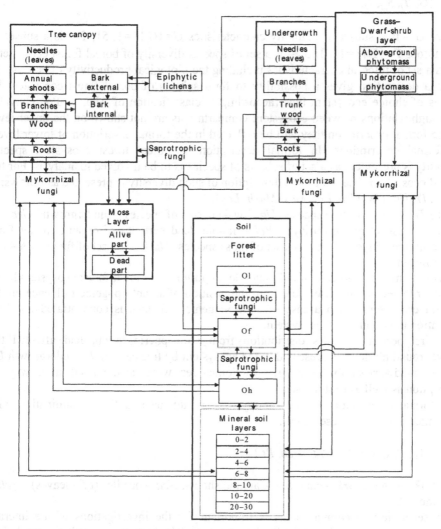

Fig. 1. The conceptual scheme of boreal forest ecosystem.

In order to create formal descriptions for further modeling the above mentioned ecosystem has been separated into 6 macro blocks. Four of them concern main blocks of vegetation: tree canopy, undergrowth, grass-dwarf-shrubs, mosses; and remaining two: forest litter and mineral soil. The conceptual scheme of described system is given in Fig. 1.

3. Connection between the conceptual scheme of boreal forest ecosystem to formal description and modeling

Formal description of boreal forest ecosystem has been presented as a set of characteristics:

$$Fji = (G_i, T_j, S_{pf}, I),$$

where G – is a subset of humidity characteristics $G=\{G_i | i =1, 5\}$; T – is a subset of trophotops $T=\{T_j | j =1, 5\}$; S_{pf} – is a set of species diversity of boreal forest ecosystem; I – is a set of integral characteristics including the biological productivity of ecosystem. Such a description gives a possibility to form subsets of boreal forest ecosystem by means of choice ecosystem type (according to classification of Alexeev-Pogrebnyak). Although relations between ecosystem compartments are not shown on the upper level of the formal description they will be reflected in the formal description of lower levels of formal descriptions. In formal description of boreal forest ecosystem species diversity is presented, as a set of subsets of species is to be involved in analysis. For the boreal ecosystem case the formal description of such diversity is presented as follows:

$S_{pf} = \{Tr, Ug, Uw, Grb, Fal, Mos, Mush, Lch\}$,

where Tr – is a set of tree species; Ug, Uw – are sets of species of undergrowth; Grb – is a set of grass–dwarf-shrub layer; Fal – fallout dead particles of organic matter from different forest layers; Mos – is a set of moss species, $Mush$ – is a set of fungi, Lch – is a set of lichen.

In order to model a process of radionuclide migration in a boreal forest ecosystem let us consider the ecosystem life cycle from the position of an antropogenic influence on the natural ecosystem. Such a case of anthropogenic influence is concentrated on ^{137}Cs migration in boreal forest ecosystem.

The first period of ^{137}Cs distribution from atmospheric fallout had affected the aboveground phytomass presented in the ecosystem by tree canopy Tr, undergrowth Ug and grass–dwarf-shrub layer Grb. Radionuclides were accumulated in a way of adsorption as well as root uptake.

The formal description of tree canopy taking into account the contamination with radionuclides can be presented as:

$Tr = \{Wd, Brk, Spr, Ned, Lf, Brch, Rt \}$,

where Wd – wood, Brk – bakr, Spr – annual shoots, Ned – needles (Lf – leaves), $Brch$ – branches, Rt – roots.

Such separation in compartments is important for the investigations of the internal processes of radionuclides redistribution in plants. The layer of grass–dwarf-shrub Grb

is presented by the aboveground phytomass and underground phytomass taking into account this layer as a whole disregarding the internal processes of redistribution.

The litter layer is presented in the model by *Ol*, *Of* and *Oh* parts but considering only the process of vertical migration of ^{137}Cs. The layer of lichens is not included into the model but formal description is given to save the openness of the approach for modeling. The layer of fungi has been presented in a model by separate types of fungi because of peculiarities of fungi types.

Mineral soil layers are divided into 10 layers with 2 cm thickness. Soil type is to be characterized by the type of natural forest ecosystem according to classification of Alexeev-Pogrebnyak.

4. Computer-aided set up and simulation for the mathematical model of radionuclides migration in boreal forest ecosystems

The mathematical model of radionuclides migration in boreal forest ecosystems is set up using the approach to the compartment ecosystem models formation [21]. The conceptual scheme of radionuclides migration in boreal forest ecosystems is a start point for working towards a model. Every compartment here adds one equation into the system of ordinary differential equation that is to agree with radionuclides migration processes.

The computer-aided nature of radionuclides migration research is presupposed both for the phase when a user sets up the model and for the phase when a user wants to analyse it. Computer-aided modeling and simulation is provided by a software complex of two components:

1. Information system "Polyn" is to store, edit, replenish and systematize radiomonitoring data, which determine computer-aided parameters matching for the mathematical model.
2. Mathematical software "DSR Open Lab 1.0" is to make an earlier planned approach for the simulation runs. It performs the eventual simulation runs for the preformed mathematical model of radionuclides migration in boreal forest ecosystems using such parameters as the input to be used, the period to be simulated, the quality of the results to be expected, etc.

Besides the usual profits of a computer-aided research process, a user can transfer easily the considered approach for the radionuclides migration in boreal forest ecosystems to an arbitrary number of other forest ecosystems due to the integration of the information monitoring data management system and the mathematical software.

A prognosis of radionuclides migration in mineral soil layers is presented on Fig. 2. Here we can see a clear trend toward considerable accumulation of the radioactivity value in the mineral soil layers at a depth of 0-2 and 2-4 centimetres. It can be explained by means of the litter behaviour for the boreal forest ecosystem. This thick layer stands as a permanent source of contamination for underlying layers.

The next figure (Fig. 3) allows to concluding that pine-tree is able to accumulate and retain big radioactivity value for a long time while taking up minerals from the great bulk of underlying layers. Ol contains a small amount of radioactivity due to the its periodical renewal with litter-fall that is characterized by a minor content of ^{137}Cs as far as the radiation disaster becomes a thing of the past.

Trends of accumulation of the radioactivity value for *Xerocomus badius* and *Russula paludosa* is shown on Fig. 4. This graph allows to conclude that the dynamics of ^{137}Cs specific activity for *Xerocomus badius* and *Russula paludosa* lies in constant growth up to years of 1996-2004 and in lessening of this value during the following period.

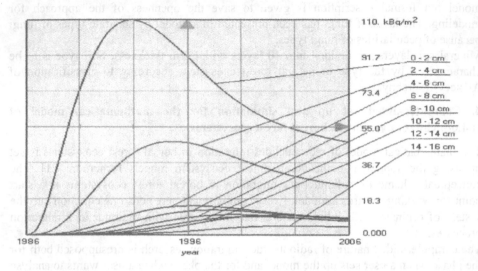

Fig. 2. Results of the research of radionuclides migration in boreal forest ecosystems: mineral soil layers.

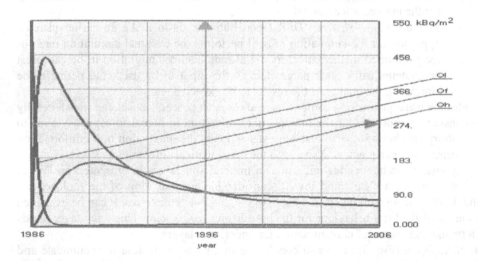

Fig. 3. Results of the research of radionuclides migration in boreal forest ecosystems: tree canopy, Ol, Of, Oh.

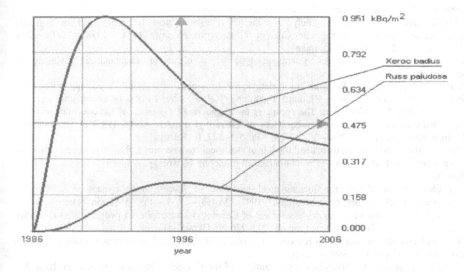

Fig. 4. Results of the research of radionuclides migration in boreal forest ecosystems: Xerocomus badius, Russula paludosa.

6. Conclusion

The software complex, considered in this work, sets up the mathematical model of radionuclides migration using formal description of the forest ecosystem. The proposed approach to the modelling process allows to automate research of radionuclides migration in the boreal forests and to transfer easily the modelling procedure to other forest ecosystems. The phase of setting up the model is reduced for a user to the defining of relations between compartments of the ecosystem. Conversion of mathematical language constructs into domain-specific ones thus and so brings modelling closer to the ecologists.

Trends of radioactivity in the compartments, taken from simulation runs of the mathematical model of radionuclides migration in the boreal ecosystems, are corresponding to the radiomonitoring data and are able to provide prognosis for radionuclides redistribution in the ecosystem. Effective management of forest resources can be reached by means of interpreting the results of the model analysis.

7. References

1. L.G. Appleby, L. Dovell, Yu.Ê. Mishra The ways of radionuclide migration in environment. Radioecology after Chernobyl Catastrophe. – Ì.: Ìyr, 1999. – 512 p. (in Russian)
2. À.Ì. Dvornik, Î.À. Zhuchenko The ¹³⁷Cs behavior in pine forests of Belorus Polissya: modeling and prediction // ÀNRI, The Journal of Radiation Ecology. – 1995. ¹ 3/4. – Pp. 56–66. (in Russian)
3. B.P. Ivchenko, L.À. Martyschenko The Information Ecology. Part 1. The risk assessment from technogenous accidents and catastrophes. The statistical interpretation of ecological monitoring. The modelling and prediction of ecological situations. – St.-Ptb.: "NorMed-Publ", 1998. – 208 p. (in Russian)
4. Ì.Ì. Kolodnytsky The typology of mathematical models of technical systems. Part 2 // Visnyk ZhITI.–

1998. – ¹ 7/Technical Sciences. – Pp. 208–218. (in Ukrainian)

5. Ì.Ì. Kolodnytsky, V.Ì. Yanchuk "Polyn" – the information system for assessment of radiation contamination of Zhitomir region after Chernobyl catastrophe // Visnyk ZhITI. – 1998. – ¹ 9/Technical Sciences. – Pp. 271–278. (in Ukrainian)

6. Ì.Ì. Kolodnytsky, V.Ì. Yanchuk Systems analysis of problems of mathematical modeling in environmental sciences. (in Ukrainian)

7. V.P. Krasnov, Z.Ì. Shelest, Ì.Ì. Orlov, Ì.Ì. Kaletnyk, S.P. Irklienko, V.Ì. Turko Padioecology of Roe Deer in Central Polissya of Ukraine. – Zhitomir: Publishing "Volyn", 1998. – 128 p. (in Ukrainian)

8. S.V. Mamikhin, L.N. Merkulova The computer investigations of dynamics of radionuclides in forest ecosystems contaminated in a consequences of Chernobyl accident (1986–1995) // The radiation biology. Radioecology. – 1996. – V. 36, ¹. 4. – Pp. 516–523. (in Russian)

9. Yu.Ì. Svirzhev About the mathematical models of biological communities and tasks of optimization and control connected with them // The mathematical modeling in biology. – Ì., 1975. – Pp. 30–53. (in Russian)

10. S.V. Chernyshenko About the mathematical modeling of biogeocenosis dynamical structure // Environmental sciences and noospherology. – 1997. – Vol. 3. – ¹ 1–2. – Pp. 65–86. (in Russian)

11. V. Yanchuk Problems of computer assessment of Chernobyl catastrophe on people illness // Visnyk ZhITI. – 1998. – ¹ 8/Technical Sciences. – P. 271–276. (in Ukrainian)

12. R. Avila Radiocesium transfer to roe deer and moose. Modeling and experimental studies. Doctor's dissertation. – Uppsala. – 1998. – 121 p.

13. Chernobyl Digest 95–98 /Interdisciplinary Bulletin of the Chernobyl Problem information. Issue 5. – Minsk, 1999. – 257 p.

14. V.P. Krasnov, Z.Ì. Shelest, Ò.V. Êurbet ¹³⁷Cs contamination of mushrooms in the Ukrainian Polessye // Abstr. Of III Congress on Radiation Research "Radiobiology, Radioecology, Radiation Safety" (Moscow, 14–17 October 1997). – Pushchino, 1997.–Vol. 2. – P. 353–354.

15. Ò.V. Kurbet The regularities of ¹³⁷Cs accumulation by edible mushrooms in forests of the Central Polessye of the Ukraine // Abstr. ² Internat. Conf. "Ecology and youth". – Gomel, 1998. – Vol. 1, Part. 2. – P. 106.

16. I. Linkov, W.R. Schell (eds.) Contaminated forests. Recent Developments in Risk Identification and Future Perspectives. Kluwer Academic Publushers, 1999. – PP. 151–160

17. I.G. Malkina-Pykh. Modeling of the dynamics in Sr-90 content in ecosystems of various geographical zones // Radiatsionnaya biologia. Radioekologia. – 1996. – V. 36, ¹ 1. – P. 112–132.

18. A.A. Orlov Regularities of technogenous radionuclide accumulation and migration in forest biogeocenosis of boreal coniferous type: research progress 1996–1999 // Chernobyl Digest 95–98 / Interdisciplinary Bulletin of the Chernobyl Problem information. Issue 5. – Minsk, 1999. – PP. 18–31.

BEHAVIOR OF Cs-137 AND Sr-90 ON FISH PONDS IN UKRAINE

E.VOLKOVA, V. BELYAEV, Z. SHIROKAYA, V. KARAPYSH
Institute of Gidrobiology, NASU, 12, Prosp.Geroyiv Stalingrada, Kiev, 04210, UKRAINE

1.   Introduction

Radiation monitoring and radioecological studies have also shown a high persistence of radionuclides in lakes located in a number of communities in the Ukraine[1, 2]. Contamination levels of fish in these lakes are very high and this pathway has become increasingly important in determining the overall dose to humans [3].

2.   Results and discussion

The region incorporates several ponds and lakes, which are located within areas of different contamination intensities. Water bodies are distributed as follows: water bodies No. 1, 2 (Narodychi rayon)-the zone of obligatory evacuation, water bodies No. 5 (Dubrovizk rayon), 6 (Zarechensk rayon), 7 and 8 (Volodimerezk rayon), 10 (Rokitnyansk rayon) - the zone of optional evacuation, ; water body No. 3 (Ovruch rayon) - the zone of strict radiation control, water bodies 4 (Novograd-Volinsk rayon), 9 (Sarni rayon) - the clean zone.

^{137}Cs levels in water bodies under study vary widely (more than 1 order of magnitude),   however, the levels do not exceed contemporary permitted levels. ^{90}Sr levels were found to be of more uniform pattern.

Levels of sediments-sorbed ^{137}Cs and ^{90}Sr were found to vary substantially - from 242 to 2470 Bq/kg and from 32 to 306 Bq/kg, respectively. At the same time there is no definite correlation between sediments' radionuclide levels and radionuclide levels in water. For example, the highest ^{137}Cs level was registered in water body No. 7, while the highest level of specific radioactivity of sediments was registered in water body No.6. While analysing radionuclide distribution between soluble and sorbed forms, it is necessary to note, that ^{137}Cs levels in water and ^{137}Cs levels of sediments are almost equal (with some prevalence of soluble forms), while ^{90}Sr is mainly represented by soluble forms (from 71 to 99%) of its overall concentrations in a water cross-section.

Bottom deposits are the prime radionuclide accumulators (this is especially true for ^{137}Cs). Levels of bottom deposits' radioactive contamination depend on multiple factors, including, *inter alia*, types of bottom deposits, bottom shapes, currents, etc. (fig. 1). High water plants belong to the most important components of aquatic ecosystems (fig. 2). We selected the most common water plants for the study.

F. Brechignac and G. Desmet (eds.), Equidosimetry, 293–298.

294

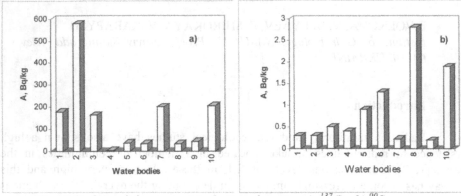

Fig. 1. Radionuclides content in bottom deposits, a) ^{137}*Cs; b)* ^{90}*Sr.*

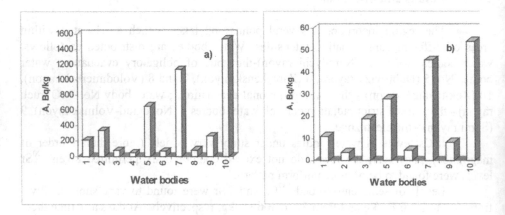

Fig. 2. Radionuclides content in high water plants, a) ^{137} *Cs; b)* ^{90} *Sr.*

^{137}Cs levels in fish of the water bodies under study vary rather significantly (fig. 3). In water body No. 1, approximately 50% of samples demonstrated ^{137}Cs levels in excess of maximal tolerable levels; in water bodies No. 10 and 7 all samples demonstrated excess ^{137}Cs levels. The highest ^{137}Cs levels were registered in the case of fish from water body No. 7 - up to 1150 Bq/kg. It is necessary to note, that in the majority of the water bodies, fish stocks are every year replenished by 1-year old fingerlings of *Cyprinus carpio* and *Carassius auratus gibelio*. We selected 2-year old samples of these species with weight from 200 to 800 g (carp) and from 50 to 120 g (crusian carp). Predator fish are rare in these water bodies. We have not registered credible differences in radioactive contamination levels of predator fish and bentophages. In water body No. 7, where fish stocks were not replenished, we sampled *Rutilus rutilus*.

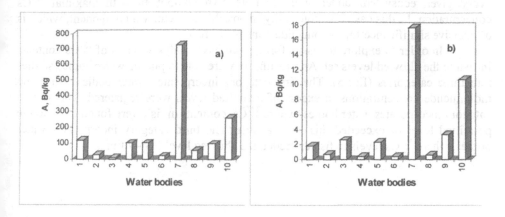

Fig. 3. Radionuclides content in fish ,a) - ^{137}Cs; b) ^{90}Sr.

^{90}Sr levels in the fish species studied vary from 0.1 to 22.8 Bq/kg and do not exceed contemporary permitted levels (35 Bq/kg). Let us now review some general trends of radioactive contamination of aquatic ecosystems studied. As fig. 4 shows, radioactivity levels of the majority of ecosystem components are determined mainly by ^{137}Cs. Water itself is an exception - overall water radioactivity is almost equally shared between ^{137}Cs and ^{90}Sr.

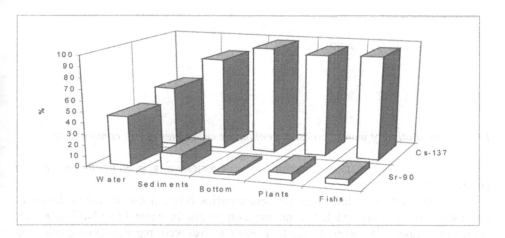

Fig. 4. Average percentages of radionuclide contents in different ecosystem components in the water bodies studied.

As analysis of the data collected shows, we have not registered cases of maximal ^{137}Cs contents in any single ecosystem component in all water bodies; instead every given ecosystem under study had its own component with maximal ^{137}Cs concentration. In this case it is practically impossible to identify a component, which is of decisive significance for radionuclide contents in fish.

In order to explore the issue further, we have analysed ratios of ^{137}Cs contents in fish to the allowed levels set. As a result, we were able to put the water bodies studies into three categories (fig. 5). The first category incorporates water bodies, where no radionuclide concentrations in excess of permitted levels were registered. The second category incorporates water bodies, where ^{137}Cs contents in fish were found close to the permitted level or exceeded this level a little. The third category incorporates water bodies, where ^{137}Cs levels in fish exceeded the allowed level 1,3 - 7 times.

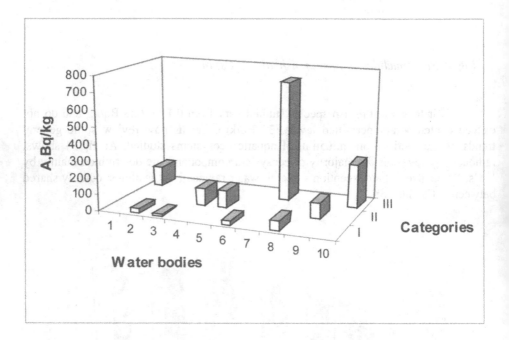

Fig. 5. Distribution of water bodies, depending on categories of fish contamination.

Then, the system of contamination categories was applied to other ecosystem components.

Now, let us review potential dependencies between contamination levels of ecosystem components and fish contamination levels. In water body No.7 (the one, where the highest ^{90}Sr levels in the fish levels in fish were registered) contamination levels of water, sediments and higher water plants also belong to the third category (fig. 6).

^{137}Cs levels in bottom sediments were rather low. In this case, it is possible to claim that ^{137}Cs is poorly sorbed by bottom sediments, as a result, the radionuclide easily contaminates other components. It is necessary to emphasize, that this water body is closed and operates as a accumulating reservoir for radionuclides, entering the water body ecosystem.

Water body No. 3 shows the opposite pattern. Radioactive contamination parameters of ecosystem components in the water body indicate that ^{137}Cs are firmly trapped by bottom sediments. Water body No. 1 shows relatively uniform contamination levels of all components, and ^{137}Cs levels in fish (the final link of trophic chains) were found to be rather high.

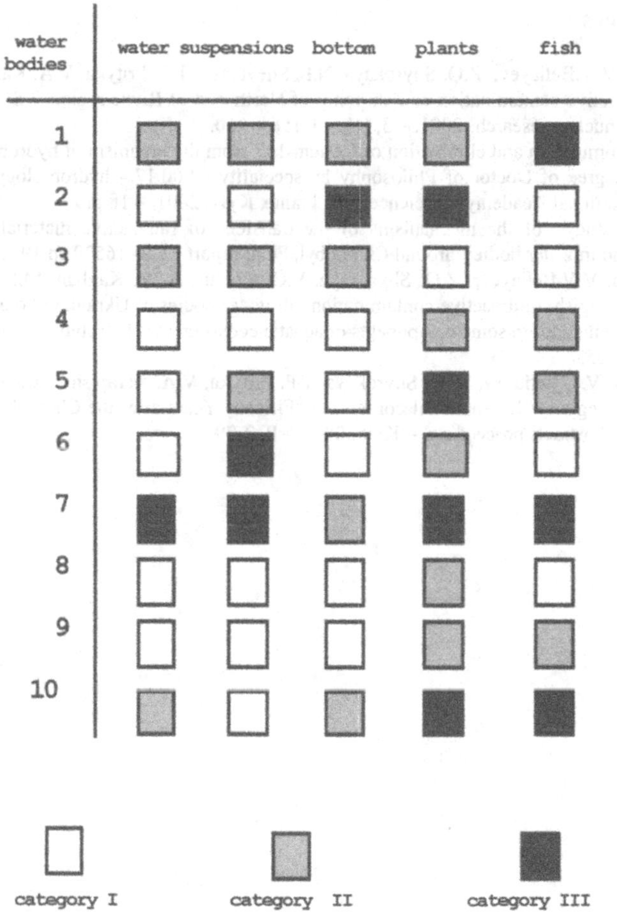

Fig. 6. Distribution of ecosystem components, depending on radioactive contamination categories.

In some cases, radionuclide contents in fish samples from the water bodies substantially exceeded permitted ^{137}Cs levels in food products (approved by the Public Health Ministry of Ukraine). Moreover, these levels substantially differ from similar parameters, observed before the Chernobyl NPP accident.

3. Conclusion

It is necessary to note, that water bodies, affected by several radionuclide accumulation factors simultaneously, might become sources of human radionuclide consumption [4, 5]. Therefore, it is necessary to review permits for commercial/amateur fishing in every particular case.

4. References

1. E. N. Volkova, V.V.Beliayev, Z.O. Shyrokaya, N.L. Shevtsova, T.P. Prityka, V.A. Karapish. Particular feature of radioactive contamination of fish ponds of North part of Rivne region // Scientific papers of the Institute for nuclear research.-2001. – 3, N1. – P.156 – 160.
2. Belyaev V. Accumulation and elimination of Cesium-137 from the organism of hydrobionts : Summary of thesis for degree of Doctor of Philosophy by speciality 03.00.17 - hydrobiology. – Institute of Hydrobioljgy, National Academy of Sciences of Ukraine, Kyiv, 2001. – 18 p.
3. Modelling and study of the mechanisms of the transfer of radioactive material from terrestrial ecosistems to and in water bodies around Chernobyl, Final report EUR- 16529 en,1996.- 184 p.
4. Ye. N. Volkova, V.V.Beliayev, Z.O. Shyrokaya, V.G. Klenus, A.Ye. Kaglian, M.I. Kuzmenko, T.P. Prityka, V.A. Karapish. Radioactive contamination of water bodies of Ukrainian Polesye and forms of presence of radionuclides in some components of aquatic ecosistems // Hydrobiological Journal.- 2000. - 36, N4. – P.50 – 65.
5. Ye. N. Volkova, V.V.Beliayev, Z.O. Shyrokaya, T.P. Prityka, V.A. Karapish. ^{137}Cs and ^{90}Sr in water bodies of Rivne region // International conference "Fifteen Years after the Chernobyl Accident. Le ssons Learned". Abstracts proceeding. - Kyiv, 2001. – P. 2-79.

GENETIC EFFECTS OF CHRONIC GAMMA-IRRADIATION AT A LOW DOSE RATE: EXPERIMENTAL STUDY ON CBA/LAC MICE

A.N. OSIPOV, A.L. ELAKOV, P.V. PUCHKOV, V.D. SYPIN
Moscow Scientific and Industrial Association "Radon", 7th Rostovsky lane 2/14, Moscow, 119121, RUSSIA;
E-mail: aosipov@radon.ru

M.D. POMERANTSEVA, L.K. RAMAIYA, V.A. SHEVCHENKO
N.I. Vavilov Institute of General Genetics, Russian Academy of Sciences Gubkin str. 3, , Moscow, 119991, RUSSIA.

1. Introduction

In contrast to high doses of ionizing radiation, the negative effect of which is beyond doubt, the question of whether the action of low doses is harmful or useful remains open. As follows from literature data [1-3] and the results of our studies [4,5], low-dose ionizing radiation induces a complex of biochemical and biophysical reactions in the animals' organism. However, it is not clear whether these changes are the result of adaptation of the organism to an increased radiation background and whether the action of low doses leads to significant genetic consequences.

In the present work we have studied the level of DNA strandbreaks in spleen lymphocytes, the percentage of normochromatic erythrocytes (NCE) with micronuclei (MN) in peripheral blood as well as the frequency of abnormal sperm heads (ASH) in mice exposed to continuous low dose-rate gamma-radiation.

2. Materials and methods

12-14 g CBA/lac mice (3-4 weeks old) purchased from Stolbovaya (Russian Academy of Medical Sciences) were used for this experiment. The animals were housed in plastic cages 7 days before irradiation. Food and sterile water were provided *ad libitum.* The control and experimental animals were maintained under identical conditions ($20\pm2°C$, $50\pm10\%$ relative humidity, 12-hr light/dark cycle).

The mice were irradiated using a UOG-1 device (course Cs-137, dose rate 0.17 cGy/day) during 80, 120, 210 and 365 days. The dose-rate of radiation was controlled with a DRG-O1T radiometer (Russia). The irradiation was 24 h a day with a short break (10-15 min) to attend to the animals. The total dose received by the animals made up 13.6, 20.4, 35.7 and 63,5 cGy respectively. Dosimetry was performed using FLi detectors TLD-100 (Sweden) and DTG-4 (Russia). The measurements were made using RE-1 (RADOS, Finland) and DTF-01 (Russia) devices.

At the 80, 120, 210 and 365th day the animals were decapitated and the spleen and testes were removed. The suspension of spleen cells in PBS with 3 mM NaN_3, and cooled to 4°C, was filtered through a nylon net. The concentration of cells was counted in a Goryaev's chamber.

F. Brechignac and G. Desmet (eds.), Equidosimetry. 299–304.

Alkaline single cell gel electrophoresis (comet assay) was carried out as described by Singh et al. [6]. According to this assay, the level of single-strand DNA breaks will be in direct proportion to the amount and migration distance of DNA after alkaline electrophoresis of single cells immobilized in agarose. The fluorescent dye Hoechst 33258 was used for DNA staining. The DNA comets were analyzed with a fluorescence microscope Lumam I-2 (LOMO, Russia). 100 comets for each slide were counted. Depending on an appearance of comets (i.e. tail length, head diameter and intensity), the comets were assigned to one of five classes from 0 to 4 (0 – undamaged, 4 – maximally damaged). This method of visual estimation was proposed by Collins et al. [7] and is considered to be suitable for studying DNA damage [8]. The results obtained with this method were confirmed with the help of a computer analytical system Kinetic Imaging, Liverpool, UK. The number of comets in each category was counted and an average index of DNA comets was calculated as

$$ACI = (0 \cdot n0 + 1 \cdot n1 + 2 \cdot n2 + 3 \cdot n3 + 4 \cdot n4)/\Sigma,$$

where n0 - n4 is the number of comets in categories 0–4 and Σ is the sum of all counted comets.

To study the induction of DNA damage in spleen lymphocytes by hydrogen peroxide the cell suspension in PBS (1×10^6 cells/ml) was incubated with 0.5 mM and 5 mM H2O2 for 30 min at 37°C. The number of single-strand DNA breaks was determined using the fluorimetric assay of DNA unwinding [9]. According to this procedure, the increase of DNA breaks is estimated by the decrease of double-strand DNA fragments after controlled alkaline unwinding of DNA of the analyzed cells.

The frequency of peripheral blood NCE with MN was counted by the conventional procedure [10]. 2000 NCE were analyzed for each animal.

The ASH frequency was analyzed on epididymis smears [11]. 200 to 500 spermatozoa were counted from each male. Statistical processing of the results of measurements was performed using Student's t-test. The samples from each animal were treated separately. Data are presented as the mean ± standard error.

3. Results and discussion

The level of single-strand DNA breaks (SSB) was estimated by the fluorimetric assay of DNA unwinding (FADU) and alkaline single cell gel electrophoresis (comet assay). It was found that beginning from the 120th day of the experiment a significant ($p < 0.05$) increase was observed in the number of SSB registered by the comet assay but not by the FADU (Table 1). This fact is further proof of a higher sensitivity of the comet assay as compared to the FADU. So, in accordance with Singh et al., the lower limit of sensitivity of a standard modification of the comet assay is ~ 3 cGy for acute irradiation (~ 30 SSB per nucleus) [12]. In addition to a high sensitivity, this method makes it possible to estimate the distribution of the total cell population by the level of DNA damage.

Figure 1 shows the distribution of the cell population by the level of DNA damage. It is seen that at the 80th day of exposure a shift for one category is observed in the distribution of spleen lymphocytes in the direction of increasing number of cells with an elevated level of DNA strandbreaks. At the 120 and 210th days of the experiment subpopulations of cells with a high level of DNA damage and apoptotic cells were registered, but their number was insignificant (~2-4 %). Thus, the results obtained indicate that a long-term low dose-rate irradiation induces an increase in the general level of DNA breaks.

It is known that in normally functioning mammalian cells there is always a background level of DNA breaks present. These breaks are always divided in two classes:
- breaks, formed upon DNA damage by free radicals resulting from cell oxygen metabolism;
- breaks, formed in the course of chromatin functioning. This group includes breaks of enzymatic nature necessary for the processes of replication, repair, transcription, condensation, decondensation of chromatin, etc.

Mammalian cells have defense mechanisms of neutralization of free radicals in which such enzymes as superoxide dismutase, catalase, glutathion peroxidase, etc. (enzymatic defence systems) and such antioxidants as vitamin E, glutathione, taurine etc. (non-enzymatic defence systems) are involved. The heterogeneity in the distribution of DNA breaks observed in our work could suggest the exhaustion of the defense antioxidant mechanisms of cells protecting from the damaging action of low levels of "by-products" of normal oxygen metabolism (O^{2-}, $OH^{\bullet}$, H_2O_2, etc.).

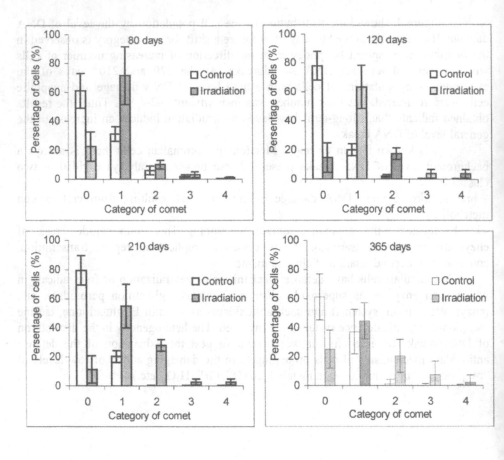

Fig. 1. The distribution of lymphocytes by the level of single-strand DNA breaks obtained in the course of analysis of alkaline DNA comets of spleen lymphocytes of mice exposed to continuous low dose-rate gamma-radiation.

To determine the antioxidant status of spleen lymphocytes of irradiated mice, we studied the indication of DNA damage in these cells by hydrogen peroxide. Hydrogen peroxide is a product of normal cell metabolism which in the presence of metal ions of variable valence (mainly Fe^{2+}) leads to the formation of a highly toxic hydroxyl radical.

A slight increase of sensitivity of spleen lymphocytes of irradiated mice to hydrogen peroxide was noted on the 80th day of the experiment (Fig. 2). This fact can be explained by the exhaustion of the antioxidant potential of the cells. Continuation of irradiation of the animals probably leads to the activation of the defence systems of spleen lymphocytes, which is expressed as a decrease of their sensitivity to the action of H_2O_2 at the 120 and 210th days of the experiment. This effect can be due to the development of adaptation processes with the accumulation of a certain dose.

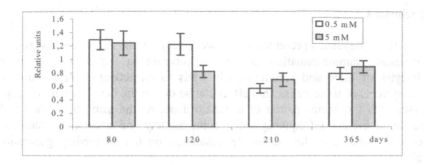

Fig.2. Alteration of the sensitivity of spleen lymphocytes of mice exposed to low dose-rate gamma-radiation under the action of hydrogen peroxide. The data are presented as the ratio of the values obtained for the experimental animals to the values for the control animals.

Thus, the increase in the level of DNA breaks registered by the method of DNA comets seems to be most likely due to the increase of breaks arising in the course of chromatin functioning rather than to the increase in the number of breaks of oxidative nature. Plausible explanations may be the following ones: activation of DNA excision repair, structural rearrangement of chromatin, induction of compensatory proliferation of lymphocytes. One cannot exclude a simultaneous occurrence of these processes.

The mutagenic effect of low dose-rate gamma-radiation was judged by the presence of peripheral blood NCE with MN. MN represent acentric chromosome fragments and individual whole chromosomes lost during mitosis. No essential changes were observed in the experimental group as compared to the control group (Table 2). On the contrary, on the 365th day of the experiment even a reduction of the percentage of NCE with MN was observed. This is probably due to the activation of the antioxidant defence mechanisms, but we have no sufficient data as yet to make a final conclusion.

The results of estimation of ASH frequency depending on the duration of irradiation are presented in Table 3. The ASH frequency in the experimental groups increased as compared to the control with the time of irradiation. At a dose of 35.7 cGy the ASH frequency significantly exceeded the control level.

A slight reduction of the testis mass as compared to the control was also observed at the same time period (210 days). Low-intensity chronic irradiation was found to activate the hypophysial-adrenal system [3]. The author believes that irradiation indirectly acts as a peculiar stress factor through the neurohumoral system inducing a chain of metabolic changes at the cell and organism levels, which is what leads to an increase in the ASH frequency.

304

4. Conclusions

Thus, the results presented in this work demonstrate that a continuous action of low dose-rate gamma-radiation induces an increase in the SSB level in spleen lymphocytes of mice and simultaneously leads to an activation of the antioxidant defence systems of these cells. This effect can be due to the development of adaptation processes with the accumulation of a certain dose. At the same time, the results of estimation of the ASH frequency cause some anxiety and necessitate studies of the influence of continuous low dose-rate irradiation on the succeeding generations of animals.

5. Acknowledgements

The work has been carried out with a partial financial support by the Russian Foundation for Basic Research (project № 99-04-49-218).

6. References

1. L.V. Slozhenikina, L.A. Fialkovskaya, I.K. Kolomiytseva Ornithine decardoxylase in organs of rats following gamma-irradiation at low dose-rate. *Intern. Journ. Radiat. Biol.* 1999. V. 75. P.195-199.
2. M.A. Smotryaeva, K.E. Kruglyakova, L.N. Shishkina et al. The structural and biochemical properties of mice blood components after low dose gamma-irradiation of different intensity. *Radiats Biol Radioecol* 1996. V. 36. P. 21-29 (in Russian).
3. M.Yu. Alesina Radiobiological effects formation in experimental animals under low dose chronic internal and external irradiation. *Intern. Journ. Radiat. Medicine.* 1999. V. 2. P.92-99.
4. A.N. Osipov, M.V. Grigoryev, V.D. Sypin et al. The Influence of Chronic Exposure to Cadmium and γ-Radiation at Low Doses on Genetic Structures of Mice. *Radiats. Biol. Radioecol.* 2000. Vol. 40(4). P. 373-377 (in Russian).
5. A.N. Osipov, V.D. Sypin, P.V. Puchkov et al. Changes in the Level of DNA-Protein Cross-Links in Spleen Lymphocytes of Mice Exposed to Low-Intensity γ-Radiation at Low Doses. *Radiats. Biol. Radioecol.* 2000. Vol. 40(5). P. 516-519 (in Russian).
6. N.P. Singh, M.T. McCoy, R.R. Tice, E.L. Schneider A simple technique for quantification of low levels of DNA damage in individual cells. *Exp. Cell. Res.* 1988. V. 175. P. 184–191.
7. A.R. Collins, A.G. Ma, S.J. Duthie The kinetics of repair of oxidative DNA damage (strand breaks and oxidised pyrimidine dimers in human cells). *Mutat. Res.* 1995. V. 336(1). P. 69–77.
8. H. Kobayashi, C. Sugiyama, Y. Morikawa et al. A comparison between manual microscopic analysis and computerized image analysis in the single cell gel electrophoresis assay. *MMS Commun.* 1995. V. 3. P. 103–115.
9. P.M. Kanter, H.S. Shwartz A fluorescence enhancement assay for cellular DNA damage. *Mol. Pharmacol.* 1982. V. 22. P.145-151.
10. R. Schlegel, J.T. MacGregor The persistence of micronuclei in peripheral blood erythrocytes: detection of chronic chromosome breakage in mice. *Mutat. Res.* 1982. V. 104. P. 367-369.
11. A.J. Wyrobec, J.A. Heddle, W.R. Bruce Chromosomal abnormalities and the morphology of male mouse sperm heads. *Can. J. Genet. Cytol.* 1975. V. 17. P. 675-681.
12. N.P. Singh, R.E. Stephens, E.L. Schneider Modifications of alkaline microgel electrophoresis for sensitive detection of DNA damage. *Int. J. Radiat. Biol.* 1994. V. 66. P. 23–28.

Part 7.

Applied Radioecology and Ecotoxicology

METHODS AND TASKS OF RADIATION MONITORING OF FOREST ECOSYSTEMS

V. KRASNOV and A. ORLOV
Polesskiy Branch of Ukrainian Scientific Research Institute of Forestry and Forest Amelioration (UkrSRIFA), pr. Mira, 38; Zhitomir, 10004, UKRAINE,
station@zt.ukrpack.net

1. Introduction

More than 15 years have passed since the Chernobyl accident changing sharply the radiation situation of an environment over a large territory. This accident also has changed way of life and consciousness of the people which were influenced by it and those who participated in liquidation of its consequences and studied biological and ecological effects of this global catastrophe. Unfortunately, now world science can't predict the remote consequences on many aspects, in particular, influence of low-dose radiation, consequences of these doses to whole biota and its separate species including medical consequences on health of man in radioactively contaminated zone. Besides last ones also another important problem has appeared i.e. the synergetic influence of pollutants of anthropogenic origin (technogenic radionuclides, heavy metals, pesticides, etc.). For this reason now when the sharpness of Chernobyl problem is essentially deadened the significance of such research increases.

Forest ecosystems are critical landscapes from the point of view of radionuclides migration in trophic chains, from its components to the man. At the same time, high values of content of radioactive elements which were accumulated in it has caused changing of methods of management of forestry as a part of industry on large territories as well as the reorganization of this industry taking into account the radiation situation. Rather extensive researches were carried out in forest ecosystems mainly after Chernobyl accident. There were studied:

- The scales, character and peculiarities of radioactive contamination of forest ecosystems of various types;
- The methodological problems of examination of forests on radioactive contamination (partially studied, with some contradictory conclusions);
- The general regularities of redistribution of radionuclides in various types of soils automorphic and hydromorphic landscapes;
- The influence of some species of mosses and lichens on rates of migration of radioactive elements in soil;
- The paths and directions of accumulation of radionuclides were determined and the quantitative evaluation of the main woodforming tree species were given;
- The evaluation of the contribution of tree canopy in radionuclides redistribution in different species composition and cenotic structure forests on the basis of forest-typological approach has been given;
- The regularities of migration of radioactive elements in various grass and dwarf-shrub species of undergrowth vegetation in various types of habitat;
- The resource evaluation of the main wild mushroom, berry and medicinal plants of forests was given from the point of view of their radioactive contamination and dose formation on local population;

F. Brechignac and G. Desmet (eds.), Equidosimetry. 307–312.

- The role of undergrowth vegetation in various types of habitats in redistribution of radionuclides in forest ecosystems;
- The specific peculiarities of accumulation of radionuclides by various species of edible mushrooms on forest-typological approach;
- The role of spring ephemeroid plants in redistribution of some radioactive isotopes in the chain "upper soil layers – plants";
- The regularities in migration of radionuclides in a system "soil - fodder plants - wild game animals" for the main ruminant game species (roe-deer, moose, wild boar);
- The regularities of migration of radioactive elements in ecosystems of ombrotrophic and mezotrophic bogs;
- The prognosis of radioactive contamination of various components of forest ecosystems on the basis of mathematical modelling.

The studying of regularities of radionuclides behaviour in forest ecosystems is a very complex and continuous research process. The main reasons of this phenomenon are differences in climatic conditions, soils and forest vegetation. Forest ecosystems are the complex of biota consisting of numerical species of plants and animals essentially differ by a singularity and specificity in different ecological conditions. The complexity of radioecological researches also is caused by existing of large spatial irregularity of radioactive contamination of forests – by complex configuration of radioactive contamination of territory as a whole. It demands precise methodological guidance of research activity.

As a whole conducted researches have allowed us to understand and to evaluate general regularities of radionuclides migration in forest ecosystems. Besides they have enabled to make up some organizational and economic measures directed on maintenance of safe working conditions of the workers of forestry and also of manufacturing of forest production which radioactive contamination would not exceed state permissible levels [1]. The conducted examination of forests of Ukraine on radioactive contamination (1986-1992) has allowed to show that on 38,6 % of investigated squares of forests in 18 regions of Ukraine density of ^{137}Cs ground deposition exceeded 1,0 Ci·km^{-2}. The highest index was in Zhitomir region (60,1 % of its total forest area), Rivne region (56,3 %), Kiev region (without 30-km exclusion zone of Chernobyl NPP) – 52,2 % of its total forest area. The generalized data of radioactive contamination of forests situated in Ukraine far from 30-km exclusion zone clearly have shown that there are areas of forests with the high levels of radiocontamination including those where forestry as a whole should be forbidden (in Ukraine – on 40,8 thousand ha), the partially limitation of use of wood production of forestry is demanded on 101,5 thousand ha and nonwood production – on 1190,5 thousand ha in Ukraine.

2. Proposed methodology

From a scientific point of view previous study of radiation situation in the forest ecosystems of Ukraine in a certain degree corresponds with purposes and tasks of radiation monitoring of forest ecosystems. However to evaluate them from the point of view that the radiation monitoring of forests should be the constituent of a state system of radiation monitoring of an environment as a whole (law of Ukraine "About a legal mode of territory that has undergone to radioactive contamination as a consequence of Chernobyl catastrophe"), and also general system of state monitoring of an environment (law of Ukraine "About protection of natural environment"), it will be necessary to recognize that unfortunately integral state system of forest monitoring in Ukraine does not exist now. Separate monitoring researches are carried out only fragmentary. Moreover the purposes and tasks of monitoring of forest ecosystems are insufficiently precisely defined, the volumes of researches are not defined enough completely, the methodological fundamentals are not supplied, the instrumental base is various but not uniform in different organizations, there is no coordination and also the main organization as well as the constant monitoring participants.

We consider that *radioecological monitoring of forest ecosystem* is the organized system of observation, control and evaluation of radionuclides migration and circulation in forests.

The system of observations, in turn, should provide spatial distribution of points of observations. Last should be placed in such way that more or less completely to describe the main radiation parameters on territory of Ukraine as a whole, all its climatic zones. It is desirable that the points of observations were located also in such a way that it would be possible to evaluate a radioecological situation determining by pollution from already existing nuclear power plant and also from the global radioactive deposition of technogenic radionuclides at any time. Taking into account essential differences of characteristics and distribution of forest ecosystems in various climatic zones of Ukraine quantity of points of observations and also their location should be different. Because radiation situation at any time and in any place is rather dynamical therefore it is necessary to take into account the main spatial, temporal and succession forest changes [2].

In our mind there is a sense in the differentiated approach to determination of volumes of monitoring researches in various regions of the country. We define the *volume of monitoring* as quantity of investigated parameters and quantity of points of observations, number of objects of researches as well as periodicity of observations.

Thus the organization of a system of state radiation monitoring of forest ecosystems should be made from units of regional monitoring which should take into account character of existing and possible radioactive contamination as well as peculiarities of forest location, their species composition and cenotic structure.

The *main goal* of radiation monitoring of forest ecosystems is information support of the state in realization of policy of maintenance of safe living conditions and protection of nature from radioactive contamination on the basis of knowledge of regularities of radionuclides migration in various forest types, detection of contemporary parameters, prediction of its volumes and possible origin of phenomena and processes caused by ionizing radiation.

So the *object* of such monitoring should be forest ecosystem as the complex natural phenomenon consisting of combination of arboreal species, shrubs, grass, dwarf-shrubs, mosses, lichens, mushrooms, various animals and microorganisms interact with each other and with environment. Thus it is possible to consider forest ecosystem as a whole on volumes of the activity of radioactive elements retained in it; on scales of its influence on radionuclides migration in forestless territories; on the values of secondary contamination of nearest areas from wood fires; on additional irradiation of the population from consumption of forest food products (wild mushrooms, berries, game). At the same time forest ecosystems are very complex and diverse therefore at certain levels of researches the objects of monitoring can serve their separate components:

- Various types of soils in different climatic zones of Ukraine at a regional level;
- Tree canopy as a basic component of forest ecosystems;
- Plant species of undergrowth: grasses and dwarf-shrubs, mosses, lichens and also mushrooms;
- Soil small animals (mainly – mezofauna) and microorganisms;
- Species of wild game.

The objects of monitoring should be selected regionally depending on problems existing on certain territory [3]. Above we have analysed in more detail objects of researches in a zone of influence of Chernobyl NPP. But around normally functioning nuclear power plant radiation monitoring of forest ecosystems should be carried out in considerably smaller volumes, mainly on some their critical components (in most common cases – species-radionuclide accumulators). Such components can serve as possible *test-objects* of radioecological forest monitoring:

- Needles and leaves of tree species;
- Mosses, lichens and mushrooms;
- Forest litter and upper mineral horizons of soils.

Their selection should conducted from the point of view that the main problems of radioecological monitoring in these conditions are:

- The control of a radiation situation during normal functioning of nuclear power plant;
- The control of levels of possible radioactive contamination during accidental and postaccidental situation on nuclear power plant;
- The control of accumulation of radionuclides in critical components of forest ecosystems during all period of work of nuclear power plant.

As a whole all these problems are aimed at maintenance of safe living conditions of the local population including forestry workers, on acceptance of fast and correct decisions on evacuation of the local population in the case of hard nuclear accident.

Closely with the choosing of objects of radioecological monitoring and its tasks concerns the main criteria and parameters on which mentioned above monitoring is carried out. On territories, which are not located near the zone of influence of Chernobyl NPP, it is correct to observe less quantity of experimental plots and measured parameters in comparison with much more contaminated territories. Such parameters should be investigated in relative clean territories as:

- Values of an exposure doze of gamma-radiation in air and on a surface of soil;
- Surface beta-contamination of soil;
- Specific activity of the main technogenic radionuclides (including those of reactor origin or atomic bomb tests) in selected critical test-objects of monitoring.

Their quantity should be considerably increased in a zone of influence of Chernobyl NPP, both at the expense of number of measured parameters and at the expense of observed experimental plots. It is expediently to study some components of forest ecosystems with periodicity proposed for the last ones:

- Vertical distribution of the main technogenic radionuclides in soil – once per two years;
- Measurement of specific activity of these radionuclides in parts and organs of tree species (needles, annual shoots, trunk wood, internal and external parts of bark, roots) – once per 3 years;
- Study of specific activity of radionuclides in aboveground and underground parts of undergrowth vegetation, in particular, in grass and dwarf-shrubs; mosses, lichens and fruitbodies of macromycetes - once per 2 years.
- Measurement of an exposure doze of gamma-radiation on soil surface – daily;
- Research of migration of radionuclides in a system "soil – fodder plants – wild game animals" – once per 2 years;
- Calculation of distribution of sum activity of the main radionuclides in forest ecosystems and values of circulation of radioactive elements in last ones – once per 3 years.
-

Thus, the radiation monitoring of forest ecosystem should solve not only mentioned above problems but also:

- Evaluation of distribution of radionuclides in components of forest ecosystems as a kinds of production of forestry;
- Study of temporal (including multiyear) circulation of radioactive elements in forest ecosystems;
- Detection of the factors of an anthropogenic and natural origin influencing on migration of radionuclides.

The radiation monitoring of forest ecosystems as the constituent of general forest monitoring can be divided at *several levels*. We already marked systems of state and regional monitoring levels. It is necessary also to mark necessity of realization of local monitoring. It can be conducted on certain relatively small territory or in separate forestry unit. In each case the purpose, tasks and test-objects of monitoring should be defined individually. Nevertheless in any case the general requirements for the higher monitoring levels should be executed to the local monitoring:

- The methodological and instrumental basis for realization of monitoring observations should be unified;
- Successiveness and continuity of realization of observations;
- The stability of criteria, test-objects and investigated parameters from the very beginning of monitoring (the last can't be chosen but can be added);
- Comparability of obtained data;
- Unification of the main computer databases and programs of accumulation and processing of obtained monitoring results and their compatibility with another special wide spread computer programs;

- Capability of application of GIS-technologies in all stages of monitoring, from data collection to its further processing and preparation of digital or graphic information.

3. Conclusion

The radiation monitoring of forest ecosystems as the constituent of general forest monitoring and state monitoring of an environment as a whole, is at the same time a constituent of a system of information support of forestry as a branch of industry. For achievement of goals of radioecological monitoring it is necessary to involve various structural subdivisions both of State Committee of Forestry of Ukraine, the Ministry of Ecology and Natural Resources and also State Committee of Nuclear Safety of Ukraine. Probably for its implementation it would be expedient to use a network of points of observation of Committee of Hydrometeorology of the country. In forestry as a branch of general industry partially specially prepared experts of local forestry units and two central forest-accounting expeditions can solve problems of radiation monitoring. From our point of view by development of the programs and methodology of observations, collecting and processing of obtained analytical data, creation of analytical materials and prognosis of development of a situation in forests should conduct special subdivisions of Ukrainian Scientific Research Institute of Forestry and Forest Amelioration (UkrSRIFA) or special State Centre of Forest Monitoring.

It is necessary to mark that for acceptance of quick decisions on management of forestry in conditions of radioactive contamination the application of GIS-technology is demanded on the basis of radiation monitoring of forest ecosystems. GIS-technology in application to purpose and main tasks of monitoring provides:

- Processing spatial and temporal monitoring data;
- The collection, updating and processing of analytical information;
- Connection digital information with a graphic data presentation;
- Modelling, prediction and evaluation of obtained results and support of administrative decisions on state level and demanded lower levels inside industry of forestry.

4. References

1. V. Derevets, Yu. Ivanov, V. Marchenko. Radiation-ecological monitoring of exclusion zone. in: Bulletin of ecological state of exclusion zone, 2, 1996. pp. 5-12 (in Ukrainian).
2. V. Krasnov. Radioecology of forests of Polissya of Ukraine. Zhitomir, Volyn publish., 1998. 112 pp. (in Ukrainian).
3. V. Zhuravel et al. Conception of foundation of agrarian-ecological informational subsystem for accompaniment of works on liquidation of consequences of Chernobyl NPP in Russia. in: Reports of 3-rd Conf on Chernobyl problem, vol. 1, part 1, Pripyat, 1992. pp. 78-87 (in Russian).

SIMILARITIES AND DIFFERENCES IN BEHAVIOR OF ^{137}Cs, ^{40}K AND ^{7}Be IN NATURAL ECOSYSTEMS

C. PAPASTEFANOU, M. MANOLOPOULOU, S. STOULOS, A. IOANNIDOU and E. GERASOPOULOS
Aristotle University of Thessaloniki, Atomic and Nuclear Physics Laboratory, Thessaloniki 54124,
GREECE

1. Abstract

Fallout-derived ^{137}Cs and naturally occurring radionuclides, such as ^{40}K and ^{40}Be in soils and grass were measured in Thessaloniki area, Northern Greece, at temperate zone (40°), in order to examine their behaviour in natural ecosystems. For long-term measurements over a 15-y period, ^{137}Cs concentrations ranged from 3.73 to 1307 Bq kg^{-1} (avg. 210.5 Bq kg^{-1}) in soils and from 0.4 to 334.9 Bq kg^{-1} (avg. 14.5 Bq kg^{-1}) for grass. Potassium-40 concentrations ranged from 141.4 to 580.2 Bq $^{kg-1}$ (avg. 224.4 Bq kg^{-1}) in soils and from 66.3 to 1480 Bq kg^{-1} (avg. 399.8 Bq kg^{-1}) for grass. Beryllium-7 concentrations ranged from 0.53 to 39.6 Bq kg^{-1} (avg. 14.4 Bq kg^{-1}) in the soils and from 2.1 to 348.0 Bq kg^{-1} (avg. 54.4 Bq kg^{-1}) for grass. Cesium-137 transfer coefficients, TF from soil to plants (grass) ranged from 0.002 to 7.42 (avg. 0.20). Potassium-40 TF values from soil to plants (grass) ranged from 0.16 to 2.42 (avg. 0.73). Beryllium-7 TF values from soil to plants (grass) ranged from 0.027 to 2.37 (avg. 0.42) 137 An ecological half-life, T_{ec} for ^{137}Cs in grassland 3 1/3 y could be derived, when TF values varied between 0.1 and 0.01 in an 11-y elapsed time since contamination (May 1986). An almost similar T_{ec} value (3 2/3 y) was derived for ^{40}K in grassland.

2. Introduction

It is known that radionuclides that occur naturally in soil are incorporated metabolically into plants. Man-made produced radionuclides introduced into soil behave in a similar manner. In addition to root uptake, direct deposition may occur on foliar surfaces, and when this happens the plants may absorb the radionuclides metabolically.

Uptake of the long-lived radionuclides by plants depends to a considerable degree on whether they remain within reach of the roots of plants and the extent to which they are chemically available. The rates at which the various radionuclides will migrate through different soils under varying conditions of pH and moisture are not well known.

Of the naturally occurring primordial radionuclides ^{40}K ($T_{1/2}$=1.28x10 9 y) is very abundant in soil as the stable K = 2.59% of earth's crust [1]. Its concentration in soils ranged between 140 and 850 Bq kg^{-1} (avg. 400 Bq kg^{-1}) [2]. Of the cosmogenic origin ^{7}Be ($T_{1/2}$ = 53.3 d), a relatively short-lived radionuclide is always present in the atmosphere. Its environmental concentration in the temperate zone is about 3 mBq.m^{-3} in surface air and 700 Bq.m^{-3} in rainwater [3]. It is deposited onto the ground through dry and wet deposition processes and is easily found in soil and plants [4]. The Chernobyl origin ^{137}Cs (T   =30.17 y) is also

F. Brechignac and G. Desmet (eds.), Equidosimetry, 313–319.
© 2005 *Springer. Printed in the Netherlands.*

found at elevated concentrations in soil and plants through the wet deposition process following heavy rainfalls occurred on May 1986 (23.9 kBq m^{-2}) [5].

Cesium, an alkaline metal, is a congener of Potassium (group I), while Beryllium belongs to group II$_a$ of the Periodic Table of the Elements. Because of its exchange capability it can substitute for Potassium where there is a lack or deficiency of the latter. So, the presence of ^{137}Cs in plants may be caused by its uptake through soil as well as Potassium uptake. The roots of plants may be unable to distinguish between chemical congeners in the 137-Cs uptake process [6]. It was found that the ^{137}Cs concentration is inversely proportional with ^{40}K concentration of Potassium content of soil [7]

This work deals with data of ^{137}Cs, ^{40}K and ^{7}Be in soil and plants (grass) which play an important role in the soil-grass-cow-milk-human pathway.

3. Experimental methods

During the period of about 15 years followed the Chernobyl accident, soil and grass samples were collected in the University Campus, Aristotle University of Thessaloniki, Greece (40° 38'N, 22° 58'E) with a very dry (precipitation free, mean annual precipitation height 424 mm for the period 1986-1997) climate. Surface soil samples were collected in the middle of each month using a rigid frame sampler. The samples were collected from a 30x30 = 900 cm^2 area and were taken to a depth of 0 to 5 cm. At that area, there was not any human activity (undisturbed area), that means, the selected area was neither cultivated and ploughed nor fertilized and neither irrigated nor mowed. The total area of this study was about 20x20 = 400 m^2, appropriately fenced. Grass (Gramineae or Poaceae, the species) samples were collected from an area of 3 m^2 (0.125 kg of grass per m^2) at the same time and place as soil sampling was carried out [5]. Cesium-137, ^{40}K and ^{7}Be activity measurements in the soil and grass samples were performed using high resolution (1.9 keV at 1.33 Mev of ^{60}Co) and high efficiency (42%) gamma spectrometer consisting of a high purity Germanium detector. The energy of gamma-rays were as follows: E$_V$ = 1461 keV (11%) for ^{40}K and E$_V$ =477 keV (10.3%) for ^{7}Be. Overall efficiency was known to an accuracy of about 12% for the geometry used (Marinelli beaker).

4. Radioactivity of soil and grass

4.1. SOIL

Cesium-137, ^{40}K and ^{7}Be concentrations of soil were carried out for over fifteen years (August 1989 – March 2002). Cesium-137 concentrations in soil varied between 3.73 and 1307 Bq kg^{-1} (avg. 210.5 Bq kg^{-1}) (Table 1). High values of Cs in soil were observed in the winter and autumn, each year. The increase was due to rainfalls and drainage as well as to the fall of autumn leaves at those periods. However, unusually high values of ^{137}Cs concentrations were measured in soil samples, particularly on October 17, 1988 (1307 Bq kg^{-1}), May 8, 1992 (809 Bq kg^{-1}), March 20, 1995 (582 Bq kg^{-1}) and October 11, 1996 (573 Bq kg^{-1}), while the ratio $^{137Cs}/^{134}$Cs at these dates was as expected for decay reason. That increase of ^{137}Cs might be attributed to soil resuspension from the surrounding area and not only by wind scavenging [8-10]. Potassium-40 concentrations in soil varied between 141.4 and 580.2 Bq kg^{-1} (avg. 224.4 Bq kg^{-1}) (Table 1). Beryllium-7 concentrations in soil varied between 0.53 and 39.6 Bq kg^{-1} (avg. 14.4 Bq kg^{-1}) (Table 1).

Table 1. Average activity concentrations of ^{40}K, ^{7}Be and ^{137}Cs in soil and grass and soil-to-plant transfer factors, TF

Radio-nuclide	Radionuclide concentration in soil[a] (Bq kg^{-1})	Radionuclide concentration in grass[a] (Bq kg^{-1})	Soil-to-plant transfer factors TP[a]
^{40}K	224.4 (141.4-580.2)	399.8 (66.3-1480.0)	0.73 (0.16-2.42)
^{7}Be	14.4 (0.53-39.6)	54.4 (2.1-348.0)	0.42 (0.027-2.37)
^{137}Cs	210.5 (3.73-1307.0)	14.5 (0.4-334.9)	0.20 (0.002-7.42)

[a]Range is given within brackets.

4.2 GRASS

Cesium-137, K and Be concentrations of grass were measured for over fifteen years (August 1986 - March 2002). Cesium-137 concentrations in grass varied between 0.4 and 334.9 Bq kg^{-1} (avg. 14.5 Bq kg^{-1}) (Table 1). There was a trend of decreasing Cs with time reflecting an effective half-life, Tgff of 40 months (3 1/3 years) [11]. Cesium-137 peaks were observed during the spring and summer, each year. This was due to the uptake by root system from the autumn leaf-litter. As fallout Cesium is incorporated in the biological cycle (soil-plant-litter-microbial utilization-soil [12], it will be environmentally persistent for a long time after the accident. Potassium-40 concentrations in grass varied from 66.3 and 1480 Bq kg^{-1} (avg. 399.8 Bq kg^{-1}) (Table1). Beryllium-7 concentrations in grass varied between 2.1 and 348.0 Bq kg^{-1} (avg. 54.4 Bq kg^{-1}) (Table 1).

5. Transfer of radionuclides from soil to plants

The concentrations of radionuclides in soils and plants can be used for the determination of soil-to-plant transfer rates (transfer coefficients between soil and plants or transfer factors) TF, i.e. the ratio

$$TF = \frac{[R_i, \ Bq \ kg^{-1} \ d.w.]_{plant}}{[R_i, \ Bq \ kg^{-1} \ d.w.]_{soil}} \qquad (1)$$

where R_i is the uptake of ith radionuclide by plants from the soil through the soil-root-trunk pathway.

In the literature this ratio is also known as relative concentration factor [6], or plant-soil concentration ratio, CR [13]. For the grassland in the region of Thessaloniki, Greece, temperate zone (40° N), the soil-to-plant transfer rate of each radionuclide R_i for ^{137}Cs, ^{40}K and ^{7}Be was estimated every month for over fifteen years (August 1986 - March 2002).

5.1. CESIUM-137

Soil-to-plant transfer rates of ^{137}Cs for grassland vs. time elapsed since contamination (May 1986) are presented in Fig.1 for the period of August 1986 through July 1997. The TF values for ^{137}Cs at that period ranged from 0.002 (December 1996) to 7.42 (February 1987) (avg. 0.20). The data of Fig.1 shows considerable scatter indicating that there is no consistent correlation between the TF values of ^{137}Cs and the time elapsed. However, an ecological half-life, T_{ec} of 3 1/3 years could be derived when the TF values varied from 0.1 to 0.01 (66.4% of the measurements). Same result was derived for the effective half-life, T_{eff} of ^{137}Cs for grass [11]. Rosen et al (1995) [14] reported that the ecological half-life of ^{137}Cs in herbage in Northern Sweden (65° N) ranged from 3 to 21 years with a mean of 7 years (latitudinal effect). Peles et al (2002) [15] reported an ecological half-life of ^{137}Cs in terrestrial plant species (*Alnus serrulata, Myrica cerifera, Salix nigra*) ranging from 4.85 to 8.35 years (avg 5.87 years), while in aquatic plant species (*Potygonwn punctatwn, Sagittaria latifolia*) averaging 6.30 years.

In a previous work, Papastefanou et al (1988) [7] reported soil-to-plant transfer rates of ^{137}Cs ranging between 0.02 and 0.2 for grass (avg. 0.07) and from 0.009 to 0.018 (avg. 0.012) for the tree deciduous leaves (wild pear trees and bush trees as brambles and briers, the species), in fairly good agreement with that determined before the Chernobyl accident in temperate zone (48°N-52°N) in Germany for grassland and ranged from 0.01 to 0.1 [16].

Eisenbud (1973) [6] reported that the relative tendency of Cesium to be concentrated from soil by crop plants ranged from 0.01 to 1, while Kathren (1984) [13] mentioned as a typical plant-to soil concentration ratio, CR for Cesium 0.01 (Source: USNRC 1977 [17]). Ban-Nai et al (1999) [18] reported an average transfer factor, TF of [137Cs] 0.02 in three root vegetables (radish, carrot and turnip).

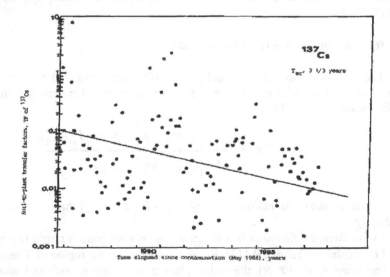

Fig. 1. Soil-to-plant transfer factors, TF, of ^{137}Cs vs. time elapsed since contamination (May 1986) for the period May 1986-July 1997.

5.2. POTASSIUM-40

The soil-to-plant transfer rates of ^{40}K for grassland varied between 0.16 and 2.42 (avg. 0.73) (Fig.2). For 85.6% of the measurements the TF values for 40K ranged between 0.1 and 1.0. There was a trend of decreasing of the soil-to-plant transfer rate of ^{40}K with time, possibly due to exchange of Potassium by Cesium, reflecting an ecological half-life, T_{ec} of 3 2/3 years of ^{40}K for grassland that could be derived for the TF values of Fig.2, pretty close to the T_{ec} obtained for ^{137}Cs for grassland. As the sampling area was never fertilized, that means, Potassium fertilizers were not used, the data also indicate a depletion of ^{40}K in the soil which took place by the contribution of grass during the elapsed time.

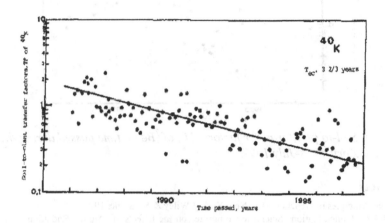

Fig. 2 Soil-to-plant transfer factors, TF, of ^{40}K vs. time passed during the period December 1986 - July 1997.

5.3. BERYLLIUM-7

The soil-to-plant transfer rates of Be for grassland varied between 0.027 and 2.37 (avg. 0.42) (Fig.3). For 70.7% of the measurements the TF values for Be ranged between 0.1 and 1.0. The data of Fig.3 shows considerable scatter indicating that there is no consistent correlation between the TF values of ^{7}Be and the time elapsed. This is a result of the continuous deposition of ^{7}Be coming from the stratosphere onto the plants and soil ground and not of the uptake by the root system of plants, because of its relatively short half-life.

6. Conclusions

The existence of naturally occurring radionuclides as well as the long-lived fallout-derived radionuclides in soil and plants might be used in determining the transfer rates of elements from soil to plants. In this study, the naturally occurring primordial radionuclides Potassium-40 rich in soil and plants and Beryllium-7 of cosmogenic origin, relatively short-lived radionuclide, were considered in association with Cesium-137 of Chernobyl origin. Soil-to-plant transfer rates, TF for grassland ranged from 0.002 to 7.42 (avg. 0.20) for ^{137}Cs, between 0.16 and 2.42

318

(avg. 0.73) for ^{40}K and from 0.027 to 2.37 (avg. 0.42) for ^{7}Be. It was found that the ecological grassland 3 1/3 y was almost similar to that derived for ^{40}K in grassland 3 2/3

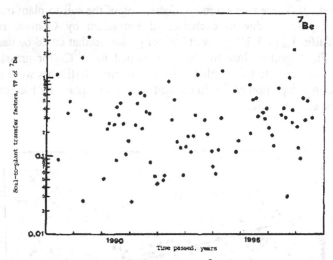

Fig. 3. Soil-to-plant transfer factors, TF, of ^{7}Be vs. time passed during the period April 1988-July 1997.

7. References

1. B.Mason. Principles of Geochemistry. New York, J.Wiley & Sons, Inc. 1996.
2. UNSCEAR - United Nations Scientific Committee on the Effects of Atomic Radiation. Sources and Effects of Ionizing Radiation. United Nations, New York. 2000.
3. UNSCEAR - United Nations Scientific Committee on the Effects of Atomic Radiation. Ionizing Radiation: Sources and Biological Effects. United Nations, New York. 1982.
4. C.Papastefanou and A.Ioannidou. Depositional fluxes and other physical characteristics of atmospheric Beryllium-7 in the temperate zones (40°N) with a dry (precipitation-free) climate. Atmosph.Environment V25A, 1991. p. 2335-2343.
5. C.Papastefanou, M.Manolopoulou and S.Charalambous. Radiation measurements and radioecological aspects of fallout from the Chernobyl accident. J.Environ. Radioactivity. V 7, 1998. pp 49-64.
6. M.Eisenbud. Environmental Radioactivity. New York, Academic Press, 2nd ed. 1973.
7. C.Papastefanou, M.Manolopoulou and S.Charalambous. Cesium-137 in soils from Chernobyl fallout. Health Phys. V55, 1988. pp 985-987.
8. J. A.Garland and K.Playford. Deposition and resuspension of radiocesium after Chernobyl. Proceedings of Seminar on Comparative Assessment of the Environmental Impact of Radionuclides Released during Three Major Nuclear Accidents: Kysthym, Windscale, Chernobyl. Luxemburg, 1-5 October 1990. VI, 1990. Report EUR 13574. p. 237-253.
9. P.J.Coughtrey, J.A.Kirton and N.G.Mitchell. Environmental distribution and transport of radionuclides in West Cumbria following the Windscale and Chernobyl Accidents. Proceedings of Seminar on a Comparative Assessment of the Environmental Impact of Radionuclides Released during Three Major Nuclear Accidents: Kysthym, Windscale, Chernobyl. Luxembourg, 1-5 October 1990. VI, 1990. Report EUR 13574. p. 473-484.
10. C.Papastefanou and M.Manolopoulou. The radioactivity of coloured rain in Thessaloniki, Greece. Sci.Total Environ. V80, 1989. p. 225-227.

11. C.Papastefanou, M.Manolopoulou, S.Stoulos and A. Ioannidou. Behavior of ^{137}Cs in the environment one decade after Chernobyl. J. Radioecology. V4, 1996. p. 9-14.
12. J.C.Ritsie, E.E.C.Clebsch and W.K. Rudolf. Distribution of fallout and natural radionuclides in litter, humus and surface mineral soil layer under natural vegetation in the Great Smoky Mountains, North Carolina Tennessee. Health Phys. V18, 1970. p. 479-489.
13. R.Kathren. Radioactivity in the environment: Sources, distribution and surveillance. Chur, Harwood Academic Publishers. 1984.
14. K.Rosen, I.Anderson and H.Lonsjo. Transfer of radiocesium from soil to vegetation and to grazing lambs in a mountain area in Northern Sweden. J.Environ. Radioactivity. V26, 1995. p. 237-257.
15. J.D.Peles, M.H.Smith and I.L.Brisbin Jr. Ecological half-life of ^{137}Cs in plants associated with a contaminated area. J.Environ. Radioactivity. V59, 2002. p. 169-178.
16. W.Kuhn, J.Handl and P.Schuller. The influence of soil parameters on ^{137}Cs uptake by plants from long-term fallout on forest clearings and grassland. Health Phys. V.44, 1984. p. 1083-1093.
17. USNRC - U.S.Nuclear Regulatory Commission. Calculation of annual doses to man from routine releases of reactor effluents for the purpose of evaluating compliance with 10 CFR Part 50, Appendix 1. Regulatory Guide 1.109. Washington, D.C. 1977.
18. T.Ban-Nai, Y.Muramatsu and k.Yanagisawa. Transfer of some selected radionuclides (Cs, Sr, Mn, Co, Zn and Ce) from soil to root vegetables. J.Radioanal.Nucl.Chem. V241, 1999. p. 529-531.

HEAVY METALS ACCUMULATION BY AGRICULTURAL CROPS GROWN ON VARIOUS TYPES OF SOIL IN POLESSJE ZONE OF UKRAINE

L. PEREPELYATNIKOVA, T. IVANOVA
Ukrainian Institute of Agricultural Radiology, Mashinostroitelej str., 7, Chabany, Kiev region, UKRAINE

1. Introduction

Environmental contamination with technogenic pollutants of different chemical origin became the actual problem of the recent century. Numerous studies carried out in various countries characterised by different soil-and-climatic conditions allowed to estimate the principal properties of pollutants, their danger for nature and human being. Estimation of pollutants toxicity promoted to elaborate approaches to standardisation of toxic substances content in biocenotic components for safe human living. But regional peculiarities of soil cover and climate, technogenic loading to ecosystems require additional studies of toxicants migration ways in food chains. Ukraine is characterised by multiplicity of landscapes and soils, created on various rock-beddings. Due to several reasons the natural content of elements in soils and other components of ecosystems varies in a wide range.

2. Technogenic environmental pollutants in Ukraine

The territory of Ukraine can be considered as unequally safe by environmental contamination. A significant part of industrial sources of heavy metals ejection into environment is located in the eastern and southern parts of the country (Donbass, Dniepropetrovsk, Zaporozhje and Nikolajev regions). In other areas the contribution of sources like transport ways and agro-amelioration of agricultural lands is more substantial. A significant part of soils is characterised by a low content of mobile species of microelements including those ones which can be considered as technogenic pollutants when their content manifold exceeds permissive levels. For instance, an extremely low content of a mobile species of zinc in soil was noted [1], soils with utterly low (no more then 0,05 mg/kg) and low (less then 0,7 mg/kg) content prevail; their shares in soil cover are 10 and 89%, respectively. Soils with middle (0,8-1,5 mg/kg) and high (>1,5 mg/kg) content of zinc occupies about 1% of arable lands area, presumably in mountainous regions of Karpaty and Zakarpatje. Deficiency of mobile copper in drained peat and peat-swamps, light soddy-podzolic, soddy soils and chernozem is estimated [1]. Prevalent part of soils needed in copper fertilisers is located in Polessje and forest-steppe zone. Soils with low copper content (less 1,5 mg/kg) are spread in Polessje zone of Zhitomir, Volyn, L'vov and Rovno regions. Particular features of the regional ecological state consist in a combination of geochemical properties per province, and soil and biocenotic components contamination with radioactive ^{137}Cs and [90]Sr. Lack of some elements, including microelements, in conditions of increased radionuclides concentrations in soils is accompanied by disruption of physiological processes in plants, by reduction of crops yield and its quality. This leads to detriment of quality of foods and damage of animal physiological state. It is important to underline that application of large-scale agrochemical countermeasures

321

F. Brechignac and G. Desmet (eds.), Equidosimetry. 321–325.
© 2005 *Springer. Printed in the Netherlands.*

aiming at the reduction of radioactive contamination during more then 15 years was accompanied by temporary changes of agrochemical soil properties and radionuclide behaviour in soil.

3. Heavy Metal (HM) behaviour in agrocenoses

Survey of heavy metals (HM) content in components of agrocenoses was carried out by UIAR specialists in 1995-2001in the Polessje zone that suffered from Chernobyl nuclear accident. Special attention was paid to copper, zinc, lead, and cadmium. Analysis of data mass is being used for following prediction and elaboration of approaches to standardisation of pollutants in components of food chains. More detailed information on heavy metals migration in the environment contaminated with radioactive elements should allow to elaborate the strategy of regulation of elements accumulation in plant production including countermeasures application.

HM migration in a soil-plant system depends on some soil characteristics as physical-and-chemical properties, organic matter content, granulometric and mineralogical composition, microbiological activity of soils, duration of pollutants existence in soil. It is assumed that multi-element technogenic contamination leads to HM interaction in a soil, their competition for adsorbing sites in soil components, inclusion into metabolism of plants, combined action on plants.

HM behaviour in typical Ukrainian Polessje soils contaminated by Chernobyl radioactive substances was studied in natural landscapes and in field experiments. Mineral (soddy-podzolic of various granulometric composition, grey, chernozems) and organic (variations of peat) soils are wide spread in the region. Low provision of soils with some elements was justified by HM clarke content in the upper 0-10 cm layer of natural landscapes. Copper content varies in a range 1,3-3,2 mg/kg, zinc one in a range 1,0-7,0 mg/kg. Lead content didn't reach limited permissible levels, but cadmium content exceeded respective norms 2 times and more (5,0-7,7 mg/kg). Differences of elements migration through a profile of various soils were found. The biggest part of copper, cadmium, lead, zinc was found in 10-20 cm layer of soddy-podzolic soil. Peat soils are characterised by heighten copper, lead and zinc concentration in 0-5 cm layer, cadmium – in 5-10 cm layer. The biggest concentration of elements in grey forest sandy loam soil was observed in 0-5 cm layer for lead and zinc, in 10-20 cm layer for cadmium.

Elements brought in a soil of experimental plots as water-soluble compounds are fixed fast in a soil-absorbing complex. Copper and zinc are bound firmly, and turn into immobile species in half a year. About 90% of lead mobile species and 30-80% of other metals transform into exchangeable ones. Therefore HM accumulation in the oat yield was not directly proportional to their content in a soil. Leguminous crops were more sensitive to excessive concentrations of HM than cereals.

HM uptake into agricultural crops depends on a soil type. On the whole it is the biggest on soddy-podzolic and peat-bog soils. Accumulation in the yield depends on species peculiarities of crops and physical-and-chemical properties of elements. For example, factors of accumulation (transfer factor –TF) of lead and zinc in oats plants on soddy-podzolic soil are significantly higher then on other soils (8-25 times). Copper and cadmium income more intensive from peat-bog soils (Table). The biggest accumulation of copper and cadmium in potato tubers was observed on peat-bog soil (1,4-7 times), lead – on grey forest (1,2-4 times), zinc –

on soddy-podzolic soil (1,2-3,0 times).

Table 1. Accumulation of heavy metals in the yield of agricultural crops in dependence on the type of soil

Type of soil	TF			
	Cu	Cd	Pb	Zn
oats				
Soddy-podzolic	1,50	0,08	0,72	4,2
Peat-bog	1,25	0,35	0,20	1,8
Chernozem typical low humic	0,17	0,10	0,008	1,34
Dark grey forest middle loam	0,17	0,05	0,028	0,50
potatoes				
Soddy-podzolic	0,20	0,08	0,006	0,34
Peat-bog	0,29	0,35	0,025	0,12
Chernozem typical low humic	0,008	0,10	0,21	0,29
Dark grey forest middle loam	0,14	0,05	0,016	0,015
cabbage				
Soddy-podzolic	0,06	0,32	0,021	0,30
Peat-bog	0,035	0,68	0,007	0,025
Chernozem typical low humic	0,04	0,08	0,015	0,15
Dark grey forest middle loam	0,03	0,07	0,01	0,04
beetroot				
Soddy-podzolic	0,33	0,32	0,006	1,02
Peat-bog	0,39	0,91	0,07	1,39
Chernozem typical low humic	0,42	0,06	0,007	0,10
Dark grey forest middle loam	0,07	0,13	0,007	0,06

The biggest accumulation of Cu, Pb and Zn by cabbage was obtained on soddy-podzolic soil (1,5-14 times more) in comparison to other soils under study, but Cd accumulation by cabbage was more intensive on peat-bog soil (2-5 times). Beetroot took up Cd, Pb and Zn more strongly from peat-bog soil (1,2-23 times), Cu – from grey forest soil (1,1-6 times).

Agricultural crops under study vary significantly for their accumulation of different elements. Comparison has shown the lowest one for cabbage and the biggest one for oats. Zinc is accumulated to the biggest extent; lead – to the lowest extent (2-3 orders of magnitude). By their capability of separate HM absorbance, crops can be ranged in the following order:

Zn: oats>beetroot> potatoes>cabbage;

Cd: beetroot > cabbage > potatoes > oat;

Cu: oat> beetroot > potatoes > cabbage;

Pb: oat> beetroot > potatoes > cabbage.

Study of caesium radionuclides transfer in the same sites under study shows that ^{137}Cs TF into meadow plants on peat-bog soil was 9,2, on soddy-podzolic soil – 3,1, on grey forest – 1.2. Factors of heavy metals accumulation in roots 2-11 times higher then in an overground part of plants. Factors of HM accumulation in roots on various types of soil vary from 1,1 to 3,6. Particularly, the parameter for cadmium wavers in a range of 0,2-1,0, for lead between 0,1-1,2, for zinc between 1,7-23,9. The special feature of lead consists in its maximum stocking up in a root system of meadow grasses at the depth 5-10 cm.

Ratio of heavy metals accumulation in different parts of plants depends on type of element. Values of ratio between metal content in roots and overground part vary for copper - 1,8-2,7, for cadmium – 2,4-11,2, for lead - 10,7-40,8, for zinc - 0,8-2,2. These data show the role of roots as geobiological barrier. Zinc and copper are retained less in comparison to other metals by roots of meadow plants. High migration capacity of zinc can be explained by its chemical properties allowing its entering into reactions of exchange, forming of soluble compounds and less immobilising by soil minerals and organic matter. From the other side zinc is the physiologically obligate microelement for green plants because of its participation in photosynthesis process and activation of some enzymes. But excessive amount of HM leads to disruption of nutritional elements balance, i.e. increase of nitrogen content, but decrease phosphorous, calcium, magnesium, iron ones. Physiological active substances content decreases also.

Application of agro-ameliorative measures correcting radioactive substances accumulation in plant production influences the heavy metals uptake in plants also. Liming of soil affected significantly HM accumulation in leguminous and cereal crops. Accumulation factor of elements by clover plants was decreased by liming, i.e. for Cu – 1,1 times, Cd – 2,3, Pb and Zn - 1,7-1,9 times. Liming of soil affected unequally the factor of accumulation by cereal grasses: for Zn and Cu it is slightly decreased, for Cd and Pb it increased a little. TF of HM to crops under study from soddy-podzolic soil changed in a different way. Cd and Zn content increased markedly after lime application in a dose one dose calculated per hydrolytic acidity.. Cd, Zn and Cu TF for cereal grasses on soddy-podzolic light loamy soil increased for Cu – 1,5 times, Cd and Zn 1,3 times. The maximum liming effect was found on peat-bog soil probably due to acidic soil solution (pH 5,2) in comparison to soddy-podzolic light loamy soil (pH 6,2).

Artificial application of HM or increased natural content of HM in radioactive contaminated soil affects ^{137}Cs TF values. So ^{137}Cs TF to clover and potatoes tubers from chernozem and to cereal grasses from peat-bog soil decreased when HM have been added. The fact can be explained by compensation of natural lack of microelements and increase of yield after HM input (Fig 1).

Multiplicity of ^{137}Cs TF increase on soddy-podzolic light loamy soil for potatoes tubers is the biggest and equal to 2,7 times; for cereal grasses and clover - 1,2 times. Radiocaesium TF increases 1,8 times for clover on peat-bog soil and for cereal grasses on chernozem.

Essential countermeasures used for reduction of ^{137}Cs accumulation by plants have not given any effect on HM accumulation. Application of mineral fertilisers in a dose $N_{60}P_{60}K_{60}$ and liming in a dose one dose calculated per hydrolytic acidity on peat-bog soil contaminated with HM increased ^{137}Cs TF to both kinds of grasses 2,1-2,6 times, while in control their efficiency reaches 1,2-3,0 times. Radiocaesium accumulation on soddy-podzolic soil contained HM increased 1,6 - 3,0 times, while in absence of HM it decreased 1,1-2,0 times.

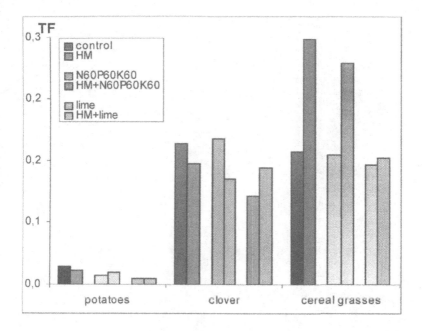

Fig. 1. Change of ^{137}Cs TF to various crops on chernozem after HM adding and countermeasures application.

So increased HM concentration in soil leads to ^{137}Cs TF increase. Moreover, HM weakens countermeasures effect for radiocaesium accumulation in plants. Probably, it can be connected with increase of radionuclide mobility due to competition of elements for sorbing sites. The largest affect of heavy metals on radiocaesium behaviour was found on the peat-bog soil, the lowest one – on chernozem due to agrochemical properties of soil. Grasses were more sensitive to interaction of technogenic pollutants in comparison to potatoes tubers.

4. References

1.   Grigorjev V.L., Plishko A.A. Microelements in soils of Ukraine. Khimizaciya selskogo hozyaistva (in Russian), 1989, 3, c. 7-8

THE *BRILLIANT BLUE* METHOD FOR WATER SOLUBLE CHERNOBYL[137]Cs BEHAVIOUR ESTIMATION IN SOILS OF SOUTH BELARUS

N. GONCHAROVA, K. KALINKEVICH, V. PUTYRSKAYA.
International Sakharov Environmental University, Dolgobrodskaya str.23, Minsk, 220009,BELARUS

A. ALBRECHT
Swiss Federal Institute of Technology, Institute of Plant Sciences, Postfach 185,Eschikon 33,CH-8315 Lindau, SWITZERLAND

1. Introduction

The radionuclides distribution in forest and agricultural soils depends on soil structure, hydromorphity, and hydrological features, in combination with root distribution, caesium mobility and characteristics. The interaction between ^{137}Cs, soil matrix and root distribution is important in radionuclide transfer processes assessment. The study of radionuclides distribution is necessary for estimating plant root uptake and for evaluating dose rate. Migration of ^{137}Cs and its forms is delineated by flow patterns in the soil. A field experiment was carried out on tilled agricultural and forest soils in the South of Belarus affected by the Chernobyl accident. The distinction in accumulation of ^{137}Cs in preferential flow and matrix zones was determined. *Brilliant Blue FCF* was used to dye flow lines and to determine the activity of Chernobyl [137]Cs (deposited in April 1986) as a function of dye presence and absence. The experiment was carried out on the soddy-podzolic soils of forest and agrocoenosis of "Sudkovo" farm, Khoiniki district, Gomel region.

2. Water flow in soils (saturated and unsaturated)

Fluxes of water in soils change their magnitudes and directions depending on spatial variability of the hydraulic properties of soils. When hydraulic properties vary continuously with position in soils, the directions of streamlines may by curved. When hydraulic properties vary discontinuously with position in soils, the streamlines may be refracted at the boundaries between regions of different hydraulic properties. These curvings and refractions of streamlines occur not only in saturated soils but also in unsaturated soils.

Thus refraction of fluxes influences several kinds of soil-hydrological processes: lateral flow of water during vertical percolation in layered soils, surface and subsurface water flow in slopes, and anisotropy of flow in saturated or unsaturated soils [1].

3. Classification of the preferential flow

Classification of preferential flows based on their phenomenological features as proposed by Kung [2-3]. Taking account of his classification, preferential flows in soils are classified into three types:
- By-passing flow

F. Brechignac and G. Desmet (eds.), Equidosimetry, 327–332.

- Fingering flow
- Funnelled flow

By-passing flow is local flow in such highly permeable zones as a macropore (right-hand side) and a crack (left-hand side) in heterogeneous soil. This flow takes place either when the macropores and cracks are open to the atmosphere or when the water pressure within the macropores and cracks are positive. When, for example, pounded water infiltrates a soil matrix whose macropores are open to the land surface, water will infiltrate the macropores faster than the soil matrix, water can infiltrates into the soil matrix across the boundary between macro- and micropores. Under certain conditions, this type of flow will occasionally be transformed into pipe flows. On the other hand, when pounded water infiltrates a soil matrix within which micropores or cracks are buried, and when the soil matrix is under suction, water will not infiltrate the buried micropores but will infiltrate only the soil matrix. In this case, micropores are not preferential for the flow of water but are obstacles to such flow.

Fingering flow takes place mainly in relatively coarse layers overlain with finer soils. Fingering flow occurs under both pounded and unsaturated condition at the land surface.

Funnelled flow is another type of partial flow accumulated along the inclined bottom of a fine layer overlaying a coarse sublayer. The pressure of funnelled flow is initially negative, and due to the accumulation of infiltrated water, increases with distance along the inclined interface between the fine top layer and the coarse sublayer. The quantity of funnelled flow also increases with distance along the inclined interface. When the pressure of funnelled flow exceeds a critical value, fingering flow will be generated in the coarse sublayer.

4. General features of Brilliant Blue FCF dye

Dyes are widely used to stain the travel path of water and solutes in soils. *Brilliant Blue FCF* – bright greenish blue food dye – effective indicator for visualizing the flow paths of water in soil. *Brilliant Blue FCF* – triphenilmethane with formula $C_{37}H_{34}N_2Na_2O_9-S_3$. Chemical structure of dye is shown on the fig.1.

Brilliant Blue FCF is mobile, distinctly visible, non-toxic, chemical stable. So it is a good reagent for field So it is a good reagent for field experiments

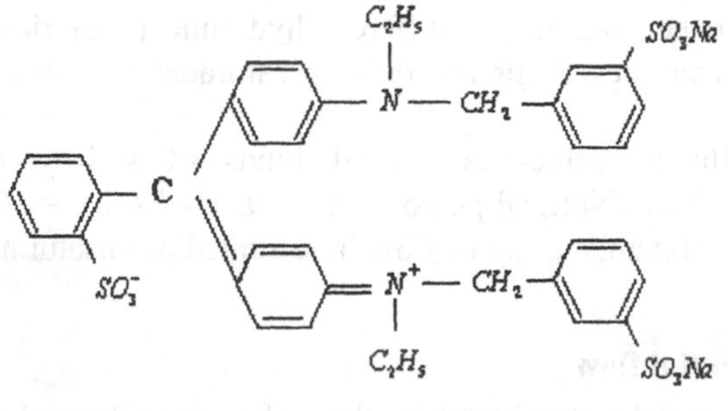

Fig. 1. Chemical structure of Brilliant Blue FCF [4].

5. Brilliant Blue FCF methods (30-km zone, South Belarus)

The experimental plot was situated on the Poles'e lowland, Sudkovo farm, v. Novosiolki, Hoiniki district. It is a monotonous waterlogged flat plain. For our experiment we took 2 plots (100 × 100 cm) with the same characteristics. Soil type is podsoluvisols on sandy loam underlying moraine. The difference between plots was in the tillage. One plot was situated on the repeated tillage field; another plot was in a forest. The soil of the second plot has been untouched since April, 1986. Plots were situated close to each other. The first plot was tilled and harvested two times per year. During our experiment (October, 2001) the field plot was under motley grass.

On forest plot we moved off all litter and cut plants at the 5 cm level. Plants were packed into plastic for further laboratory analysis.

Brilliant Blue FCF concentration – 1 g per 1 l of water. Irrigation was 6 times every 15 minutes with 5 litres of dye.

6. Vertical profile observation and soil sampling

We cut the soil every 20 cm on the 25 cm depth. Every slice was indicated by Latin letters (Fig.2). First slice - $\grave{A}_1$, second - $\hat{A}_1$, next 20 cm– slice $\tilde{N}_1$, then – D_1 and E_1. Thus, we obtained 5 vertical slices: A_1, B_1, C_1, D_1, E_1. Another 5 vertical slices were obtained at a depht of 20 cm: A_2, B_2, C_2, D_2, E_2.

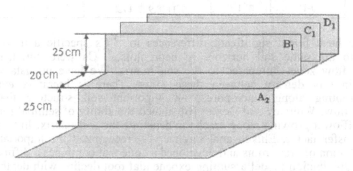

Fig. 2. Vertical slices.

All samples of plants and soils were air-dried in laboratory till constant weight. Then samples were weighted (repeated 3 times) and their radioactivity measured (Bq/kg). The root density has been calculated from those root measurements, which included handpicking. The average activity at each depth is based on soil measurements and estimated distribution of individual samples (Table 1, 2).

7. Discussion

The fast passage of the tracer solution through the repacked monolith and the appearance of colour and radionuclide tracers indicate a type of bypass flow, which is generally termed fingering flow [5].

Table 1. ^{137}Cs specific activity distribution in Sudkovo farm agrocoenosis (October, 2001)

Slice	Depth, cm	^{11}№ spot	Specific activity ^{137}Cs, Bq/kg	
			M	PF
A$_1$	0-10	1	993,1 ± 22,1	1000,8 ± 21,8
		2	927,7 ± 16,7	254,4 ± 12,5
		3	1508,3 ± 19,4	698,8 ± 16,1
	20-25	4	0 ± 9,5	0,4 ± 5,6
B$_1$	20-25	1	2,9 ± 6,4	14,8 ± 3,5
B$_2$	25-40	1	0 ± 6,0	1,9 ± 5,7
C$_1$	0-10	1	898,6 ± 13,3	1340,0 ± 25,1
	10-25	2	981,2 ± 15,3	731,7 ± 14,8
C$_2$	25-40	1	19,8 ± 4,2	0 ± 6,3
D$_1$	0-20	1	920,1 ± 15,2	946,9 ± 13,0
	0-10	2	1068,6 ± 15,1	367,6 ± 8,3
D$_2$	25-40	1	3,7 ± 4,6	0 ± 5,1
E$_1$	0-10	1	1178,9 ± 17,2	n.d.

We have found significant differences in ^{137}Cs specific activity in zones belonging to the same soil horizon. In both plots, ^{137}Cs were enriched in dyed preferential flow zones compared to the soil matrix. We have no data yet on the difference in root density between these zones. There is literature evidence on enhanced rooting along macropores, an important soil structure that triggers preferential flow. Whiteley and Dexter [6] studied the ability of seminal roots of pea, rape and safflower growing along cracks to penetrate the soil matrix. In their analysis of ^{137}Cs transfer factor, Ehlken and Kirchner [7] recognized the importance of the depth distribution of grass roots and developed a code that combines a diffusion model with a root distribution model assuming exponential root decline with depth. With the incorporation of root architecture, they were able to reproduce the change in transfer factors as a function of time to a much better degree than with a model based on meteorological results (soil moisture).

8. Conclusions

During our experiment, we were not able to carry out root and ^{137}Cs tracing to such an extent that both could be fully associated with the different flow compartments within the soil. This is related to the choice of Brilliant Blue as a flow tracer during the irrigation experiment. Brilliant Blue is relatively mobile The distribution of surface applied ^{137}Ñs in the soil profile is delineated by flow patterns in the soil and therefore by soil structure. Tracer experiments in the forest and in the field have shown that ^{137}Cs concentrations between zones of preferential flow (several hundreds Bq/kg) and matrix (in cases below the detection limit) dramatically. In soils without bioturbation, such as acidic forest soils these differences prevail even after 16 years of application

(Chernobyl NPP). In permanent meadows with high biological activity, reduced aggregate stability and fast turnaround of organic matter, the horizontal differences fade away with time, whereas vertical variation prevails. In the ploughed layer of agricultural fields both the horizontal and vertical distinction vanishes. Roots develop their architecture also depending on soil structure, zone of enhanced bioaccumulation therefore develop.

Table 2. ^{137}Cs specific activity distribution in Sudkovo farm forest coenosis (October, 2001)

slice	Depth, cm	¹ spot	Specific activity ^{137}Cs, Bq/kg	
			M	PF
A₁	0-10	1	4,6 ± 6,4	19,9 ± 9,3
		2	4,6 ± 6,4	27,2 ± 6,9
		3	4,6 ± 6,4	9,9 ± 6,0
	20-25	4	2,4 ± 5,5	0 ± 7,2
B₁	0-10	1	209,3 ± 15,3	87,2 ± 11,2
	0-20	2	66,2 ± 10,8	12,2 ± 6,4
		3	66,2 ± 10,8	270,6 ± 10,4
B₂	25-40	1	2,7 ± 4,7	4,8 ± 5,4
		2	6,1 ± 4,3	18,1 ± 5,6
C₁	0-20	1	302,1 ± 17,7	17,1 ± 7,6
		2	301,9 ± 8,5	13,5 ± 6,8
		3	23,9 ± 9,0	14,3 ± 6,7
C₂	25-40	1	6,2 ± 4,2	8,9 ± 5,1
		2	2,8 ± 4,3	4,6 ± 4,6
	>50	2	n. d.	0,7 ± 3,1
D₁	0-10	2	n.d.	54,7 ± 9,1
	0-20	1	35,6 ± 5,6	24,3 ± 5,7
D₂	25-40	1	0 ± 4,3	3,8 ± 3,6
E₁	0-10	2	59,6 ± 5,6	110,3 ± 8,2
	0-20	1	10,9 ± 4,7	63,5 ± 5,4
E₂	25-40		0 ± 5,2	n.d.

9. Acknowledgements

The Swiss National Science Foundation (SNSF) funded SCOPES JRP.

10. References

1. T .Miyazaki Water flow in soils. in: Marcel Dekker,1993 Inc. New York   pp.93- 95.
2. K-J.S. Kung.Preferential flow in a sandy vadose soil: 1. Field   observations. in: Geoderma, 1990, 46, pp. 51–58.
3. K-J.S. Kung,Preferential flow in sandy vadose zone 1. Mechanism and implications, in: Geoderma, 1990, 46, pp. 59-71.
4. M.Flury, H. Fluhler H. Tracer characteristics of Brilliant Blue FCF.in: Soil Science Society of

332

America Journal,1995, 59(1), pp. 22–27.

5. A. Albrecht., U. Schultze, M. Leidgens, H. Fluhler, E. Frossard. Incorporated soil structure and root distribution into plant uptake models for radionuclides: toward a more physically based transfer model. in: J. Environmental Radioactivity,2001(submitted for publication).

6. D. Whiteley, A. Dexter. Behaviour of roots in cracks between soil peds. in: Plant and Soil, 1983.74.pp 153-162.

7. S. Ehlken, G.Kirchner Seasonal variations in soil-to-grass transfer of fallout strontium and cesium and of potassium in North German soils. In Journal of Environmental Radioacctivity, 1996, 33(2), pp.147-181.

AQUATIC ECOSYSTEMS WITHIN THE CHERNOBYL NPP EXCLUSION ZONE: THE LATEST DATA ON RADIONUCLIDE CONTAMINATION AND ABSORBED DOSE FOR HYDROBIONTS

D.I. GUDKOV, M.I. KUZMENKO
Institute of Hydrobiology, NASU, Geroyev Stalingrada Ave. 12, Kiev, 04210, UKRAINE,
digudkov@svitonline.com

V.V. DEREVETS, A.B. NAZAROV
State Specialised Scientific Enterprise "Ecocentre" of the Ministry of Ukraine on the Emergency and Affairs of Population Protection Against the Consequences of the Chernobyl Catastrophe, Kirova Str. 17, Chernobyl, 07270, UKRAINE

1. Radionuclides in components of aquatic ecosystems

The territories of the Chernobyl NPP (ChNPP) exclusion zone are characterised by significant heterogeneity of radionuclide contamination, which is significantly reflected by the radioactive substances contents in aquatic ecosystem components. Primarily this is due to the composition and the dynamics of radionuclide emissions into the environment as a result of accident in 1986, as well as to the subsequent processes of radioactive substances transformation and biogeochemical migration in the soils of a catchment basin and bottom sediments of reservoirs. Relatively low contents of radioactive substances are found in the river ecosystems. Due to high water change rate the river bottom sediments have undergone decontamination processes (especially during floods and periods of high water) and over the years that passed since the accident have ceased to play the essential role as a secondary source of water contamination. The main sources of radionuclides in rivers are currently the washout from the catchment basin, the inflow from more contaminated water bodies, as well as the groundwater. On the other hand, the closed reservoirs, and in particular the lakes in the inner exclusion zone, have considerably higher levels of radioactive contamination caused by limited water change and by relatively high concentration of radionuclides deposited in the bottom sediments. Therefore, for the majority of standing reservoirs the level of radionuclide content is determined mainly by the rates of mobile radionuclide forms exchange between bottom sediment and water, as well as by the external wash-out from the catchment basin.

Our research was carried out during 1998–2001 on Azbuchin Lake, Yanovsky (Pripyatsky) Backwater, the cooling pond of the ChNPP, the lakes of the left-bank flood plain of Pripyat River – Glubokoye Lake and Dalekoye-1 Lake and also on Uzh River and Pripyat River. The sampling station on Uzh River is situated near the river mouth (Cherevach village), on Pripyat River – near the town Chernobyl (Figure 1).

F. Brechignac and G. Desmet (eds.), Equidosimetry, 333–341.

Fig. 1. Map of reservoirs within the Chernobyl NPP exclusion zone.

The radionuclide content in biological tissues was measured for 28 higher aquatic plant species, 6 species of molluscs and 18 species of fish. The results of the radionuclide content measurements in hydrobionts are expressed in Bq·kg^{-1} of wet weight at natural humidity. The tendency of the aquatic organisms to accumulate radionuclides, traditionally expressed as the concentration factor (CF), which is determined by calculating the ratio of the specific activity of radionuclides in tissue to the average annual content (for molluscs and fish) or to the average content in the environment water during the vegetation period (for higher aquatic plants). The estimation of the absorbed dose rate for hydrobionts was carried out according to the method described in literature reference [1].

1.1. WATER AND BOTTOM SEDIMENT

The highest radionuclide activity in water among the studied objects was found in the Azbuchin Lake. During 1998–2001, the average content of ^{90}Sr and ^{137}Cs in lake water reached 120–190 and 18–43 Bq·l^{-1} respectively. The radionuclide contamination density values found in the lake bottom sediments for ^{90}Sr, ^{137}Cs, $^{238+239+240}$Pu and ^{241}Am averaged at 6.70, 11.50, 0.24 and 0.22 TBq·km^{-2} respectively, with the maximum values of 33.30, 14.40, 1.10 and 0.29 TBq·km^{-2}.

The ^{90}Sr and ^{137}Cs content in the water of Glubokoye Lake come to 99–120 and 13–14 Bq·l^{-1} respectively. The average values of contamination density in the bottom sediments by ^{90}Sr, ^{137}Cs, $^{238+239+240}$Pu and ^{241}Am in 1998 were 2.6, 5.6, 0.07 and 0.06 TBq·km^{-2}, with the maximum values being 10.0, 13.7, 0.22 and 0.23 TBq·km^{-2} respectively. In Dalekoye-1 Lake the average content of ^{90}Sr and ^{137}Cs in the research period reached 82.5 and 11.8 Bq·l^{-1} respectively. The maximum value of radionuclide contamination density in the bottom sediments by ^{90}Sr in 1999 was 18.9, by ^{137}Cs – 15.2, by $^{238+239+240}$Pu – 0.6 and by ^{241}Am – 0.4 TBq·km^{-2}. The average values were, accordingly, 4.0, 3.1, 0.08 and 0.08 TBq·km^{-2}.

The average specific activity values for ^{90}Sr and ^{137}Cs in water of Yanovsky Backwater for the period 1998–2001 were 75.2 and 5.6 Bq·l^{-1} respectively. The radionuclide contamination of the bottom sediments of reservoir is extremely heterogeneous, which is obviously caused by the non-uniform character of the nuclear fall-out and by the absence of wind-induced turbulence in deep water. The average content of ^{90}Sr, ^{137}Cs, $^{238+239+240}$Pu and ^{241}Am in bottom sediments was, respectively, 16.3, 14.8, 0.4 and 0.3 TBq·km^{-2}. At the same time, within the bounds of silt sediment deposition, some sites with abnormally high density of contamination by ^{90}Sr, ^{137}Cs and $^{238+239+240}$Pu (307.1, 251.6 and 5.3 TBq·km^{-2} respectively, which is 20 times higher than the average values in the backwater), were found.

The cooling pond of the ChNPP has undergone the highest radionuclide contamination in comparison with other reservoirs of exclusion zone. In the course of time, after the cessation of radioactive emissions into the atmosphere and due to disintegration of short-lived isotopes, ^{90}Sr and ^{137}Cs have become the main radioactive contaminants of the cooling pond water. During 1998–2001, the specific activity of ^{90}Sr in the cooling pond water was found to be within the range of 1.7–1.9, with the range being 2.7–3.1 Bq·l^{-1} for ^{137}Cs. The heterogeneity of the bottom sediment contamination in the cooling pond is currently determined by the nature of the silt accumulation processes. The height of silt layers at the depth of over 11 m (for up to 35 per cent of the bottom area) reaches up to 100 cm., with the density of contamination by ^{137}Cs at 18.5–133.2 TBq·km^{-2}. The bottom at the depth of 3–11 m consists of primary soils, which are covered, with a 1–6 cm layer of silt, with the contamination density by ^{137}Cs in the range of 1.5–5.9 TBq·km^{-2} [2].

Uzh River is the main tributary of Pripyat River within the exclusion zone and the runoff of its inflow covers the southern part of a zone with a rather low level of ^{90}Sr contamination. The contents of ^{90}Sr and ^{137}Cs in water of Uzh river during 1998–2001 averaged at 0.11 and 0.31 Bq·l^{-1}. The specific activity of ^{90}Sr and ^{137}Cs in water of Pripyat River during 1998–2001 was found in the ranges of 0.22–0.50 and 0.11–0.14 Bq·l^{-1} respectively. The radionuclide content in the bottom sediment of main riverbed sites of Uzh River and Pripyat River is currently only slightly in excess of the pre-accident levels. Considerably higher activity in the bottom sediments is still observed in the backwaters, the old riverbeds and other slow-running sites of the rivers.

1.2. HIGHER AQUATIC PLANTS

The radionuclide contents in higher aquatic plants (macrophytes) in the studied water bodies were largely determined by the nature of radionuclide contamination of the water objects and nearby territories, as well as by the hydrochemical regime in the reservoirs. The latter affects the forms of radionuclides in the reservoirs, thus affecting the level of their bioavailability to the hydrobionts. The specific activity data for the main radionuclides in macrophyte tissues are shown in Table 1.

The patterns of ^{90}Sr and ^{137}Cs accumulation have been shown to be species-specific. Among the species with relatively high ^{137}Cs content are the helophytes (air-water plants) of genus *Carex, Phragmites australis, Glyceria maxima, Typha angustifolia*, as well as strictly water plant species *Myriophyllum spicatum* and *Stratiotes aloides*. The low values of ^{137}Cs activity in all reservoirs were found in the representatives of family Nymphaeaceae – *Nuphar lutea* and *Nymphaea candida* as well as *Hydrocharis morsus-ranae*. Relatively high content of ^{90}Sr was shown by the species of genus *Potamogeton*. Obviously this is related to this plant's tendency to accumulate large quantities of calcium (which is not washed off during standard sampling) on its surface during photosynthesis. At the same time, calcium carbonate that is removed from the plant could contain 7–20 times more radioactive strontium than the plant tissue [3]. Thus, *Potamogeton* species makes a good prospective radioecological monitoring object as a specific accumulator of ^{90}Sr.

Table 1. Content of radionuclides in macrophytes (1998–2001), Bq kg^{-1} w. w.

Water reservoir	^{90}Sr			^{137}Cs		
	max	min	average	max	min	average
Glubokoye Lake	14060	67	2212	36470	1215	8730
Dalekoye-1 Lake	5100	200	1808	19470	1167	5170
Azbuchin Lake	24210	730	4895	23860	220	2025
Cooling pond	1600	36	218	3125	310	1379
Yanovsky Backwater	2200	110	779	1702	38	814
Uzh River	234	3	10	185	4	42
Pripyat River	357	5	15	164	5	30

The analysis of the ^{90}Sr and ^{137}Cs content at various vegetative stages has revealed seasonal dynamics of radionuclide accumulation by macrophytes. The majority of higher aquatic plant species showed the increased content and CF of ^{90}Sr and ^{137}Cs at the peak of vegetation (end of July – August) in comparison with the spring and autumn periods.

The content of radionuclides $^{238+239+240}$Pu and ^{241}Am in higher aquatic plants of the left-bank flood plain of Pripyat River was found, respectively, in the ranges of 1–66 (11) Bq·kg^{-1} with CF – 24–4175 and 1–45 (11) Bq·kg^{-1} with CF – 83–7458. *Typha angustifolia* showed the highest CF, which was 5–7 times higher than the average CF values for other studied plant species. That allows considering this species as a specific accumulator of transuranic elements in reservoir conditions within the ChNPP exclusion zone.

1.3. FRESHWATER MOLLUSCS

Freshwater molluscs are often considered as bioindicators of radionuclide contamination of water objects. These invertebrates accumulate practically all the radionuclides found in water and, due to their high biomass, molluscs play an important part in bioaccumulation processes and radionuclide redistribution in aquatic ecosystems. The ^{137}Cs and ^{90}Sr content in molluscs of reservoirs of the ChNPP exclusion zone are shown in Table 2.

The highest CF for both ^{90}Sr and ^{137}Cs were found in bivalve molluscs *Dreissena polymorpha* and *Unio pictorum,* which are the most active filtrators. The highest CF for ^{90}Sr was noted in *Dreissena polymorpha* – in excess of 1100, while for ^{137}Cs the highest CF (about 500) was found in the tissues of *Unio pictorum.* Considerably lower CF's were determined for the gastropod species *Lymnaea stagnalis*, *Planorbarius corneus* and *Viviparus viviparus.*

The lowest CF for both ^{90}Sr and ^{137}Cs were found in *Lymnaea stagnalis* (440 and 137 respectively). Whereas the differences in ^{90}Sr CF for gastropods could be explained by their shell morphological structure and its specific weight, the distinctions in value of ^{137}Cs CF are related to the functional ecology and feeding modes of these invertebrates.

Table 2. Content of radionuclides in molluscs (1998–2001), Bq·kg^{-1} w. w.

Water reservoir	^{90}Sr			^{137}Cs		
	max	min	average	max	min	average
Glubokoye Lake	170300	39770	61830	27150	3156	9067
Dalekoye-1 Lake	62830	17123	32310	2410	847	1523
Azbuchin Lake	87670	34010	51120	4750	2704	3620
Cooling pond	2500	1600	2133	2900	830	1450
Yanovsky Backwater	13350	7720	10350	590	430	510
Uzh River	216	98	177	30	14	25
Pripyat River	101	77	86	61	31	35

The average contents of transuranic elements ^{238}Pu and $^{239+240}$Pu in mollusc tissues in Glubokoye Lake and Dalekoye-1 Lake were as follows: the lowest value was determined for *Lymnaea stagnalis* – 0.1 and 0.2 Bq·kg^{-1} respectively in Dalekoye-1 Lake, 2.7 and 6.4 in Glubokoye Lake. The highest content was determined for *Stagnicolia palustris* from Glubokoye Lake – 14 and 36 Bq·kg^{-1} respectively. The highest activity among gastropods was shown in *Planorbarius corneus* – 1 and 2 Bq·kg^{-1} respectively from Dalekoye-1 Lake; 25 and 53 Bq·kg^{-1} in Glubokoye Lake. *Dreissena polymorpha* from the cooling pond of the ChNPP showed ^{238}Pu and $^{239+240}$Pu contents of 3 and 6 Bq·kg^{-1} respectively.

The contents of ^{241}Am in *Lymnaea stagnalis* tissue was the lowest – in the range of 4–30 (15) Bq·kg^{-1} in Dalekoye-1 Lake and 6–51 (27) Bq·kg^{-1} in Glubokoye Lake. For *Stagnicolia palustris* from Glubokoye Lake the value was about about 75 Bq·kg^{-1}. The highest value was found in *Planorbarius corneus* – 18–29 (24) Bq·kg^{-1} in Dalekoye-1 Lake and 80–310 (170) Bq·kg^{-1} in Glubokoye Lake. The content of ^{241}Am in *Dreissena polymorpha* tissue from the cooling pond of the ChNPP was at the level of 8 Bq·kg^{-1}.

1.4. FISH

The fish species that are found at the upper levels of the food webs may also constitute a part of human diet and, therefore, are of a particular interest in radioecological research of water ecosystems. The comparative contents of ^{90}Sr and ^{137}Cs in fish of the exclusion zone reservoirs are represented in Table 3.

The concentration of transuranic elements ^{238}Pu, $^{239+240}$Pu and ^{241}Am was measured in fish of Glubokoye Lake and Dalekoye-1 Lake. The activity of ^{238}Pu in fish tissue was found in the range of 0.4–0.5 (0.4) Bq·kg^{-1} with CF – 72–98 (83), $^{239+240}$Pu – 0.7–0.9 (0.8) Bq·kg^{-1} with CF – 68–87 (75) and ^{241}Am – 2.2–10.0 (6.2) Bq·kg^{-1} with CF – 367–1667 (1028).

The contents of ^{90}Sr and ^{137}Cs radionuclides in lake fish of the left-bank flood plain of Pripyat River in all cases considerably exceeded maximum permissible level (MPL), according to the standards accepted in Ukraine for fish production: for ^{90}Sr on average 146 times higher (MPL – 35 Bq·kg^{-1}), for ^{137}Cs – 134 times (MPL – 150 Bq·kg^{-1}). The highest measured values were 373 and 180 times in excess of MPL. The content of ^{90}Sr in fish of the cooling pond practically in all caught specimens also exceeded MPL (on average 8 times higher), with the highest registered values being 43 times higher than MPL. The ^{137}Cs contents in all cases also considerably exceeded MPL – on average 33 times higher, with highest registered values exceeding MPL 84 times.

Table 3. Content of radionuclides in fish (1998–2001), Bq·kg^{-1} w. w.

Water reservoir	^{90}Sr			^{137}Cs		
	max	min	average	max	min	average
Glubokoye Lake	3300	660	2582	11000	5200	8520
Dalekoye-1 Lake	13060	410	5103	27020	16110	20030
Cooling pond	1518	18	272	13260	1855	4990
Uzh River	7	1	3	75	3	44
Pripyat River	87	9	23	287	32	109

Despite the fact that the average values of the radionuclide contents in fish of Pripyat River did not exceed MPL, the cases of excess ^{90}Sr and ^{137}Cs concentrations for the research period have constituted about 15 per cent of the total studied individuals. The highest values of ^{90}Sr and ^{137}Cs contents exceeded MPL by a factor of two. No values in excess of MPL for the research period were found in Uzh River.

2. Dose rate for hydrobionts

The values of the absorbed dose for hydrobionts from reservoirs of the ChNPP exclusion zone were found to be in the range from 1.8E–03 to 3.4 Gy·y^{-1}. The highest value was found for hydrobionts from lakes within embankment territory on the left-bank flood plain of Pripyat River, the lowest – for specimens from the running water objects – Uzh River and Pripyat River (Table 4). The ratio of external and internal doses varied considerably for hydrobionts from different reservoirs and depended on the contents of γ-emitting radionuclides in the littoral zone bottom sediment, as well as in soils close to the riverbank. Thus, in Glubokoye Lake, which contains a so-called abnormal contamination strip at the shoreline border, about 95% of absorbed dose results from external exposure and only about 5% – from internal exposure to radionuclides incorporated in tissue (Figure 2). The similar ratio is observed for the exclusion zone rivers – Uzh River and Pripyat River; however, in these objects the ratio is related to the high flow rate as well as to the relatively low radionuclide content in water and, consequently, in hydrobiont tissues.

Table 4. Diapasons of absorbed dose for hydrobionts of littoral zone from different water objects within the sampling sites, Gy·year^{-1}

	Dose	Uzh River	Pripyat River	Cooling pond of the ChNPP	Yanovsky Backwater	Azbuchin Lake	Dalekoye-1 Lake	Glubokoye Lake
Internal dose from incorporated radionuclides	^{90}Sr	9.9E–06 / 1.8E–04	1.5E–05 / 8.6E–05	2.2E–04 / 2.1E–03	7.7E–04 / 8.1E–04	4.9E–03 / 5.1E–02	1.8E–03 / 3.2E–02	2.2E–03 / 6.1E–02
	^{137}Cs	1.0E–04 / 1.8E–04	1.2E–04 / 4.5E–04	5.7E–03 / 2.0E–02	3.2E–03 / 4.2E–02	8.3E–03 / 1.5E–02	6.2E–03 / 8.2E–03	3.6E–02 / 4.9E–02
	^{238}Pu	x* / x	x / x	5.2E–06 / 7.9E–05	5.4E–05 / 2.5E–04	4.2E–05 / 3.3E–04	6.5E–05 / 3.7E–04	1.5E–04 / 1.3E–03
	$^{239+240}$Pu	x / x	x / x	8.3E–05 / 1.6E–04	1.3E–04 / 5.9E–04	1.3E–04 / 7.2E–04	1.6E–04 / 7.9E–04	3.2E–04 / 2.3E–03
	^{241}Am	x / x	x / x	2.3E–05 / 3.6E–04	4.5E–05 / 6.3E–04	6.7E–05 / 7.5E–04	9.7E–05 / 8.0E–04	7.2E–04 / 2.5E–03
	Total	1.1E–04 / 3.6E–04	1.4E–04 / 5.4E–04	6.0E–03 / 2.3E–02	4.2E–03 / 4.4E–02	1.3E–02 / 6.8E–02	8.5E–03 / 4.2E–02	3.9E–02 / 1.2E–01
External water dose	^{90}Sr	4.9E–08 / 9.8E–08	6.8E–08 / 1.5E–07	5.2E–07 / 5.8E–07	2.1E–05 / 2.8E–05	4.4E–05 / 7.0E–05	2.8E–05 / 3.3E–05	3.2E–05 / 4.0E–05
	^{137}Cs	2.7E–07 / 3.8E–07	2.9E–07 / 4.3E–07	7.2E–06 / 8.3E–06	1.3E–05 / 1.6E–05	4.5E–05 / 1.3E–04	2.7E–05 / 4.0E–05	3.8E–05 / 4.5E–05
	Total	3.2E–07 / 4.8E–07	3.6E–07 / 5.8E–07	7.7E–06 / 8.9E–06	3.4E–05 / 4.4E–05	8.9E–05 / 2.0E–04	5.5E–05 / 7.3E–05	7.0E–05 / 9.5E–05
External γ-dose		1.7E–03 / 2.9E–03	2.3E–03 / 3.6E–03	7.4E–03 / 7.9E–03	3.3E–03 / 6.2E–03	4.8E–03 / 1.2E–02	4.3E–02 / 5.0E–02	1.5–3.3
Total absorbed dose		1.8E–03 / 3.3E–03	2.4E–03 / 4.1E–03	1.3E–02 / 3.1E–02	7.5E–03 / 5.0E–02	1.8E–02 / 8.0E–02	5.2E–02 / 9.2E–02	1.6–3.4

Note: * – the measurements were not carried out

In Azbuchin Lake and Yanovsky Backwater, at a relatively low external radiation dose, the main contribution to the absorbed dose is made by radionuclides incorporated in hydrobiont tissues. It is linked to the high radionuclide content in water and at the same time to the low contamination level of the bottom sediment within littoral zone and the soils of nearby areas (with sandy soils showing low levels of radionuclide fixation). In this respect, the cooling pond of the ChNPP is in a mid-way position.

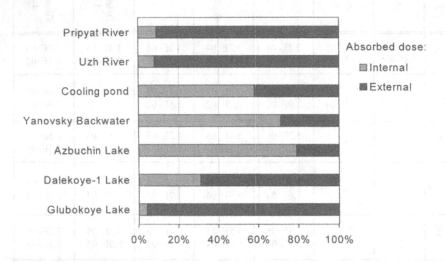

Fig. 2. Ratio of internal and external absorbed dose for hydrobionts from different water objects within the Chernobyl NPP exclusion zone.

According to the classification by G. Polikarpov [4, 5], the studied littoral sites of Uzh River and Pripyat River can be assigned to the radiation safety zone; the sampling stations of Azbuchin Lake, Yanovsky Backwater, the ChNPP cooling pond and Dalekoye-1 Lake – to zones of physiological and ecological disguise, which are also approaching the ecosystem effect zone (Glubokoye Lake), where reduction in aquatic organisms numbers and loss of radiosensitive species can be observed. Populations of benthos organisms from the studied water objects cannot, however, be assigned to the above mentioned zones as, at the known levels of bottom sediment contamination in non-flowing reservoirs within exclusion zone, the absorbed dose for these hydrobionts could be considerably higher.

3. References

1. Amiro B.D. Radiological Dose Conversion Factors for Generic Non-Human Biota Used for Screening Potential Ecological Impacts // Journal of Environmental Radioactivity. – 1997. – Vol. 35, No 1, 37–51
2. Kanivets V.V., Voytsekhovich Î.V. Current state of radionuclide contamination of the bottom sediment complex of the cooling pond of the ChNPP and perspectives of its deactivation (rehabilitation) after ChNPP decommission // Abstract of the Int. Conf. "Radioactivity at Nuclear Tests and Accidents". – SPb.: Gidrometeoizdat, 2000, 128 (in Russian).
3. Laynerte Ì.P., Seysuma Z.Ê. The role of littoral-water plants of freshwater reservoirs in accumulation of Ca and Sr (stable and radioactive) // Higher aquatic plants and littoral-water plants: Abstracts of the 1st All-Union Conference (Borok, September 7–9, 1977). Borok: 1977, 117–119 (in Russian).
4. Polikarpov G.G. (Invited paper by ASIFSPCR from IAEA, Monaco). Effects of ionizing radiation upon aquatic organisms (Chronic irradiation) // Atti della Giornata sul Tema "Alcuni Aspetti di

341

Radioecologia". XX Congresso Nazionale, Associazione Italiana di Fisica Sanitaria e Protezione contro le Radiazioni, Bologna, 1977. – Parma Poligrafici, 1978, 25–46.
5. Polikarpov G.G. Effects of nuclear and non-nuclear pollutants on marine ecosystems // Marine Pollution. Proc. of a Symp. held in Monaco, 5–9 October 1998. IAEA-TECDOC-1094. IAEA-SM-354/22. International Atomic Energy Agency, July 1999, 38–43.

ROLE OF HIGHER PLANTS IN THE REDIDSTRIBUTION OF RADIONUCLIDES IN WATER ECOSYSTEMS

Z. SHYROKAYA Z.O., Ye.VOLKOVA, V. BELIAYEV,
V. KARAPISH, I. IVANOVA
*Institute of Hydrobiology, National Academy of Sciences, Kiev,
UKRAINE*

1. Introduction

Higher water plants actively participate in biological cycles and energy balance of water bodies and thereby play an important role in functions of freshwater ecosystems. Macrophytes comprise approximately one-third of the total biological productivity of water bodies [5]. In accumulation of organic matter in water ecosystems, they are second only to microorganisms, including microalgae.

Macrophytes intensively accumulate radionuclides. Their role in self-purification of freshwater ecosystems depends on their abilities to accumulate, to fix, and to withdraw radionuclides from the turnover in reservoirs for a prolonged period. The objectives of this work to analyze of a role of highest aqueous plants in redistribution of radionuclides in hydroecosystems, which are in conditions of radiocontamination and to identify higher water plant cenoses that make the greatest contribution to water self-purification from radionuclides.

2. Materials and methods

Shallow regions covered with vegetation of the Kyiv, Kanev and Kakhovka reservoirs were studied from 1986 to 1998. Aerial photo and video shooting were used to describe vegetation and to estimate the resources of dry biomass produced by the major dominant macrophyte species. Specific radioactivity was estimated by gamma-spectrometry and by the standard radiochemical methods [1, 3]. The specific radioactivity of the vegetation was calculated from their air-dried weight (Bq/kg).

The amount of radionuclides accumulated by plants of the Kyiv, Kanev and Êàkhovka reservoirs was obtained as a product of their average annual content in a given species and phytomass produced by the species over the vegetation season. Shallow regions covered with vegetation of the Kyiv, Kanev and Êàkhovka reservoirs were studied from 1986 to 1998.

3. Results and discussion

Until 1986, the major regularities as forming vegetation were determined by intensive bogging in the Kyiv and Kakhovka reservoirs, especially in their upper regions. This slightly increased regions occupied by plants and caused the formation of monodominant communities of cat's-tail *Typha angustifolia* and bur reed *Phragmites australis* (associations *Typhetum angustifoliae* and *Phragmitetum australis*). Regions covered with these communities increased through expansion to inundated meadows in the flood plain and the second terrace. In downstream shallows, regions occupied with vegetation were smaller because of the absence of surviving and temporary plants.

F. Brechignac and G. Desmet (eds.), Equidosimetry, 343–345.
© 2005 *Springer. Printed in the Netherlands.*

The Chernobyl meltdown caused dramatic transformation of vegetation in the Kyiv, Kanev and Kakhovka reservoirs. Data of long-term observations allowed us to establish the major regularities at its formation after 1986.

At present, higher water plants mostly occupy shallow areas in the upper and central regions of the Kyiv, Kanev and Kakhovka reservoirs. The spatial distribution of plants is closely associated with the composition of benthic sediment, depth, local water currents, and sheltering of a habitat.

Aeroaquatic plants communities — *Typha angustifolia L.* and *Phragmites australis (Cav.) Trin. ex Steud.* (More than 70 % from common reserve) occupied large areas in the Kyiv and Kakhovka reservoirs.

Plants with floating leaves in Kyiv, Kanev and Kakhovka reservoirs occupied vast areas. From 1986 to 1993, the areas occupied by plants with floating leaves and submerged species, is possibly markedly greater in Kanev reservoir due to of continuously increasing anthropogenic effect.

The wide distribution of association *Ceratophylletum demersi* was determined by increasing anthropogenic eutrophication of the reservoirs of the Dnieper cascade. Cenoses of the association were abundant in all Dnieper reservoirs; over the recent years, their development was the most significant in shallow regions of the Kanev reservoir. Growth of these plants required benthic sediment enriched in organic matter. These cenoses are most frequently found and have maximal biomass in shallow areas with slightly alkaline water and lime-containing benthic sediment [2].

Growing plants assimilate water and soil radionuclides and thereby withdraw them from turnover for a long time, thus playing an important role in purification. Associations of higher aquatic plants vary in their involvement in this process. The amount of radionuclides assimilated from water by aquatic plants depends on species, phytomass produced, and ecological conditions of its production. Cenoses of associations of the class *Phragmiti* (aeroaquatic plants) withdraw more radionuclides and for a longer period than those of the classes *Lemnetea* and *Potametea*.

Most characteristic species in association of the aeroaquatic plants are perennial; i.e., their above-ground organs die at the end of the vegetation period. Radionuclides accumulated in this phytomass are dissolved in water and deposited in benthic sediment.

The life of underground organs is more prolonged: more than three years in Phragmites australis and Schoenoplectus lacustris, 18-22 months in *Typha angustifolia* and *Sparganium simplex*, and 18-24 months in *Glyceria maxima* [4]. Radionuclides accumulated in rhizome biomass are transported in deeper layers of benthic sediment and thus buried underground (everyone year 1/3 parts of an assemblage of roots dead).

The proportions of the root weight and spring weight is a relatively stable characteristic of a species [4], though a slight variation can be induced by aging and a number of ecological factors. These proportion were 84.2 and 15.8%, respectively, in *Phragmites australis* and 54 and 46%, respectively, in Typha angustifolia.

In 1993 the total radionuclides reserve of plants was 1100 ^{90}Sr and 2500 MBq ^{137}Cs in the Kanev reservoir.

In 1992, the total radionuclides reserve (content) of ^{90}Sr and ^{137}Cs in above-ground/underground in phytomass were 1787/2723 and 11332/16830 MBq, respectively, in *Typha angustifolia* and 853/4165 and 7867/41373 MBq, respectively, in *Phragmites australis* from the Kiev reservoir.

The radionuclides reserve of plants was 1100 ^{90}Sr and 2500 MBq ^{137}Cs in the Kanev reservoir.

It is stated that distribution ^{90}Sr and ^{137}Cs between above-ground and underground parts airwater plants — *Typha angustifolia L.* and *Phragmites australis (Cav.) Trin. ex Steud.* and *Glyceria maxima L.* is conditioned phase of development of each species.

4. Conclusions

- Character of radionuclides ^{90}Sr and ^{137}Cs for their distribution between over-ground and underground parts phytomass of the airwater plants — *Typha angustifolia* and *Phragmites australis* and *Glyceria maxima* depend on their phase of development. ^{137}Cs were mostly accumulated in the underground parts phytomass of the airwater plants — *Typha angustifolia, Phragmites australis* and *Glyceria maxima L.*

- From 1986 to 1993 we estimated the amount of radionuclides (about 22800 MBq ^{90}Sr and 323000 MBq ^{137}Cs) withdrawn from water and bottom sediments by die airwater organs *Typha angustifolia* in the Kiev reservoir.

- The result obtained provided for a better understanding of role of highest aquatic plants in the redistribution of radionuclides in hydroecosystems. In addition, they offer a possibility of associations of higher aquatic plants for water purification from radiocontamination.

- It is determined that Communities with emergent airwater plants take the most effective part in processes of self purification of basins from radioactive contamination. Cenoses formed by association of airwater plants — *Typha angustifolia* and *Phragmites australis* played the major role in water self-purification from radionuclides.

5. References

1. Golfman, A.Ya. and Kalmykov, L. V., determination of Radiactive Cesium by the Ferrocianide Method, Radiokhimia, 1963, 4, ¹ 10,107-109.
2. Dubyna, D.V., Stoiko, S.M., Sytnik, K.M. et al., Macrophytes as Indicator of Changes in the Natural Environment, Naukova Dumka, Kiev, 1993.
3. Lavrucina, A.K., Malysheva, T.V. and Pavlotskaya, F.I., Radiochemical Analysis, Akad. Nauk SSSR, Moscow, 1963.
4. Lukina, L. K. and Smirnova, N.N., Physiology of Higher Water Plants, Naukova Dumka, Kiev, 1989.
5. Wetzel, R.G., A Comparative Study of the Primary Productivity of Higher Water Plants, Periphyton and Phytoplankton in a Large Shallow Lake, Int. Rev. Hydrobiol., 1964, 49, 1-61.

PROBLEMS OF THE RADIATION SAFETY ON MILITARY OBJECTS OF UKRAINE

A. KACHINSKIY
National Institute of Strategic Researches, Pirogova street 7à, Kyiv, 252030, UKRAINE

V. KOVALEVSKIY
The General Staff of Ukrainian Armed Forces, Povitroflotskiy Pr., 6, Kyiv, 03168, UKRAINE

1. Abstract

The present radiation situation in Ukraine officially refers as difficult. Radiation safety includes a component which is formed by military activity. This component is very important. Special attention to this radiation safety component is necessary because of significant specificity of military activity. In the article separate problems of radiating safety on military objects are considered. Also ways of the decision of problems to Armed Forces of Ukraine are discussed in the article.

2. A condition of radiation safety

Preservation of the environment is a global problem. This problem covers the whole world, infringes interests of all countries and peoples. Ecological failures and accidents in the separate countries began to affect the ecological condition of the countries of the whole worlds. In this connection the relation of mankind to environmental problems does not depend on borders at length.

The present ecological situation in Ukraine is in a condition of crisis. This fact is marked in official documents. This condition was formed during a long period because of neglect by objective laws of the development and restoration of a resource complex of Ukraine. The radiation situation is a dominant component of ecological stress in Ukraine. The radiation situation is determined officially as difficult [4] and as a situation which contains an even greater amount of significant problems.

Among the problems connected to radiation safety: the problem of liquidation the consequences of the Chernobyl accident occupies an important independent place; the problems of the exclusion zone around the Chernobyl atomic power station; problems of "Shelter" object in which dangerous radioactive substances and nuclear materials with total activity about 20 million Ci [4] are rather concentrated problems.

The significant part of problems is connected with a high level of development of branches and means which determine a level of radiating safety. Ukraine has about fifteen working blocks of atomic power stations. These blocks provide one of the biggest nuclear power programs. Ukraine takes the eighth place in the world and the fifth place in Europe after France, Great Britain, Russia and Germany in capacity of nuclear blocks. Research reactors are located in Kiev and Sevastopol. Ukraine is the

F. Brechignac and G. Desmet (eds.), Equidosimetry, 347–359.

country in which use of sources of ionization radiations is advanced. About eight thousand enterprises, organizations and scientific institutes use over hundred thousand sources of ionization radiation and radioisotope devices [4, 5].

There is a problem of use and burial place of radioactive waste products. Nuclear power plants are the main place of accumulation of radioactive waste products on which their initial processing and time storage [4] is carried out.

In Ukraine there are five enterprises for extraction and processing of radioactive ore. These enterprises are situated in the Dnepropetrovsk, Nikolaev and Kirovograd provinces. Almost all waste products of uranium ores which are formed during extraction and processing (stores of mine breeds, liquid dumps and gaseous emissions), are sources of environmental contamination. The basic danger is big with regard to the volumes with stores of waste products and the radioactive materials concentrated in them [4].

Radiation safety of Ukraine is formed, first of all, under influence of the enterprises for manufacture of fuel for the nuclear stations (the industry obtaining uranium; the industry processing uranium; atomic power stations), "UKRYTIE" object and Chernobyl problems. Compound radiation safety which is created by military activity, is important, and demands special attention. It is caused by significant specificity of military activity. There are a lot of potentially dangerous military objects and some kinds of military activity which have potential ability of substantial increase of a collective doze of an irradiation of the population. It is possible in adverse conditions, for example, in case of failure.

For the years of existence of Armed Forces of Ukraine such incidents did not occur. Absence of significant radiation incidents in Armed Forces of Ukraine does not mean that military activity was safe. Absence of significant radiation incidents is consequence of appropriate level of the organization of system of safety. The example for this is the work on liquidation of nuclear weapons.

3. Performance of works on nuclear disarmament

Ukraine has received in the inheritance from Soviet Union the third of the size of the nuclear arsenal in the world. It consists of 176 starting complexes of strategic rockets. From them are 130 rockets SS-19 (on Soviet classification RS-18) and 46 rockets SS-24 (on Soviet classification RS-22). In addition to strategic rockets in Ukraine were fourteen heavy strategic bombers and a significant arsenal of tactical nuclear weapons [7, 8].

The fighting part of rocket SS-19 contains 6 warheads. The fighting part of rocket SS-24 contains 10 warheads. The total of warheads on strategic rockets was equalled to 1240 units. 600 warheads were on cruise missiles for big range.

Warheads incorporate highly dangerous substances for their ecological behaviour. It is known that on each warhead about 3 kg of plutonium or 15 kg of enriched uranium is present that has created considerable accumulation of dangerous materials. For example, plutonium Pu-239 (a half-life period of 24065 years) is one of the most toxic elements. One ten-thousandth part of a gram of plutonium, in case of a hit in an organism of the person, will provoke a lethal consequence [7, 8].

All activities, which were connected to the liquidation of the strategic nuclear arsenal of Ukraine, were ecologically dangerous and were carried out during a long period (figure 1, [18]). The application of the nominal rules and technologies of safety guaranteed exception of threat of an unauthorized atomic explosion. However the possibility of start offing of emergencies in atypical situations could not be eliminated. For example, in emergency situations, at errors or violations of technology, there is a potential of explosion of explosive material in warheads. Such an explosion, fire or transport accident can produce a radiation contamination on a distance of several kilometres

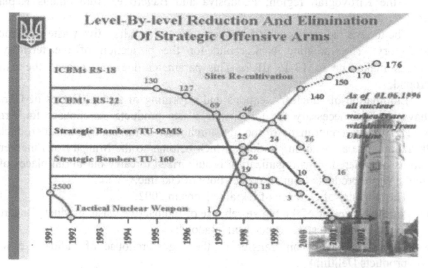

Fig. 1. Reduction and elimination of strategic and offensive arms.

Unfortunately, cases of dangerous failures happened in world practice. In January, 1985 in Germany, there was an explosion of a rocket "Pershing-2". In December, 1987 there was a non-authorized start-up of a research rocket of NASA. In September, 1986 there was an explosion of a rocket in the shaft of the Soviet submarine in the area of the Bermuda triangle. In April, 1990 there was an explosion of the accelerator of a "Titan - 4" rocket on the Military - Air Forces of USA basis. Cases of destruction of rockets SS-20 took place at overturning mobile launchers. The Joint Staff of Armed Forces SNG informed, that in USA during 1950 - 1989, 178 incidents have arisen with nuclear weapons. 43 refusals were accompanied by non-nuclear explosion of explosive which was in warheads. Approximately 170 similar cases happened in Great Britain [7]. Safety of a nuclear arsenal of Ukraine caused concern prior to the beginning of works on liquidation of the nuclear weapon. The characteristic name of articles of that time was V. Litovkina's article - "On rocket shafts of Ukraine: the second Chernobyl" [9]. The successful fulfilment of all complexes of activities on nuclear disarming was especially considerable on a background of such publications. The Armed Forces of Ukraine have acquired experience with a security system at fulfilment of activities of elimination of nuclear weapons and experience of unprecedented moving of nuclear ammunition.

4. The manipulation with radioactive waste products and sources ionization radiations

As a heritage from Armed Forces of Soviet Union one more problem has been identified. Such problem is the existence of a series of large radioactive waste storehouses. These storehouses are located close to Deljtin the Ivano-Frankovsk region; Cybulovo the Kirovograd region; Feodosiya and Bagerove, Autonomous Republic Crimea.

The today's radiating condition of territory of the radioactive waste storehouses is satisfactory. The radiating background for the protection of the territory of storehouses makes up to 13-15 µR/h. This parameter does not exceed the natural background.

Liquidation of nuclear weapons and reforming of Armed Forces of Ukraine [10] have made unnecessary the radioactive waste products storehouses for Armed Forces. The question on transfer of the big storehouses of radioactive waste products to the Ukrainian state association "Radon", which belongs to the Ministry of Emergency, therefore is considered. This organization is authorized to carry out burial places of all radioactive waste products from all the territory of Ukraine.

The Ministry of Defence of Ukraine plans in 2002:

- to finish transfer of the big storehouse of radioactive waste products Bagerove, to the Ukrainian state association "Radon";
- to continue work on transfer of the big storehouse of radioactive waste products Deljtin;
- to develop proposals, concerning increase of a level of radiation safety;
- to improve the organization of realization in Armed Forces of Ukraine of the State inventory of radioactive waste products and radiating sources.

Recycling of radioactive waste products is the most actual task. It speaks about the presence of a significant amount of radiation sources. After export from Ukraine of nuclear ammunition those radiations sources in Armed Forces of Ukraine were estimated to remain at about 75 thousand of them. Overwhelmingly, the majority of these sources are closed low-power γ-sources. The basic kind of use of radiation of sources is their application for the control of serviceability of dosimetric devices. Under certain conditions these sources may create a threat of pollution of the territory.

In Armed Forces of Ukraine the State an inventory of radioactive waste and radiations sources is carried out. Work is carried out according to requirements of the reference of Program of use of radioactive waste [6]. The results have allowed to obtain the information on the quantity of radioactive waste outside of the big storehouses of radioactive waste. Specification of structure of radioactive waste and radiations sources is carried out. The tendency of reduction of quantity of radiations sources and radioactive waste became obvious (Table 1).

Table 1. Presence radiations sources and radioactive waste in Armed Forces of Ukraine

№	Amount of organizations which have RS and RW, units.	Amount of RS, units.	Activity of RS, Bq	RS which are used		RS which pass in group RW	
				Units	Activity of RS, Bq	Units	Activity of RS, Bq
Zone of service of State association RADON, Kyiv							
In total	139	8598	5,74E+14	7862	5,36E+14	429	2,80E+13
Zone of service of State association RADON, Dnipropetrovs'k							
In total	107	18331	6,09E+14	1194	5,36E+14	27	2,11E+12
Zone of service of State association RADON, Lviv							
In total	284	16380	8,13E+12	16282	9,70E+16	78	3,80E+11
Zone of service of State association RADON, Odessa							
In total	166	17812	7,43E+13	3892	6,21E+11	3623	2,14E+13
Zone of service of State association RADON, Harkiv							
In total	20	1959	2,40E+12	1901	9,77E+11	16	5,25E+11
Result	716	63080	1,27E+15	31131	9,81E+16	4173	5,24E+13

RS: radiations sources; RW: radioactive waste.

Control of the condition of operation and storage of radiation sources and radioactive waste is executed simultaneously by the State inventory of radioactive waste. As a whole, the level of radiating safety corresponds to requirements of the present legislation.

5. Maintenance of radiation safety in case of a difficult radiation situation

Radiation safety is a component of ecological safety. The modern legislation of Ukraine obliges commanders of all levels, without exception, to solve questions of ecological safety. Therefore all officials in all kinds of armed Forces of Ukraine provide a necessary level of radiation safety. In addition, the service of ecological safety of Armed Forces of Ukraine carries out the control of completeness and correctness of observance of ecological laws.

The Ministry of Defence has entered a system of expedite and periodic reports. Their purpose is well-timed reacting to ecological threats by the military serving and environment. The expedite reports are done in case of detection of considerable violations of the ecological legislation, or in case of detection of pollutions of a natural medium. The periodic reports go into controls with a frequency of once a week.

The controls organs analyse the obtained information, organize and conduct measures with the purpose of avoidance of extraordinary situations in crisis cases. The purpose of activity varies in that case, when the extraordinary situation has taken place. In this case the purpose is the minimization of negative consequences.

The most indicative examples of such actions are works carried out for the liquidation of an extreme situation in Artemovsk area and works carried out for maintenance of radiation safety of the Ukrainian contingent of peace-making forces in

Kosovo. Works in Kosovo were carried out in connection with application on the Balkan of weapons with depletion uranium.

5.1. LIQUIDATION OF AN EXTREME SITUATION IN ARTEMOWSK AREA

The extreme situation in Artemovsk area was fixed after revealing radiation dangerous subjects (further in the text - RDS) on an object which in past time belonged to the Ministry of Defence of Ukraine.

Documents, which specified that RDS belong to Armed Forces of Ukraine, were not found. Therefore for participation of Ministry of Defence in liquidation of this extreme situation, a special decision of the Cabinet of ministers of Ukraine (date of decision making - 14.11.2000, №15159/29) was adopted.

For participation in works on liquidation of the specified extreme situation in Artemovsk area a special working group of experts of the Ministry of Defence was created.

The extraordinary situation was a consequence of the detection of an unidentified RDS on the terrain of a former ammunition warehouse of Air Forces. In figure 2, the RDS lie in zones of accumulation.

The operators searched RDS and dangerous segments on the terrain. This search up was executed by spectrometric complex of field VEKTOR (red fingers, figure 4). The complex VEKTOR is made by the Ukrainian Institute of Ecological and Medical systems. By complex VEKTOR the operators have conducted field examination of the whole terrain. The contaminated terrain has occupied about 1 percent of the area, on which there could be RDS. The labs made a refinement of outcomes of field measurements only on the contaminated terrain. Such procedure of measurements has allowed to analyze in lab the only "contaminated" objects and to not analyze samples from clean places. Usage of a complex VEKTOR has given considerable economies of time and means. The measurements by a field complex VECTOR have compared to laboratory researches. The matching has shown coincidence of outcomes of these methods. The high performance of the field method has been confirmed.

Search of RDS and potentially dangerous sites of district were made with the help of a field γ-spectrometer VECTOR complex (red pointers, figure 4) manufactured by the Ukrainian research-and-production firm, Institute of Ecological and Medical systems. Specification of field measurements was made in laboratory conditions. Such organization of the work has allowed to analyze in laboratory only polluted objects and to not consider samples from clean places. Comparison of results of measurements by a field complex the VECTOR and laboratory researches has shown concurrence of results of these methods. High efficiency of a field method has proved to be correct.

The field and laboratory methods have given identical results on the structure of radiation sources. Radiation sources contain Ra-226 and Th-232. The researches of air, dust and aerosols have not detected any α-radiation. RDS were dangerous as a radiation source. In addition, in connection with presence Ra-226 and Th-232, RDS created threat of chemical pollution of district. Pollution might be more than maximum concentration limit (MCL).

According to the legislation of Ukraine there were situations which corresponded to attributes of extreme situations:

10504 - Failures with radioactive unnecessary products which are not connected to nuclear stations;

10401 - Surplus in the soil of harmful substances.

In addition in the Ministry of Defence of Ukraine for performance of liquidation of an extreme situation were involved the Ministry of Ukraine on questions of extreme situations and the Ukrainian state association "Radon". All works were carried out in interaction with the Ministry of ecology of Ukraine, Donetsk regional sanitary epidemiological station and local State administration.

Fig. 2. An accumulation zone of Radiation Dangerous Subjects (RDS)

The organization of the interdepartmental interaction and the general management of liquidation of an extreme situation were carried out by the Ministry of Ukraine concerning the questions of extreme situations.

Experts of Armed Forces of Ukraine have already done most part of works on search and withdrawal of RDS.

From seven working groups which have executed this work, five groups were composed out of experts of the Ministry of Defence of Ukraine. The Ministry of Defence made separate group of servicemen for the most difficult works.

60 military men and 8 special cars of the Ministry of Defence of Ukraine have taken part in the liquidation of the extreme situation. The working group consists of 15 sappers, 13 experts in dose measurements, 3 military ecologists, a military doctor and group of servicemen. Most part of the working group consist of NBC protection experts of the Armed Forces.

The territory of a former ammunition warehouse has an area of 700 000 m2. In this territory a search for RDS and explosive subjects were carried out. Figure 3 represents these works: yellow flags mark the places of collecting RDS.

RDS of a weight to 6165 kg were handed over the State association "Radon" in the Kharkov area after the termination of works.

Fig. 3. Searching for Radiation Dangerous Subjects (RDS) and explosive subjects.

Security measures of works were chosen in view of absence of the necessary information. The degree of danger, the characteristic and condition of RDS were unknown. Therefore experts worked with RDS with observance of the greatest possible norms of safety. Experts worked protected by leather clothes - in overalls of filtering type, rubber gloves and stockings. When there was dust threat, experts put on respirators. Daily, after the termination of works, experts carried out a deactivation of their means of protection. A separate group carried out medical and a radiation control.
During work the external irradiation dose to the staff has not exceeded 7 mRem (at an annual rate limit of 500 mRem). Cases of diseases of staff, or the local population, connected to liquidation of an extreme situation, were not discovered.

The basic conclusions of the carried out work were:
- the legislation has shown its capacity in practice;
- the central authorities have shown ability to organize interaction of various executors;
- the executed works have confirmed expediency of the decision on creation of incorporated working group for liquidation of extreme situations;
- the executed works have confirmed expediency of use γ-spectrometric complex in the field;
- the executed works have allowed drawing conclusions on improvement of the organization of works and equipment of working group by dosimetric devices.

Fig. 4. Field γ-spectrometric complex VECTOR.

5.2. SOLUTION OF PROBLEMS OF A RADIATION SAFETY IN KOSOVO

During the air raids of the NATO in former Yugoslavia in 1999 significant harm was brought to the environment of this country . First of all, pollution has taken place from application of the weapons on chemical enterprises and other industrial targets. The facts of deterioration of health of a certain circle of people are connected with the deterioration of an ecological situation. Many experts connect deterioration of health to application of depleted uranium ammunition. A manual of the Armed Forces of Ukraine regarding the decision on radiating inspection of military men, engineering, property, places of basing, routes of patrolling in a zone of the responsibility of the Ukrainian contingent of peace-making forces in Kosovo was accepted. Under the decision of Minister of Defence of Ukraine in the period from January 23 till January, 27, 2001, radiating inspection in Kosovo was carried out. Check was carried out by working group of NBC protection experts of the armies. The chief of the NBC protection of armed Forces of Ukraine general - lieutenant V. Litvak headed the working group.

Only with depleted uranium ammunition, which one applied on Balkan, there were shells of calibre of 30 mms. These shells came from the "A-10". armed airplanes of of USA air force. Officers of NATO of the working group have transferred documents which contained the list of objects on which, ammunition with depleted uranium was applied. It has allowed conducting examination and parts of such objects. For realization of radiometric examination were used:

1. Field γ-spectrometric complex VECTOR (red fingers, figure 4). The complex was applied for the measurement of the strength of the absorbed dose of γ-radiation on the surface of objects, for the definition of the type of radionuclides and for the measurement of spectra of radionuclides. The calibration of the device was done with the help of monitoring an ionising radiation source of Cs-137. The time for calibrating and measurement was not less than 120 seconds.

2. Automated complex SKRINER (figure 5) was effected by the Ukrainian research-and-production corporation Institute of Ecological and Medical systems. The complex was used for an estimation of the pollution by radionuclides of the organism of persons, for measurement of spectra of radionuclides in a body of person and on the condition of the servicemen. The calibration of the device was made on a monitoring ionizing radiation source Cs-137.

3. For measurement of the strength of of γ-radiation the absorbed dose the nominal radiometer by IMD-12 and radiometer DKS-01 SELVIS of effecting of the Ukrainian research-and-production corporation SPARING-VIST were used.

The exploratory exploitation of a perspective complex for personal dosimetry constructed by the French corporation MGP Instruments has been conducted as a follow-up. In figures 4 and 5 the service-men carrying personal dosimeters DMC2000 (blue fingers), and in figure 6 the unit for reading the information coming from personal dosimeters (blue finger) are shown. The measurement of strength of the absorbed dose of γ-radiation was conducted by arranging the detector directly on the surface of the objects. The measurement of a background spectrum was conducted on uncontaminated terrain of the same type as the terrain that was investigated.

The duration of the measurement of a spectrum of γ-radiation took not less than 300 seconds. In a number of cases the duration of measurement of a spectrum was augmented up to 600 and 800 seconds. Spacing interval of the detector up to object of measurement was 10-15 mms and was separately fixed during the action.

Calibration of a complex SKRINER and measurement of a background spectrum of a were conducted in absence of the patients. The duration of measurement of spectra of radionuclides in a body of the person and on the form of clothes of the servicemen took not less than 120 seconds, in a number of cases it was augmented to about 360 seconds.

During the operation medical examinations of about 350 servicemen serving as special subdivision of Ministry of Internal Affairs of Ukraine (UNMIC) are conducted and their radiation exposure measured.

The diseases, connected with irradiation, were not detected. Also at militaries and on their clothes no excess of the radioactive nuclides norms is revealed. Also potable water and foodstuffs which are consumed by militaries, do not contain radionuclides.

Fig. 5. Automated complex SKRINER.

The overall objective function maximizing is shown up and distribution of limits of interval and gradients in the grate is possible in case cost of injection of successive [9, 10].

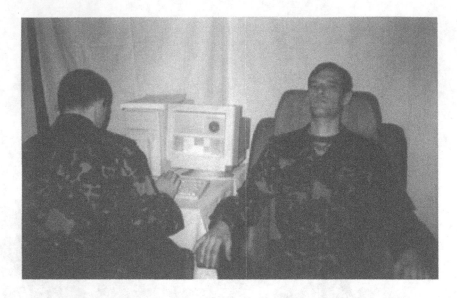

Fig. 6. Reading the information coming from personal dosimeters.

The whole Ukrainian contingent of peace-making forces participated in the observance of safety at performing their tasks in districts where weapons with depleted uranium were used.

Inspection of cars has revealed the presence Cs-137. The maximal pollution made 482.3 Bq/m2. It is below minimally allowable activity on workplaces under requirements of the Ukrainian standards. With all divisions deactivation of the engineering gear was carried out. All cars were deactivated.

Inspection of places of aviation impacts in the zone under the responsibility of the Ukrainian peace-making contingent and in places of performance of tasks has revealed pollution radionuclides Cs-137, Th-232 and Ra-226.

The basic conclusions of the carried out work were:
- threats to the Ukrainian peace-making contingent it is not revealed;
- the zone of a disposition of the Ukrainian peace-making contingent has no territories polluted with the depleted uranium;
- the decision on creation of a uniform complex of the equipment for radiation monitoring and the control was correct;
- VECTOR, SKRINER and DMC 2000 complexes have confirmed high efficiency;
- the obtained experience allows improving of radiating safety system.

6. Conclusions

The overall objective of sanitary-and-hygienic normalization is studying and substantiation of limits of intensity and duration of actions on people in conditions of impact of harmful substances [9, 12].

The considered types of military activity confirm an urgency of application of methods of sanitary-and-hygienic normalization with the purpose:

- choice of parameters on which the establishment of the fact of occurrence of a new source of pollution is possible;
- studying and substantiation of the boundary values, the chosen parameters which authentically confirm presence of pollution.

References

1. The Law of Ukraine about the protection of the population and territory from the extreme situations of technogenic and natural character (in Ukrainian)
2. The Law of Ukraine about the area of extreme ecological situation (in Ukrainian)
3. Decree of the Parliament of Ukraine about the main directions of the government policy of Ukraine in the branch of the environmental protection, use of natural resources and securing of ecological safety, 05.03.98 № 188/98-BP (in Ukrainian)
4. Decree of the Parliament of Ukraine about the conception of the state regulation of the safety and management of nuclear branch of Ukraine, 25.01.94 № 3871-XII (in Ukrainian)
5. Decree of the Cabinet of Ministers of Ukraine about the approval Composite programme the radioactive wastes treatment of 29.04.96 № 480 (with changes, made by Decree of the Cabinet of Ministers of Ukraine 05.04.99 № 542 (in Ukrainian)
6. Decree of the President of Ukraine 23.02.2000 about the primary measures of the implementation of the Message of the President of Ukraine to the Parliament of Ukraine "Ukraine, entrance into XXI century. Strategy of economic and social development for 2000-2004" (in Ukrainian)
7. Charter about the particular partnership between Ukraine and Organisation of the North Atlantic Treaty, 09.07.97, Madrid (Spain) (in Ukrainian)
8. Yu. Izrael. Ecology and control of environment. Leningrad, Hydrometeoizdat, 1984 (in Russian)
9. A. Kachinsky. Risk concept in the context of ecological security of Ukraine. National institute of strategic research. Vol. 14, Kyiv, 1993 (in Ukrainian)
10. A. Kachinsky. Modern problems of ecological security of Ukraine. National institute of strategic research. Vol. 33, Kyiv, 1994 (in Ukrainian)
11. A. Kachinsky,O. Lavrinenko. War in Yugoslavia as a technogenic ecological catastrophe. Strategichna Panorama, Vol. 1-2, 2000 (in Ukrainian)
12. A. Kachinsky, O. Nakonechny. Stability of ecosystems and problem of standardization in the ecological security of Ukraine. National institute of strategic research. Series "Ecological security", Vol. 1, Kyiv, 1996 (in Ukrainian)
13. A. Kachinsky, G.Khmil'. Ecological security of Ukraine: analysis, estimation and state policy. National institute of strategic research. Series "Ecological security", Vol. 3, Kyiv, 1997 (in Ukrainian)
14. S.Zgourets. Start-1, or steeplechase. Narodna Armiya, 23.12.1999 (in Ukrainian)
15. V. Kovalevsky. Concept of evaluation ecological stay the military sites. Proceedings. Defence technologies, Vol.1, Kyiv 1998 (in Ukrainian)
16. V. Litovkin. Second Chernobyl matures at the missile shafts (?) of Ukraine. Izvestia, 16.2.1993 (in Russian)
17. O. Saltykov. Ecological standardization. Problems and perspectives. Ecologia, - 1989, #3 (in Russian)
18. The State programme of the Ukrainian Armed Forces reform and development until 200
19. V. Kovalevskyi, Radiological control and monitoring on the military installation sites in Ukraine. General approaches, SCOPE-RADSITE 12-14.11.98, "Radioactivity from Military Installation sites and Effects on Population Health", Brussels, Belgium
20. E. Prohach. The ecology of nuclear disarmament. (scientific report). The World of Security VIII, International Institute on Global and Regional Security, Kyiv, 1995

LANDSCAPE CRITICALITY INDEXES FOR THE DIFFERENT POLLUTANTS

N. GRYTSYUK
Ukrainian Institute of Agricultural Radiology,
7, Mashynobudivnykiv St. Chabany, Kyiv prov. 255207 UKRAINE
root@inrad..kiev.ua

V. DAVYDCHUK
Institute of Geography, National Academy of Sciences,
44 Volodymyrska St. 01034 Kyiv, UKRAINE
chornob@geogr.freenet.kiev.ua

1. Introduction

Landscape can be defined as a multidimensional and multileveled spatial structure, which unites systematically such natural components as an earth crust with its lithology and relief, ground water, air, vegetation cower and animal population, as well as a soil, which is a result of their interaction. Landscape plays a role for an environment and can be characterized by a number of geochemical parameters, which reflect pathways, barriers, fluxes and balances of redistribution for the different substances, including pollutants. So both a primary and secondary contamination field is forming and evolving within the landscape, under influence of the landscape factors.

This permits use of the landscape approach for the monitoring of different pollutants and further evolution of the ecological situation, for the identification of the ecologically vulnerable landscapes.

Generally the landscape approach in ecotoxicology means a substantiation of the influence of the the sum of natural factors, which evolve themselves (relief, lithology, soil and vegetation cover, geochemical pathways and barriers), on the migration and accumulation of the pollutants, and identification of some areas, which are critical by the combination of these factors.

2. Landscape criticality factors

From this point of view, the criticality of landscape is a result of co-influence of the negative natural factors, which cause the pollutant accumulation at the geochemical barriers, or their intensive biogenic migration and accumulation by the biomass. As the landscape is positioned at the initial link of the trophic/food chain, its criticality reflects the potential risk of the pollutant bioavailability.

Because of chemical similarity of the number of technogenic pollutants (radionuclides and heavy metals etc.), the same methodological approach can be used in ecotoxicology. The ecotoxicological criticality of the landscapes can be defined via indexes of the pollutant balances for the elements of landscape structure and intensity of their biogenic migration.

F. Brechignac and G. Desmet (eds.), Equidosimetry, 361–368.

3. Radiological criticality indexes

For the long-term evaluations and predictions of the pollutants behaviour, to considering the intensity of their horizontal migration and spontaneous decay are of special importance. The experimental data confirm very low velocity of the radionuclides re-distribution between landscape elements by the washing-off at the local level. The ^{137}Cs and ^{90}Sr annual transfer between geochemically adjacent landscapes is about 1% [3], which is less than their losses by physical decay. At least, it means: the consideration of the horizontal migration input/output in the mid-term evaluations is desirable. Moreover, this is essential for the long-living pollutants behaviour forecasting and migration ability evaluation.

The Chornobyl experience increasingly shows the migration processes intensity from the automorphous landscapes to the hydromorphous superaquatic ones, and this is in good correlation with the ^{137}Cs transfer factor meanings for grass and wood [1, 4].

The balance of the radionuclides between different elements of the landscape structure, which are positioned at the different locations of the relief (plateau, slope, pediment or depression), reflects the ability of the ecosystem to evacuate or to collect the pollutants [3]. Both intensity and sign of the pollutant migration varies, following the combination of the landscape factors. Depending on the sign of the migration, the pollutant balance for an areal can be defined as:

- negative (green, predomination of the pollutant evacuation);
- positive (orange, predomination of its accumulation);
- neutral (yellow), where processes of the evacuation and accumulation are balanced in equilibrium (Figure 1).

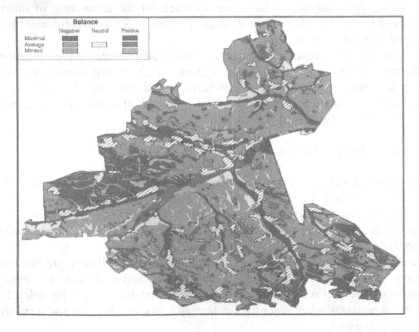

Fig. 1. ^{137}Cs *balances at the landscape areals of the Narodychi district.*

Intensity of the migration varies in the different landscapes, and areals with a strong, intermediate or low evacuation/accumulation of the pollutants reflect the thin local variations of the soil forming deposit lithology, soil, and vegetation cover. Generally, most intensive redistribution of the radionuclides and other pollutants by surface migration are expected at the steep open slope surfaces, which connect autonomous and hydromorphic landscape areals. The mostly intensive accumulation is observed at the slope pediments and the bottoms of depressions, which reflects a role of mechanical geochemical barriers in the accumulation of the surface-migrating pollutants.

The approach used makes it possible to mapping the balance evaluations at the basis of a landscape map. Figure 1 shows intensity and balances of the redistribution of ^{137}Cs or other corresponding pollutant by the washing-off/accumulation for the Narodychi district of Zhytomyr province at the qualitative level. Using appropriate experimental data, this can be transformed into quantitative evaluation, and such a quantitative balance index is one of the landscape criticality indexes.

The phytocoenoses as a biogeochemical barrier consume the pollutant movable forms from soil and ground water and disable that in the phytomass, more or less, depending of their radiological or ecotoxicological capacity and criticality. The transfer factor (TF) is an integral quantitative index, which reflects co-influence of the geochemical, geophysical, and biological factors on the intensity of the biogenic migration, and it can be used for the evaluation of the ecotoxicological criticality (Table 1 and Figure 2).

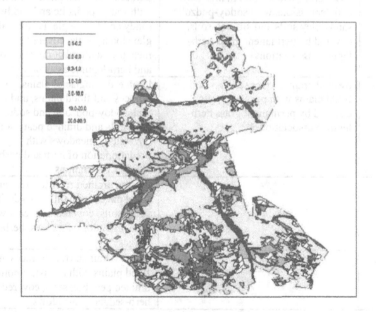

Fig. 2. Means of Tf ^{137}Cs for the natural and cultivated grass associations (Narodychi district, agricultural lands withdrawn).

Table 1. Means of the ^{137}Cs TF to the phytomass of the natural and cultivated grass at the different landscape areals

Means of TF Bq·kg^{-1}/kBk·m^{-2}	Soils and landscape areals under cultivated grass associations	Soils and landscape areals under natural grass associations
0,1 - 0,2	Moraine-fluvioglacial plains and river terraces with soddy-podzolic loamy soils, covered by permanent grass associations with predomination of Dactylis glomerata L.	-
0,2 - 0,3	Moraine-fluvioglacial plains, river terraces and flood plains with soddy-podzolic dusty-sandy and loamy gleic soils, covered by permanent grass associations with predomination of herbs	Moraine-fluvioglacial plains and river terraces with soddy-podzolic loamy soils, covered by meadows with predomination of herbs and legumes
0,3 - 1,0	Moraine-fluvioglacial plains, river terraces, flood plains, and dells with soddy-podzolic and soddy dusty-sandy and loamy gleyic soils, covered by the permanent grass (herb) associations	Moraine-fluvioglacial plains, river terraces and flood plains with soddy-podzolic dusty-sandy and loamy gleic soils, covered by the meadows with predomination of xerophytous herbs
1,0 - 3,0	Lowered moraine-fluvioglacial plains, river terraces and flood plains, dells and depressions with soddy-podzolic sandy gley soils and silt bog soils, covered by permanent grass herb-legume associations	Moraine-fluvioglacial plains, river terraces and flood plains, and dells with soddy-podzolic and soddy dusty-sandy and loamy gleyic soils, alluvial gleic loamy soils, covered by meadows with predomination of herb and herb-legume associations
3,0 - 10,0	Lowered drained river terraces and flood plains with peat bog soils, covered by permanent grass herb-legume associations	Moraine-fluvioglacial plains, river terraces and flood plains, and dells with soddy-podzolic and soddy sandy gley soils and drained peaty soils, covered by meadows with predomination of herb and herb-legume associations
10,0 - 20	-	Lowered drained river terraces and flood plains with (alluvial) drained peaty soils, covered by meadows with predomination of grasses, herbs and sedges
20,0 - 80...	-	Lowered drained river terraces and flood plains with (alluvial) non-drained peat bog soils, covered by herb-sedgeous meadows

Results of studies of the ^{137}Cs soil-plant transfer factor dependence upon landscape structure of the area show that such integral characteristic of biogenic radionuclide migration as TF correlates strongly with the landscape structure of the territory. Consequent analysis and summarising of data allowed to obtaining the following table as an algorithm for the legend to TF map elaboration on the base of the landscape approach.

4. Some evaluations and applications

Data presented in the Table 1 show, that the lowest criticality is specified for automorphous landscapes with low TF values, their meanings mount to 0,1-0,2 for sown grasses, and 0,2-0,3 for natural meadow coenoses. They were estimated for soddy-podzolic sandy loamy and loamy non-gleic soils of the river terraces and moraine-fluvioglacial plains. These soils of the Northern Ukrainian Woodland are characterised by the highest natural fertility and favourable water-physical properties. Moreover, these soils are typical for landscape areals occupying relatively raised, well drained locations with relatively good conditions of surface washing-off and additive potential of self-cleaning from radionuclides due to horizontal outflow during a long-term period. These areals as low critical ones with regard to radiological aspect could be considered as mostly attractive ones for the rehabilitation activity.

Average TF values equal to 0,2-1,0 for sown grasses and 0,3-3,0 for natural meadows are usual to soddy-podzolic sandy loam and loamy gleic soils, soddy-podzolic sandy soils gleic to different extent and alluvial soddy gleic loamy soils. This group includes soils significantly differed by the fertility and their position in landscape. The third group includes soils with highest values of TF, as peaty gleic, peat bog and alluvial peat bog soils. Transfer factor values in these soils mount to 3,0-10,0 for sown grasses and 10,0-80 for meadow coenosis. Alluvial soddy gleic sandy soils draw near to organogenic alluvial ones by high TF values. So, the highest level of radiological criticacy is characteristic for landscape areals in low located territories of hydromorphic type. Level of hydromorphism (automorphism) is determined not only by the type of soil, but also by the location in a relief, its hydrological regime, and the type of plant association. By whole natural properties these areals are considered to be little suitable for use even in conditions of low radioactive contamination.

Extent of soil cultivation influences significantly on ^{137}Cs plant to soils transfer. At the same landscape conditions ^{137}Cs TF for sown grasses on cultivated soils and in natural coenosis on non-cultivated ones can differ more than two orders of magnitude.

Data obtained show that the same radiological criticality level can be observed for landscape areals with types of soils and plant association sufficiently different, formed in various types of relief, but similar by the edaphic (grooving) conditions - fertility and humidity of soils. This fact shows, that the landscape influence due to completeness of natural factors is very important, hence it is very difficult to reveal the mostly important one. The last one concerns meadow coenoses especially, characterised by wide ecological amplitude.

The next step, predetermined by estimated regularity between TF and landscape structure of territory, became the cartographic interpretation of the factor values space distribution on the base of the landscape map. On the table data are presented above the map of TF for the abandoned agricultural lands of Narodychi district (Figure 2).

The map characterizes, at the new level, patterns of the spatial distribution of the TF meanings for the natural and cultivated grass associations. It was used for solving a number of the radiological tasks, both scientific and practical, which are related to the factors of the biogenic migration of the radionuclides. Among them there are the evaluation and forecast of the potential contamination of the agricultural products for the abandoned lands of the Narodychi district, to support the decision making in the rehabilitation of these lands. By the overlay the TF map and ^{137}Cs soil contamination map, the maps of potential ^{137}Cs contamination for hay, and therefore maps for the milk and meat (Figures 3, 4, 5) were generated.

Being based on the actual experimental data, TF mean maps can be used correctly within at least nearest years, if hydro-termic (climatic) conditions are appropriate. For the further perspective, the map will be useful, if the actual TF meanings would be corrected at the base of the TF evolution forecast.

As the grass belongs to the easily contaminated phytomass fractions, the grass TF maps are very useful for the evaluations of the criticality the agricultural lands under any crops.

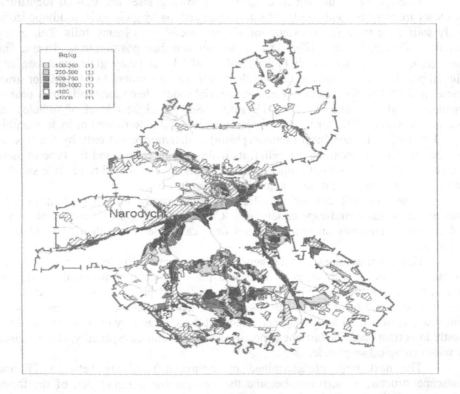

Fig. 3. Forecast of the natural and sowed grass hay by ^{137}Cs contamination.

Fig. 4. Forecast of the milk by ^{137}Cs contamination.

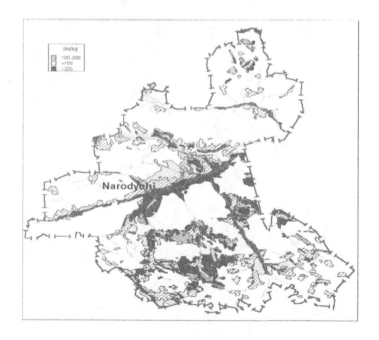

Fig. 5. Forecast of the meat by ^{137}Cs contamination.

5. Conclusion

The effect of the landscape indication permits a reliable inter-component cross-identification within the landscape. Thus, the landscape approach to the evaluation the criticality indexes can be used for the radioecological evaluations, even if the environmental characteristics of the territory affected are limited or non-reliable. The approach seems to be useful also in the general ecotoxicology, as the behaviour characteristics of any pollutant of interest can be easily adopted.

6. References

1.   Grytsyuk N. Dependence of Cs[137] transfer factor for grass on the landscape structure of territory. Visnyk Agrarnych Nauk,. ¹ 4, 2001, p.97-99 (in Ukrainian)
2.   Davydchuk V., Cs-137 balance in the Chornobyl exclusion zone. Ukrainian Geographic Journal, 1996, ¹1, p.39-44.
3.   Davydchuk V., Zaroudna R. et al. Landscapes of Chernobyl zone and their estimation on the radionuclides migration conditions. Kiev, Naukova Dumka, 1994, 112 p. (in Russian).
4.   Sorokina L. On accumulation of Cs[137] by phytocomponents of forest ecosystems depending on edaphic conditions. Ukrainian Geographic Journal, 1996, ¹1, pp. 44-48(in Ukrainian)

ECOSYSTEMS OF 30-KM ZONES OF KHMELNYTSKY AND RIVNE NPP: ESTIMATION OF MIGRATION CONDITIONS OF THE RADIONUCLIDES AND OTHER TECHNICAL POLLUTANTS

L. MALYSHEVA
Center of Ecological Monitoring of Ukraine, Glushkov Prospekt 2, korp.5 Kiev, 02127, UKRAINE

L. SOROKINA
Institute of Geography, NASU, Volodymirska str., 44, Kiev, 01034, UKRAINE

A. GALAGAN, S. GAYDAY, A. GRACHEV, O. GODYNA,
S. DEMYANENKO, S. KARBOVSKAYA, R. MALENKOV, A. NOSON
Center of Ecological Monitoring of Ukraine Glushkov Prospekt 2, korp.5 Kiev, 02127, UKRAINE

1. Introduction

Khmelnytsky and Rivne NPP are two Ukrainian nuclear power stations, where new blocks are built. Their working has to compensate the closing of Chornobyl NPP. Increase of stations capacity by means of the new blocks exploitation requires more detailed estimation of their influence upon the environment. The 30-km zones of radiological control around Khmelnytsky and Rivne NPP were explored. The aim of our investigation was to estimate landscape premises of back reactions of nature upon the NPP's probable negative pressure.

The *main research tasks* are the following:
- to describe the 30-km zones soil cover;
- to analyse the landuse structure;
- to carry on an investigation of landscape and landscape-geochemical structure;
- to estimate landscape and landscape-geochemical conditions of ^{137}Cs and ^{90}Sr migration.

The *methodological base* of the investigation is the use of a landscape approach. It enables to analyse the ecosystems as a genetically unit formation. Investigation of landscape-geochemical structure is also necessary to estimate Khmelnitsky and Rivne NPP influence on environment.
A territorially indivisible image of the transfer of the pollutants is possible by means of using the knowledge about landscape and landscape-geochemical conditions of the migration process coming about in the ecosystems.

The *main investigation methods are*: expedition to landscape and landscape-geochemical exploring; using data of the geochemical analysis; landscape mapping; using GIS-technologies to treat pervious data base, making special thematic and complex maps.

F. Brechignac and G. Desmet (eds.), Equidosimetry, 369–376.

The *base information* to decide the main tasks is the following:

- The results of the expedition of landscape and landscape-geochemical exploring (above 500 complex observation points for the 30-km zone of each NPP)
- Geochemical analyses data (by the results of the exploring expedition)
- Analysis data of the physico-chemical and water-physical soil properties which were made in the years 1980-1990 (the results of State soils exploring of Ukraine)

- Publications and materials about the geological and geomorphology structure, soil and vegetation cover of the explored territory; topographic maps (scales 1:25 000; 1:100 000).

2. Landscape structure of the exploring polygons

The significant nature diversity of the *Khmelnitsky NPP 30-km zone* territory reflects upon its landscape and landscape-geochemical structure, upon the diversity of the water-physical soil properties.

The NPP (its 30-km zone) is situated in the West-Ukrainian province of the Lesostepp zone; it includes the parts of the Volynska and Pivnichnopodilska high plains, Mizochsky chain of hills, Maloye Polissya (forested lowland). The first, second and third of them are the Lesostepp regions of the loess denudated-accumulated and denudated-structured high plains and more low-loess accumulated plains. The top levels of Lesostepp regions is from 220-320 *m*. The main soil types there are grey forest soils, chornozem podzolic and typical chornozem, and the main soilforming rock – loess loam.

Maloye Polissya is the area with outwash-alluvial and terrace landscapes of the Mixed forest zone. Soddy-podzolic sandy soils are representative to the Maloye Polissya. This is the lowest part of the explored territory (average high level – about 200 *m*).

The *Rivne NPP 30-km zone* has a less contrasted landscape structure. This is a part of Volynske Polissya (forested lowland). Chalk-rock plays an essential role in the geological structure of the territory. It is mainly covered with antropogenic sediments. The landscapes of mixed forest zone, namely morainic ridges, moraine-outwash plains, terraces and alluvial plains are typical for the explored part of the Volynske Polissya.

The central part of the Rivne NPP 30-km zone is the highest relief level of the territory (average spot heights 170 – 200 *m*). There is Volynska morainic ridge from West to East of the 30-km zone. Soddy-podzolic sandy soils and oak-pine forests are representing the landscapes of morainic ridges. The ecosystems of the North part of the Rivne NPP 30-km zone are low sandy terraces and alluvial plains (150 – 165 *m*) densely with a swamp forests and meadows. The South part of the 30-km zone is the area where ecosystems of the outwash accumulative and terrace plains (average spot heights 165 – 175 *m*) are predominating. Soddy-podzolic, sod and sod carbonate dusty-sandy and loamy-sandy soils are representing the territory. The good fertility of the soils allows to productively making use of the territory for agriculture.

3. Landscape-geochemical conditions of the territory: investigation methods; characteristic of the polygons explored

The aim of the landscape-geochemical mapping is to study the geochemical background of the chemical elements (natural substances and pollutants) migration. Maps of such kind are used to analyse and estimate the territory ecological condition, the factors of the pervious and secondary pollutants distribution, the space particularity of geochemical influence on the biota and the health of people [2, 3]. The series of factor maps and complex maps shows landscape-geochemical migration conditions of the explored territory.

Landscape information is the base for these maps. Thematic classifiers are worked out and MAPINFO-operations (*Thematic Mapping, SQL Query*) are used for the mapping.

Chemical elements, which have *concentration Clarks* more than a one were determined. *Clark* is an average content (in %) of an element in Earth's crust. *Concentration Clarks* is a correlation between *Clark* of the element in soil at present place and the average *Clark* of the element in Earth's crust. Isolines maps of these elements distribution were made. Chemical elements, which have content (in %) in soil at present place more than its average content in Earth's crust were determined. Isolines maps of these elements distribution were made.

Landscape-geochemical map contains information about classes, types, families and species of Landscape-Geochemical Systems (LGS). Type of the LGS determines "stable" geochemical background. Class of the LGS determines the "mobile" geochemical background; subclass, family and specie are showing physico-chemical properties of LGS. These properties have a principal role for the formation of the chemical elements migration and accumulation conditions [4].

The particularity of geochemical and biogeochemical processes reflectsthe complicated landscape structure of the *Khmelnytsky NPP 30-km zone* territory According to our investigation the ecosystems of the Khmelnitsky NPP 30-km zone are composed of 14 geochemical classes by typeforming elements (table 1).

Forest-steppe landscapes on the calcareous loess rocks abound at the Khmelnitsky NPP 30-km zone territory. This is why calcium is one of the most important typeforming elements here. Ecosystems of the calcium class (Ca^{2+}) compose 12% of the territory; more then 1/3 of the territory being composed by ecosystems of the transitional acid calcium class (H^+-Ca^{2+}). Such geochemical classes are characteristic for ecosystems, which have weak acid soil reaction and not a deep occurrence of carbonate in soil horizons. There are ecosystems of the acid calcium| acid, acid-gleyic class (H^+-Ca^{2+}| H^+, H^+-Fe^{2+}) representing about 13% of the territory. The formation of these landscape-geochemical conditions is connected to the high level of the ground water in depressive landforms.

LGS of the acid (H^+), acid, acid gleyic (H^+, H^+-Fe^{2+}) and acid gleyic (H^+-Fe^{2+}) classes are abounding mainly in the central part of the Khmelnytsky NPP 30-km zone, in the Male Polissya. It is a typical landscape-geochemical feature of the outwash-alluvial plains.

Landscape- geochemical diversity of the *Rivne NPP 30-km zone* territory is less than the territory around Khmelnitsky NPP. The ecosystems of the Rivne NPP 30-km zone compose only 6 geochemical classes by typeforming elements.

Typical for forested lowland's ecosystems classes of migration is the abundance here of acidity: acid class (H^+) – 23%, acid, acid gleyic class (H^+, H^+-

Fe^{2+}) – 22% and acid gleyic class (H^+-Fe^{2+}) – 50% of the territory. This geochemical situation is developing in the territory were poor sandy soils and bog soils are predominant. Cretaceous sediments do not significantly influence on the geochemical background. Hardly any part of the territory contains Ca among the LGS typeforming chemical elements. This landscape-geochemical specifity can be explained by a lot of parameters Quaternary has deposited over the Cretaceous rocks here.

Table 1. Geochemical classes of the ecosystems (by typeforming elements) of the Khmelnytsky and Rivne NPP 30-km zones territories

	Geochemical classes	Chemical indexes	% of the territory	
			Khmelnytsky NPP	Rivne NPP
1	Acid	H^+	5,6	23,6
2	Acid, acid gleyic	H^+, H^+-Fe^{2+}	3,5	23,0
3	Acid gleyic	H^+-Fe^{2+}	12,2	50,0
4	Acid calcium	H^+- Ca^{2+}	35,2	0,5
5	Acid calcium | acid, acid gleyic	H^+-Ca^{2+}|H^+, H^+-Fe^{2+}	13,2	-
6	Acid calcium | acid gleyic	H^+- Ca^{2+}|H^+-Fe^{2+}	3,6	-
7	Calcium	Ca^{2+}	12,0	-
8	Calcium | acid, acid gleyic	Ca^{2+}|H^+, H^+-Fe^{2+}	1,2	0,4
9	Calcium | acid gleyic,	Ca^{2+}|H^+-Fe^{2+}	5,0	2,5
10	Calcium carbonate	Ca^{2+}-CO_3^{2-}	3,7	-
11	Calcium-carbonate | acid, acid gleyic	Ca^{2+}-CO_3^{2-}|H^+, H^+-Fe^{2+}	0,5	-
12	Calcium carbonate | acid gleyic	Ca^{2+}-CO_3^{2-}|H^+-Fe^{2+}	4,0	-
13	Salt	Ca^{2+}-SO_4^{2-}	0,1	-
14	Salt | acid gleyic	Ca^{2+}-SO_4^{2-} |H^+-Fe^{2+}	0,2	-

4. Landscape-geochemical barriers of the exploring polygons

The places of the mostly probable accumulation of the natural and technological elements are shown on the Landscape-Geochemical Barriers (LGB) maps. Contrasting landscape-geochemical migration conditions is the main cause of the different kinds and types of the landscape-geochemical barriers. The ecosystem characteristics, which are causing a severe change of the migration conditions of the chemical elements, and their space situation, were defined.

Landscape-geochemical map and analysing of LGB-forming factors is the base for landscape-geochemical barriers mapping. LGB of the exploring territories were classified according theirs forms (linear and sheet barriers), transfer mechanism (diffusing and infiltrating barriers), genesis (natural and technological).

Dependence on the formation mechanism and main kind of migration of each LGB type is to divided in classes: mechanical, physico-chemical, biogeochemical.

Landscape and landscape-geochemical particularities of the territories around Khmelnytsky and Rivne NPP show that conditions of chemical elements transfer and conditions of the ecosystem's secondary pollution are different here.

5. Estimation of migration conditions of the radionuclides and other pollutants

One of the principal tasks of the investigation is the complex estimation of the ecosystems as a scene of radionuclides transfer by water and biogenic migration.

Research experience [1] shows the advantages of using a landscape approach to decide on the problems devoted to forecasting of the radionuclides behaviour in the environment. Pollutants join in natural processes of the cycle of matter. Therefore geochemical conditions and migration processes intensity depend on the structure and dynamics of the contaminated territory landscapes. Landscape approach allows to making a complete picture of the potential capacity of ecosystems to carrying out, to transporting and to accumulating the different pollutants.

There are the following types of migration: mechanical, physico-chemical, biogenic and technical. The first of them is the result of the effect of mechanic, hydrodynamic and gravity laws (rockfalls, slumps, erosion processes etc.) Physico-chemical migration is submitted to laws of interaction of matter and energy. Its main means are diffusion, dissolving, sorption, desorption and so on. Biogenic migration is a result of activity of living organisms and biochemical processes after their death. Transference of matter and transformation of energy occurs by means of photosynthesis, respiration, biogenic absorption of soil mineral substations, and mineralization of organic substation. Technogenic migration is a result of human activity (contamination of environment).

Some landscape characteristics that determine the particularity of chemical elements' migration can be distinguished.

The landforms are the main criterion of the surface wash-out forming and water mechanic transference of matter. Among all kinds of migration, water mechanics and water physico-chemical migration define the main particularities of the chemical elements transference by radial and lateral drifts. Therefore most active carrying out of substances by water sheet is being observed on the slopes. The active carrying out of substances from the river flats is occurring seasonally during the floods. Transit, that is to say transfer and partial accumulation of the substances, characterizes the ecosystems of the inclined planes, of the erosion net and some other systems. Accumulation of substance is occurring on the low planes, low terraces, in depressions.

Soils characteristics are important criteria for the geochemical transference and surface washing-out of material. The fixation capacity of a soil is depending on its texture. Quantity of the fixed pollutants in soils depends on quantity of the fine clay minerals (fraction $<0,01$ mm). Correlation between the fraction content of the soil and fixing capacity of soil has been established.

Washing out of soils is an indicator of the deluvial process intensity. The consequence of the deluvial process is destruction of the soil cover and transport of the technogenic pollutants across the soil surface. Therefore the potential to carrying out ^{90}Sr and ^{137}Cs on the slopes was classified according to the degree of the soil washed out.

Soil gleying is an indicator of regime of soil humidity. Gleying is an important factor of ^{90}Sr and ^{137}Cs mobility, which determine their migration conditions in ecosystems. Changing of the soil humidity (degree of gleying) shows changing of the conditions of the chemical elements dissolving and possibility of their redistribution by water physico-chemical migration.

Quantity of the organic matter – the not-decomposing and the one in the stage of decomposition affects the radionuclide behaviour in the peat soils essentially. As a result of the plant remains decomposition, the aggressive organic acids are being exudated. The acids increase the radionuclides mobility and aid their migration activity. When, however, remaining plants have been decomposed, the radionuclides accumulated by them are involved in migration floods. In such a way the cycle of these elements in system "soil – plant" is established. That is why only ecosystems with a peat soil that had formed in the conditions of wash-out regime (river plans, small flat-bottom valleys) were classified as zones of active carrying out of ^{90}Sr and ^{137}Cs. The ecosystems with a peat soil where wash-out regime is absent (low swamp terraces, flat-bottom depressions) were classified as zones of accumulation of these radionuclides.

Actual vegetation is an important biological and mechanical factor for the radionuclides transfer.

Some part of contamination accumulates in wood (in perennial plants) and remains here during a long time (the trees' lifetime). The quality of the radionuclides (mainly ^{137}Cs and ^{90}Sr) which migrate on the territory is increased by its fixation on the phytomass. The landscape predisposition of the radionuclides (Cs^{137}) accumulation by the main forestmaking kinds of trees in the territory of the Rivne NPP 30-km zone was analysed. Potential plant capacity to bioaccumulating a substances from soils is defined by the transfer coefficient. Combining landscape and edaphical approaches is the theoretical base to study the space differences of the migration intensity in the system "soil - plant" [5]. Landscape map and forest map is the cartographic base to analyse the potential capability of the forest ecosystems to catch Cs^{137} in wood.

Besides, actual vegetation cover (situation of forests, meadows, plough-lands) is a factor of mechanical transfer too. Intensity of pollutants' surface wash-out and flow decreases through stabilising the soil surface by roots.

Landscape and landscape-geochemical characteristics, information about relief, actual vegetation and landuse of the territory were used to make a classification of ecosystems for the estimation of the radionuclides migration conditions. According the discussed criteria of the classification, the maps "Carrying out, transport and accumulation of ^{90}Sr and ^{137}Cs" are made by using the landscape map as a cartographic base by means of GIS-tools (Figure 1). The maps show the location of ecosystems with about 20 different types of migration conditions.

Fig. 1 Carrying out, transport and accumulation of ^{90}Sr and ^{137}Cs in the ecosystems of 30-km zone of Rivne NPP (fragment):
1 – 3 (green colours) - carrying out (increasing of the process intensity from 1 to 3); 4 – 6 (dark-green colours) - carrying out from river plans ecosystems during a flood; 7 (blue colour) - accumulation and seasonal carrying out from river plans' ecosystems during a flood; 8 – 10 (yellow colours) - transport; 11 – 18 (pink and red colours) - accumulation; 19 – 21 (yellow-green colours) - primary accumulation and following intensive carrying out.

6. Conclusions

Knowledge about natural laws of substance migration in the environment, expedition and experimental data of landscape structure and landscape-geochemical peculiarities of the territory make it possible to forecast the conditions of ^{137}Cs and ^{90}Sr migration in ecosystems. Our investigation corroborated once more the sensibility of using a landscape base for these prognoses.

The main landscape and landscape-geochemical characteristics, which have to be taken into consideration, are: landforms, soil texture, washing out of soils, soil gleying, quantity of organic matter, regime of humidity of ecosystems, actual vegetation, edaphical conditions.

The maps "Carrying out, transport and accumulation of ^{90}Sr and ^{137}Cs" show ecosystems with a different transfer conditions, including areas of active water- and biogenic accumulation of the contaminants. These potential zones of contamination were recommended to be included into a radioecology monitoring net of the NPP 30-km zones.

The results of this investigation can be used for the estimation of migration conditions of radionuclides, other technical pollutants and their complexes as well.

7. References

1. Davydchuk V.S., Zarudnaya R.F., Mikcheli S.V. et al. Landscapes of Chornobyl zone and their estimation by migration conditions of the radionuclides. K., Naukova Dumka, 1994. 112pp. (in Russian).
2. Malysheva L.L. Landscape-geochemical mapping of ecological states of territories. Environmental Geochemistry. 3th International Symposium, 12-15 September, 1994. – Krakow, 1994
3. Malysheva L.L. Landscape-geochemical estimation of ecological state of territories. – Kiev: Edditing Centre "Kiev University", 1997. - 264 p. (in Ukrainian).
4. Perelman A. Geochemistry of Landscapes. M., Vysshaya Shkola, 1975, 342 pp. (in Russian).
5. Sorokina L.Yu. About accumulation of ^{137}Cs by phytocomponents of the forested ecosystems depending the edaphic conditions. Ukrainian Geographical Journal. – 1996, ¹ 1. pp. 44-48. (in Ukrainian).

Part 8.

Possibility of Standardization of Radionuclides and Chemotoxicants

RADIONUCLIDES ^{137}Cs AND ^{60}Co UPTAKE BY FRESHWATER AND MARINE MICROALGAE *CHLORELLA, NAVICULA, PHAEODACTYLUM*

M. ŠVADLENKOVÁ
University of South Bohemia, Dept. of Physics, Èeské Budìjovice, CZECH REP.

J. LUKAVSKÝ
Institute of Botany, Academy of Sciences of Czech Republic, Tøeboò, CZECH REP.

J. KVÍDEROVÁ
University of South Bohemia, Dept. of Botany, Èeské Budìjovice, CZECH REP.

1. Introduction

Water plants accumulate to a large extent radionuclides and heavy metals [1, 2]. This fact is used among others for surface water pollution monitoring purposes as well.

It is proved [1, 5] that the radioisotopes of some physiological elements, such as ^{32}P, and furthermore the radioisotopes of those elements such as Co, Y, Fe, Ce, Zr are largely accumulated by water plants. These radionuclides concentration in plants is up to four orders higher than in the ambient water milieu. The radioisotopes of elements, Cs, Cr, Sr, I, S, Ca, are much less accumulated .

Generally, water plants, which have a large effective surface with the predominant part submerged in water, have a big sorption capacity. These are for example water mosses and algae. Here for example, radiocaesium is most accumulated by the microalgae *Oocistis elliptica* and *Oedogonium sp.*, by the filamentous algae *Cladophora sp.* and by the moss *Fontinalis antipyretica*; a big accumulation capacity has the duckweed *Lemna minor*.

The nature of radionuclide uptake from water by plant can be passive (non-metabolic) or active (metabolic).

Diffusion, convective flow, physical adsorption and chemical sorption are passive processes. Radioisotopes such as Na, Cl, Ca, Co, Sr, Cs, Ce are partly fixed at certain plants with the help of these processes [6]. Experiments with ^{137}Cs, have shown that the part of it which was fixed on the plant surface is 14% for *Zanichellia sp.*, 46% for *Potamogeton sp.*, 15% for *Myriophyllum sp.*, 27% for *Spirogyra sp.*, 44% for *Cladophora sp.* and 36% for *Chara sp.* [7]. Except for *Potamogeton sp.*, ^{137}Cs is adsorbed mainly on algae.

The active processes are complex actions the course of which depends, apart from others, on the presence of enzymes. The kinetics of the active transfer can be seen if the modified Langmuire equation developed by Michaelis and Menten is used for the enzymes kinetics [8, 9].

For water plants (as well as for water animals) the uptake capacity of the certain radionuclide depends on a range of factors. The plant species and its vegetation period, the type of radionuclide and its physical-chemical speciation, (radio)activity concentration in water and its time course, mineral composition and

F. Brechignac and G. Desmet (eds.), Equidosimetry, 379–387.

pH of water, time of contamination, presence or absence of other plants or animals, stagnancy or move of water, water temperature belong among the most important.

The concentration factor (CF) is the quantity, which characterizes the amount of a radionuclide accumulation by water plant under steady state conditions. It is defined as a quotient of the radionuclide concentration in the organism or tissue and of the radionuclide concentration in the ambient water. CF gives the evidence neither about the uptake and elimination kinetics of radionuclide nor the character of uptake processes. Nevertheless, CF is a useful indicator for the biota's suitability for biomonitoring or bioindication of the surface water radioactive pollution. Mathematical models of the radionuclides transfer from water to plants are developed to provide more precise assessment of water pollution [10, 11, 12].

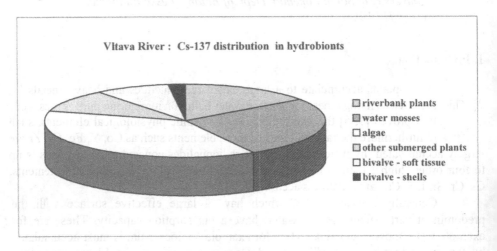

Fig. 1. Distribution of the ^{137}Cs activity concentration in hydrobionts from the Vltava River (Czech Rep.) before the running of NPP.

Hydrophytes are also used for the assessment of the radionuclides activity in the hydrosphere before and after starting-up the nuclear power plants operation. In the Czech and Slovak countries the NPP Bohunice, Dukovany and Temelin were involved in [13, 14, 15]. The algae and the mosses have proved to be suitable for this biomonitoring.

For example, in the years 1996-2000, before the start of the PWR type nuclear power plant in Temelin (South Bohemia, Czech Rep.), the monitoring of the Vltava River, where the liquid radioactive effluents will be discharged, was accomplished („point zero" assessment). The biomonitoring of the hydrosphere was made by means of water plants and bivalves. The representation of the ^{137}Cs activity concentration in hydrobionts from the Vltava River is shown in Figure 1. It can be seen that ^{137}Cs is concentrate most of all in micro-algae mixture and in water mosses while least of all in bivalve shells. The results from the year 2001 - the first year after the running of the NPP Temelín - are similar to those of the point zero assessment.

Considering a more precise interpretation of measured values of radionuclides activity concentrations in water plants and for deeper understanding of radionuclides uptake mechanisms by individual plant species as well it is suitable to study the chosen phenomena at laboratory conditions.

In this contribution, some results of the laboratory experiments of the radionuclides uptake and elimination by the freshwater and saltwater algae are presented. The radionuclides, ^{60}Co and ^{137}Cs, that are relatively important from the point of view of the nuclear industry environmental impact are chosen.

2. Materials and methods

Algae and cultivation

The experiments were pursued in batch cultures with the freshwater diatom *Navicula* (KÜTZ.) GRUN, strain KOVÁÈIK 1978/9, the freshwater chlorococal algae *Chlorella kessleri* FOTT & NOVÁK. strain LARG/1 and on the marine diatom *Phaeodactylum tricornutum* BOHLIN. Strain COUGH 632. The algae were supplied by the Culture Collection of Algal Laboratory of the Institute of Botany of Trebon (Czech Republic).

The saltwater diatom *Phaeodactylum tricornutum* is one of the most used algal species in the marine biotest because of its easy cultivation [16]. *Navicula atomus* is the representative of freshwater diatomae. *Chlorella kessleri* is the representative of common cocal algae and it is often used in the biotests.

These algae species have a different shape but comparable dimensions. This is important when comparing the capability of the algae to adsorb radionuclides on their surface for individual species.

The freshwater and saltwater algae were cultivated in solution (medium) Zehnder [17] and in the artificial seawater after ISO 10253 [18]. The experiments on the algae were then pursued in volumes of 1 litre with solutions of the same composition, which was used for algae cultivation. The solutions preparation is given in Tables 1 (freshwater) and 2 (saltwater).

Table 1. Freshwater solution preparation

Into 1 000 ml dist. H2O:	NaNO3-467mg; Ca(NO3)2.4H2O-59mg; K2HPO4-31mg; MgSO4.7H2O-25mg; Na2CO3-21mg; Fe-EDTA stock sol.-10ml; Microelements after GAFFRON-0.08ml.
Into 500ml of dist.H2O.	Fe-EDTA stock solution: 5ml of 0.1 N solution of FeCl3.6H2O in 0,1N HCl; 5ml of 0.1 N solute. Na2-EDTA;
GAFFRON's microelements: Into 100ml dist. H2O:	H3BO3-310mg; MnSO4.4H2O-223mg; Na2Vo4.2H2O-3.3mg; (NH4)6Mo7O24.4H2O-8.8mg; KBr-11.9mg; KJ-8.3mg; ZnSO4.7H2O-28.7mg; Cd(NO3)2.4H2O-15.4mg; Co(NO3)2.H2O-14.6mg; CuSO4.5H2O-12.5mg; NiSO4(NH4)2SO4.6H2O-19.8mg; Cr(NO3)3.7H2O-3.7mg; V2O4(SO4)3.16H2O -3.5mg; Al2(SO4)3K2SO4.24H2O-47.4mg.

Table 2. Artificial seawater preparation

Into 1 000 ml dist. H2O:	NaCl-22g; MgCl2.6H2O-9.7g; Na2SO4(anhydrous)-3.7g; CaCl2(anhydrous)-1.0g; KCl-0.65g; NaHCO3-0.20g; salts of H3BO3-0.023g; nutrient stock solution No.1-15ml; nutrient stock solution No. 2- 0.5ml; nutrient stock solution No. 3 - 1ml.
Stock solution No.1: into 1 000 ml dist. H2O:	FeCl3.6H2O-48mg;MnCl2.4H2O-144mg; ZnSO4.7H2O-45mg; SuSO4.5H2O-0.157mg; CoCl2.6H2O-0.404mg; H3BO3-1140mg; Na2EDTA-1000mg.
Stock solution No.2: into 1 000 ml dist. H2O:	Thiamine hydrochloride-50mg; Biotin-0.01mg; Vitamin B12 (cyanocobalamin)-0.10mg.
Stock solution No.3: into 1 000 ml dist. H2O:	K3PO4-3g; NaNO3-50.0g; Na2SiO3.5H2O-14.9g.

For each experimental culture the equations of the optical density conversion to dry weight of the algae biomass ($g.mL^{-1}$) and to the number of cells (mL^{-1}) were set.

The algal suspension was put into polyethylene bottles with supply of an air and CO_2 mixture. The flow rate of CO_2 was regulated to maintaining the pH value of suspension at a constant value. The mean pH values of the individual batch cultures during the experiment are given in the table 3.

The radionuclides uptake by algae

The experimental suspensions were contaminated in one-shot way with the radionuclides, ^{60}Co and ^{137}Cs, in the form of chlorides. After the contamination, three parallel samples of the suspension were taken from each experimental bottle. The activities of the unfiltered suspension, the filtered suspension and filters with algae were measured. Polycarbonate membrane filters Sartorius of the pore diameter of 0,45 µm were used for the filtration, which was made through two filters placed one on the other. The filter underneath served for the correction of radionuclides self-adsorption on the filter. After the filtration the individual filters were mineralised directly in the measuring vessel.

Table 3. Mean pH values of the individual batch cultures (experiment duration period in days is indicated in parenthesis).

Culture	mean pH	stand. dev.
Navicula atomus	6,760 (14 d)	0,133
Chlorella kessleri	6,891 (18 d)	0,132
Phaeodactylum tricornutum	6,733 (10 d)	0,181

At every sampling, the values of pH and the suspension optical density were measured. The pH values were measured by means of pH meter WTW 330i with gel electrolyte electrode Sentix 41-3. The optical density at 750 nm was measured by Titertek Uniscan II for samples in the plate of well volume of 3 mL. During the experiment the samples for bacteriological test were taken and fixed.

The radionuclides elimination from algae

About ten days later when radionuclides were sorbed, the algae were gently filtered on the membrane filter and then washed from the filter into the inactive medium. In this way the medium was periodically changed for a fresh one during the elimination experiment.

At given time intervals, three parallel samples of suspension were taken and filtered for the algae activity measurement. It was found that the self-adsorption on the filters was negligible in that case.

The pH value and the optical density were measured at every sampling.

3. Results and discussion

The radionuclides uptake by algae

The time course of the radionuclides uptake by individual algal species is shown on Figures 2, 4 and 5. The corresponding decrease of radioactivity in the filtered medium is also depicted on these figures. The separation of the volume activity between the medium and the algae relatively to the initial volume activity in the batches is represented on these figures. The part of activity adsorbed on the surface of experimental vessels was deduced from the time course of suspension (medium with algae) activity. The estimated values are 5-10% for ^{137}Cs and 10 -50% for ^{60}Co (the upper limit is valid for the batch with *Chlorella kessleri*).

The values of concentration factor, CF, were estimated in steady state, about 7 days after the time of contamination of the algal suspension in batch. They are shown in Table 4. The concentration factor is calculated according to the following formula:

$$CF = \frac{\text{activity in algae (Bq/g dry weight)}}{\text{activity in filtered medium (Bq/mL)}}.$$

Table 4. The estimation of the concentration factors CF for the radionuclides, ^{60}Co and ^{137}Cs in three algal species.

Algal species	CF	
	Cs-137	Co-60
Navicula atomus	300	85000
Chlorella kessleri	1300	3200
Phaeodactylum tricornutum	30	30

The volume activity in algae relatively to the unitary optical density of the algal suspension is shown in figure 6. It can be seen that the uptake varies significantly both for the type of radionuclide and for the algal species. For the freshwater diatom *Navicula atomus* the uptake of cobalt is substantially higher than the uptake of caesium. In the case of cocal algae *Chlorella kessleri* the difference in the uptake of these two radionuclides is much less evident. For the saltwater diatom located in the ISO 10253 medium, the difference of the uptake for two radionuclides were not observed and the quantity of radionuclides concentrated by this algae species was very small.

The cobalt uptake of *Navicula atomus* is much higher than the one of *Chlorella kessleri*. The uptake of *Phaeodactylum tricornutum* in ISO 10253 medium is one to two order and two to three orders of magnitude less for caesium and cobalt, respectively, than the uptake of two freshwater algal species.

The radionuclides elimination from algae

The elimination kinetics of radionuclides, ^{60}Co and ^{137}Cs, is shown in Figure 3 for the algal species *Navicula atomus* and *Chlorella kessleri*. Because of the very weak radionuclide uptake by *Phaeodactylum tricornutum*, the elimination kinetics was not measured in frame of this experiment. For *Chlorella kessleri*, the activity of both radionuclides decreased in one hour to 30% of the initial value and than remained practically constant. At *Navicula atomus*, the decrease is slower than at the *Chlorella kessleri*. The activity decreased to 50% in about 1 and 4 days for ^{137}Cs and ^{60}Co, resp. After about 12 days the initial activity of *Navicula atomus* fell to 15% and 40% for ^{137}Cs and ^{60}Co, respectively.

Conversion curves

The conversion curves were calculated for each culture. The summary of curves is shown in Table 5. The regression coefficients $R^2 > 0.99$.

$$OD_{750} = (a+c*x) / (1+b*x),$$

OD_{750} = optical density at 750 nm,
x = dry matter [g.ml^{-1}] or cell number [10^6.ml^{-1}]

Table 5. Parameters of conversion curve for three measured algal species.

Algal species		Dry matter [g.ml^{-1}]			Cell number [10^6.ml^{-1}]		
		a	b	c	a	b	c
Phaeodactylum tricornutum		-0,06054	4443,65	6232,19	-0,06054	0,20966	0,2940 5
Navicula atomus		0,01827	595,682	2623,32	0,01827	0,05383	0,2370 8
Chlorella kessleri	OD<1.3	0,02499	3533,07	7411,53	0,02499	0,17118	0,3591 0
	OD>1.3	0,36052	854,726	3523,40	0,36052	0,04141	0,1707 1

Figures 2-6 : actual values -- o -- Cs-137 , -- ■ -- Co-60

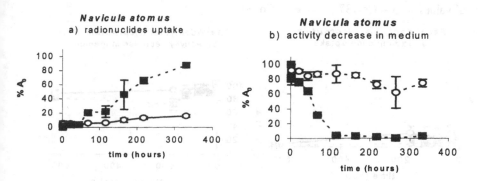

Fig. 2. a) The time course of accumulation of the radionuclides, ^{137}Cs and ^{60}Co, in algae Navicula atomus. b) The relative decrease of activity in the filtered medium. A_0 is the initial radionuclide volume activity in the batch.

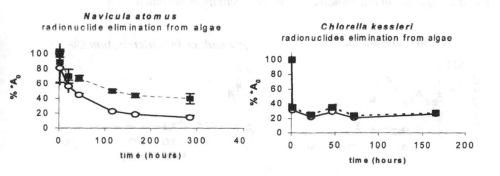

Fig. 3. The elimination of radionuclides from the algae Navicula atomus and Chlorella kessleri into the non-active medium. $^*A_0$ is the initial radionuclide activity in algae.

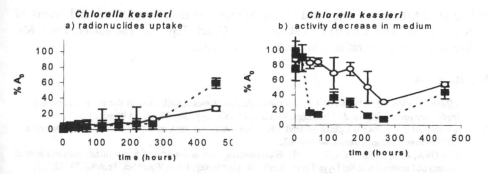

Fig. 4. a) The time course of accumulation of the radionuclides, ^{137}Cs and ^{60}Co, in algae Chlorella kessleri. b) The relative decrease of activity in the filtered medium. A_0 is the initial radionuclide volume activity in the batch.

386

Actual values -- o -- Cs-137 , -- ■ -- Co-60

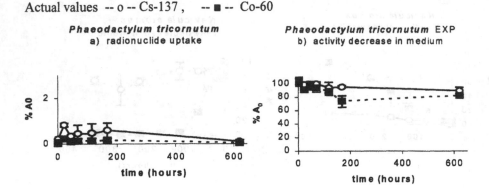

Fig. 5. a) The time course of accumulation of the radionuclides, ^{137}Cs and ^{60}Co, in algae Phaeodactylum tricornutum. b) The relative decrease of activity in the filtered medium. A_0 is the initial radionuclide volume activity in the batch.

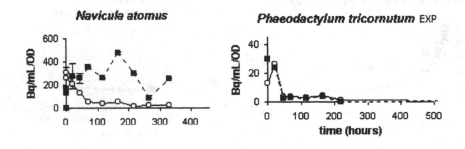

Fig. 6. The uptake of radionuclides, Co-60 (--■--) and Cs-137 (□o□), by two algae species. OD = optical density of algal suspension measured at 750 nm.

4. Acknowledgement

This project ME 419 is supported by the Ministry of Education, Youth and Sports of Czech Republic, and K 60 051 14 of Acad. Sci. Czech Republic. The authors thank Mr. K. Svadlenka for cooperation with the data processing.

5. References

1. TIMOFEYEVA-RESOVSKAYA,E.A. (1963) Radioisotopes distribution in the main components of freshwater reservoirs. Trudy Uralskogo filiala AN SSSR, vyp. 30., 77 pp. (in Russian)
2. VYMAZAL, J. (1990a): Uptake of lead, chromium, cadmium and cobalt by *Cladophora glomerata*. – Bull.Environ. Contam.Toxicol. **44**: 468-472.
3. ŽÁKOVÁ, Z. & KOÈKOVÁ, E. (1998): Biomonitoring and assessment of heavy metal contamination of streams and reservoirs in the Dyje/Thaya river basin, Czech Republic – Water Sci. Technol. **39**: 225-232.
4. FOULQUIER, L. & BAUDIN-JAULENT, Y. (1992) Impact Radioécologique de l'Accident de Tchernobyl sur les Ecosystèmes Aquatiques. CEC - Radiation protection 58, UIR, No. 90-ET-014.

5. VANDERPLOEG, H.A. *et al.* (1975) Bioaccumulation Factors for Radionuclides in Freshwater Biota. Oak Ridge Nat. Lab., Environ. Sci. Div. Public. No. 783, ORNL-5002, 222 pp.
6. BARINOV, G.B. (1970) Dynamic characteristics of the radioecological concentration process. In: Marine radioecology. Kiev, Naukova Dumka, p. 69. (In Russian)
7. GLOYNA, E.F. et al. (1966) Transport of radionuclides in model water. In: Disposal of Radioactive Water into Seas, Oceans and Surface Waters, Proc. Symp. Vienna, IAEA, p. 11-32.
8. SUTCLIFFE, J.F. (1964) Mineral Salts Absorption in Plants. (ruský pøeklad) Moskva, Mir, 221 pp.
9. THORNLEY, J.H.M. (1976) Mathematical Models in Plant Physiology. London, Academic Press, 318 pp.
10. KULIKOV, N.V. & TCHEBOTINA, M.Ya. Radioecology of Freshwater Biosystems. AN SSSR, Ural´s Dept., Sverdlovsk 1988 (in Russian)
11. RICE, D.L. (1984) A simple mass transport model for metal uptake by marine macroalgae growing at different rates. J. Exper. Marine Biol. Ecol. 82: 175-182.
12. SVADLENKOVA, M. -DVORAK, Z. -SLAVIK, O. (1989) A mathematical model for ^{137}Cs uptake and release by filamentous algae. Internat. Revue der Gesamten Hydrobiologie, **74** (4): 461-469.
13. SLÁVIK, O. - OBDRŽÁLEK, M. - JANEÈKA, S. (1983) Occurrence and behaviour of the radionuclides in hydrosphere of the NPP Bohunice environment (in Slovak). Research Report VÚJE Jaslovské Bohunice 1983. 42 pp. (in Slovak)
14. STANÌK, Z. (1988) Impact of the Nuclear Power Plant Dukovany on the River Jihlava. Vater Menagement **8**: 214-217. (in Czech)
15. ŠVADLENKOVÁ, M. & HANSLIK, E.: Biomonitoring of the Vltava River Radioactivity. In: Proceedings of the int. conf. ECORAD 2001: Radioécologic - ecotoxicologic des milieux continentaux et estuariens. France, Aix-en-Provence, 3-7 Sept 2001 (in press)
16. LUKAVSKÝ, J. & SIMMER, J. (2001): Microprocedure for the standard marine algal bioassay (ISO 10253). Algological Studies 101: 137-147.
17. STAUB, R.(1961): Untersuchungen an der Blaualge *Oscillatoria rubescens* DC. - Schweiz. Z. Hydrol. **23**: 83-198.
18. ISO 10253 (1995): Water quality – Marine algal growth inhibition test with *Skeletonema costatum* and *Phaeodactylum tricornutum*. 8 pp.

PRINCIPAL COMPONENT ANALYSIS OF CHRONIC INFLUENCE OF LOW-DOSES OF IONIZING RADIATION AND CADMIUM ON ORGANISMS

V. VOITSITSKY, S. HIZHNYAK, O. KYSIL
N. KUCHERENKO
National Taras Shevchenko University of Kyiv, Biological Faculty, 64
Volodymyrska street, 01033 Kyiv, UKRAINE
E-mail: vmv@biocc.univ.kiev.ua

A. KURASHOV, *National Taras Shevchenko University of Kyiv,*
Radiophysical Faculty, 64 Volodymyrska street, 01033 Kyiv, UKRAINE
E-mail: vnk@mail.univ.kiev.ua

L. BEZDROBNAYA, *Institute for Nuclear Research, National Academy*
of Sciences of Ukraine, 47 Nauki Ave., 03680 Kyiv, UKRAINE

1. Introduction

Exceptional variety of different physical and chemical actions on biota outlines the topical environmental problems resulting from combined biological influence of external agents on organism [1, 2]. The reaction of a biological object on ionised radiation and chemical agents over the area of low dose effects (LDE) becomes highly versatile and ambiguous. The study of organism status by conventional sequential statistical analysis of individual indices as compared with their control levels by variation model does not provide a way of estimating its manifold reactions. Adequate valuation of organism reaction on such complex influence may be done only by manifold correlation of various biological indices. The most perfect method to solve a problem is provided by the principal component analysis (PCA) [3, 4]. With this method, it becomes possible:

- to evaluate latent but objective regularity in the control mechanism of the processes in the biological object under consideration;
- to obtain essential information by process description with minimum number (much less than the number of initial indices) of resulting appreciable variables;
- to reveal and study statistical dependence of biological status on external influences from whence one can draw a valid conclusion about the effectiveness of agent action and predict the process evolution.

The objective of this paper consists in the investigation of LDE under combined chronic action of low-dose ionizing radiation and cadmium ions on the basis of morphologic and biochemical indices of various rats' organs by PCA method.

2. Materials and methods

The white underbred rats (180-200g) were divided into 10 groups (6-8 animals in each group) according to the experimental conditions: control (1),

F. Brechignac and G. Desmet (eds.), Equidosimetry, 389–402.
© 2005 *Springer. Printed in the Netherlands.*

received 0,01mg/l of CdCl$_2$ solution during 42 days (2), 84 days (5), 145 days (8); subjected to γ-irradiation during 42 days (3), 84 days (6), 145 days (9); subjected to combined influence of irradiation and CdCl$_2$ during 42 days (4), 84 days (7), 145 days (10). Animals were chronically irradiated by γ-quanta from ^{60}Co (dose rate 0,72 cGy/per twenty-four hours) at installation "ETALON" (Scientific centre of INP, NAS of Ukraine) and decapitated on the 42, 84 and 145 days after gaining the total absorbed doze of 0,3; 0,6 and 1,0 Gy, respectively. The objects of the research were blood, liver and intestinal parts (duodenum and jejunum) according to their high sensitivity to cadmium effect along with the differences in their radiosensitivity. Cytophotometric researches were carried out by light microscopy [5]. Enzyme activities were determined by routine procedures [6, 7]. Lipid peroxidation products were detected as described in [8]. The lipids content was assayed by enzyme methods according to [9]. The measured variables are grouped in Table 1. Hereinafter only the parameter 43 numbers are used for short.

3. Main features of PCA model in LDE estimation

The general condition of a biological object is qualitatively characterized by a great set of indices, which account for the dose-dependent action of external agents on the organism. The conventional approach to the problem based on the sequential analysis of the finding (a correlation between every individual index and its value in control group) does not ensure an unambiguous interpretation of experimental results. The adequate estimation of the combined action of small dose of radiation and xenobiotics must be based on radically different integral vital signs, which correctly determine the effect of the action. It should be mentioned that the procedures presently used (see, for example, [2, 10]) cannot be tolerated just for this reason. Consequently, the problem of the mathematical analysis of LDE under the combined action of several external agents is the development of proper multi-factor and informative estimation, which is valid far from lethal doses. We propose for this purpose the multi-factor PCA-based model, which was effectively used when studying the acute damaging action of ionizing radiation and cadmium ions on the organism of rats [11].

The PCA model is used in the analysis of a great many interdependent characteristics of the objects under consideration with regard to their structure and associations [3]. According to this model, the set of observed variables $\mathbf{x} = \{x_1, x_2, ..., x_N\}$ is represented as a linear combination of some hypothetic variables, principal components, which most closely match their correlation. The ultimate aim of the analysis consists in the determination of such transformed data $\mathbf{a} = \{a_1, a_2, ..., a_M\}$, $M < N$, which maximize statistical compatibility of observed correlation with number M of used variables a_i as few as possible.

Table 1. Biological indices used for an estimation of ionizing radiation and cadmium influence on organism

	Enterocyte height		27	Content of malondialdehyde (small intestine)
1	Duodenum			
2	Jejunum		28	Rate of MDA accumulation in small intestine
	Enterocyte nuclear area			
3	Duodenum			**Lipid content (liver)**
4	Jejunum		29	Triacylgliceroles
5	**Hepatocyte nuclear diameter**		30	Phospholipids
	Mitotic index		31	Cholesterol
6	Duodenum			
7	Jejunum			**Lipid content (small intestine)**
	Cell quantity in kripta			
8	Duodenum		32	Triacylgliceroles
9	Jejunum		33	Phospholipids
	Enzyme activity in liver		34	Cholesterol
10	Alkaline phosphatase			**Lipid content (enterocyte plasma membrane)**
11	γ-Glutamyltransferase			
12	Alanine aminotransferase		35	Triacylgliceroles
	Enzyme activity in blood serum		36	Phospholipids
13	Alkaline phosphatase		37	Cholesterol
14	γ-Glutamyltransferase			**Lipid content (blood serum)**
15	Alanine aminotransferase		38	Triacylgliceroles
	Enzyme activity in small intestine		39	Phospholipids
16	Alkaline phosphatase		40	Cholesterol
17	γ-Glutamyltransferase		41	High density lipoproteins
18	Alanine aminotransferase		42	Low density lipoproteins
	Content of sulfhydryl groups		43	Very low density lipoproteins
19	Liver			**Antioxidative activity**
20	Small intestine		44	Liver
	Content of lipid peroxidation products		45	Small intestine
21	Diene conjugates (liver)			**Na⁺,K⁺–ATPase activity :**
22	Diene conjugates (small intestine)		46	Liver
23	Schiff bases (liver)		47	Small intestine
24	Schiff bases (small intestine)			**Mg²⁺–ATPase activity:**
25	Content of malondialdehyde (liver)		48	Liver
26	Rate of MDA accumulation in liver		49	Small intestine

Technically, the problem of PCA consists in the following. In an early stage of investigation, wherever possible, the researcher has to collect all available data $\mathbf{x}^{(k)} = \left\{ x_n^{(k)} \right\}_{n=1}^{N}$, which are presumably reflect the effects of external influence on the system (k). Furthermore, the linear transformation of primary data $\mathbf{U} = \left\{ u_{kj} \right\}$ can be found, started from the assumption of more concise description of this multivariate sample by means of a few informative secondary quantities. Strictly speaking, such condition does not provide the only choice of transformation U, because any reasonable objective function is allowed. However, in many instances the optimal criterion consists in the requirement of mutual non-correlatedness of transformed variables. This case is an equivalent to the primary condition of maximum exhaustion of observed correlation by minimum set of used variables z_i, and it directly leads to the PCA. The desired transformation U can be found uniquely: it is a discreet analogue of decorrelating Karhunen-Loeve transform of random processes [12]. This unitary transformation $\mathbf{U} = \left\{ \mathbf{z}_n \right\}_{n=1}^{N}$ isdefined by eigenvalue equation for correlation matrix $\mathbf{G} = \left[\left\langle x_i x_j \right\rangle \right]_{i,j=1}^{N}$:

$$\mathbf{G}\mathbf{z}_n = \lambda_n \mathbf{z}_n \tag{1}$$

In principal component representation matrix G becomes diagonal, $\widetilde{\mathbf{G}} = \mathbf{U}^+ \mathbf{G} \mathbf{U} = \mathrm{diag}\left\{ \left[\lambda_n \right]_{n=1}^{N} \right\}$, and the average contribution of every component z_n into the whole data set $\left\{ \mathbf{x}^{(k)} \right\}$ is determined by the correspondent eigenvalue λ_n. This permits to select the bounded subset of the most significant components $\left\{ \mathbf{z}_n \right\}_{n=1}^{M}$, $M < N$, so that the projection on them of the whole sample $\left\{ \mathbf{x}^{(k)} \right\}$ ensures the required level of accountable correlations.

It is convenient to treat the results of the analysis in geometrical representation. For this purpose every data vector is represented in the M-dimensional sample subspace (usually $M \leq 3$). In this subspace the vector coordinates a_n, are the coefficients of its expansion in the eigenvectors z_n:

$$\mathbf{x} = \sum_{n=1}^{N} a_n \mathbf{z}_n \tag{2}$$

A point, the totality of which forms the scatter diagram of objects under consideration, represents every object vector x in this sign space. By this means certain compact regions of scatter diagram may join together the representatives of the population, which have the same or closely related characteristics. On this interpretation the adequate diagnostics of object state consists in proper classification (space decomposition) of point pattern in sign space.

4. General approach to the quantitative analysis of LDE under combined exposure

Basically, the quantitative estimation of the response of biological objects on the external influence is determined by "dose-response relationship" [1]. Conventional approach to this problem is based on dose-dependent survival value, that is, on the probability $P_A(D_A)$ of the event, when given biological object survives under the dose D_A of the external agent A. Based on the information criterion, the information capacity of this event is chosen as its quantitative estimation:

$$E(A) = -\ln P_A(D_A) \qquad (3)$$

With this definition, combined influence of the agents A and B can be determined as follows. If agents A and B act independently, the total effect of their joint action is equal to the sum of their individual effects, as it follows from the additive property of information:

$$P_{add}(A; B) = P_A(D_A) \cdot P_B(D_B)$$
$$E_{add}(A; B) = E(A) + E(B) \qquad (4)$$

In such a way, the combined action of the agents A and B one can characterize by the coefficient of synergy K_s [2]:

$$K_s(A; B) = \frac{E(A; B)}{E_{add}(A; B)} = \frac{E(A; B)}{E(A) + E(B)} \qquad (5)$$

Basically, three situations are possible if the ionizing radiation R and any other external agent A are acting simultaneously:
1. $K_s(R, A) < 1$. Agent A reduces the biological action of radiation, so that the inhibition, biological antagonism occurs in this case.
2. $K_s(R, A) = 1$. Agent A does not modify survival value of radiation, – the result of combined action is equal to the direct sum of their independent effects. This is the case of additive action.
3. $K_s(R, A) > 1$. Agent A increases the radiation influence. Combined effect is greater than the sum of independent effects. This situation is known as the synergistic effect, synergy.

It should be pointed out that the assessments based on the coefficient K_s are unusable in LDE study, first of all because of the fact, that the survival probability $P_A(D_A)$ in Eq.3 is estimated as the experimental survival rate:

$$P_A(D_A) \sim 1 - \frac{N_A(D_A)}{N}, \qquad (6)$$

where N is the total number of objects in finite sample of laboratory animals, $N_A(D_A)$ is the number of objects, which have been lost under the influence of the absorbed dose D_A. As usually under the small absorbed dose $N_A(D_A) \approx 0$, the information content of the effect is close to zero.

Nevertheless, in practice there are techniques found, which are directed to quantitative indices different from that of Eq.3. In fact, according to Eq.5 the coefficient K_s basically can be defined by every biological index (effect), if it satisfies two main conditions: 1) effect is dose-dependent, and 2) when external agents act independently, the total effect is additive. Such approach was used, as an example, in Ref. [2], where mutual synergy interaction was estimated for combined action of radiation and hyperthermia in terms other than survival value (such as enzyme activity). However, this criterion is not wholly satisfactory in LDE study, as it operates with any individual biological index, which could not be an effective statistical estimator of object state under these conditions.

Now that we have presented the features of qualitative classification of complex systems by a PCA-based model, we can use it for correct quantitative estimation of combined influence of external agents, as well.

The general concept of quantitative analysis in non-additive interaction of small-dose influences is as follows. Let us assume that the influence of some external agents A, B, ..., presumably consists in the variations of any index of object state x_i in reference to their values in control group $x_i^{(0)}$, which can be defined experimentally. In other words, the sample value $\tilde{x}_i$ can be represented as:

$$\tilde{x}_i(D_A, D_B,...) = x_i^{(0)} + \delta x_i(D_A, D_B,...) \tag{7}$$

Here $\delta x_i(D_A, D_B,...)$ is an increment of index x_i for the object (object group) under consideration, and D_A, D_B, ..., are the absorbed doses of corresponding agents. Hence, δx_i determines object's reaction on the influence, and it can be accepted as an effect of action:

$$E^{(x_i)}(A, B,...) \equiv \delta x_i(D_A, D_B,...). \tag{8}$$

Furthermore, if value $\tilde{x}_i$ under combined influence of some agents (say, A and B) will be a direct sum of $x_i^{(0)}$ and increments $\delta x_i(D_A)$, $\delta x_i(D_B)$:

$$\tilde{x}_i(D_A, D_B) = x_i^{(0)} + \delta x_i(D_A) + \delta x_i(D_B), \tag{9}$$

it may be concluded that these agents have an additive nature in relation to parameter x_i over a range D_A, D_B of absorbed doses . To the contrary, if the combined action leads to interrelated responses:

$$\delta x_i(D_A, D_B) \neq \delta x_i(D_A) + \delta x_i(D_B), \tag{10}$$

we can conclude that the interaction is non-additive (synergistic, antagonistic) in relation to x_i. Corresponding quantitative estimation of the combined action of the agents A and B is determined by the coefficient of synergy:

$$K_s^{(x_i)}(A; B) = \frac{E^{(x_i)}(A; B)}{E^{(x_i)}(A) + E^{(x_i)}(B)} \equiv \frac{\delta x_i(D_A, D_B)}{\delta x_i(D_A) + \delta x_i(D_B)} \tag{11}$$

Unfortunately, such estimations are usually ineffective in LDE-analysis because of their statistical ambiguity. Moreover, a variety of individual parameters (indices) do not reflect influence effectively and univalently (unically): for the most part quantitative and qualitative changes of system attributes are determined by their totality only.

As a consequence of these features, one can construct an integral analog of the estimation (11) by means of PCA. As we shall see subsequently, in most cases such approach is correct and effective. More importantly, it is especially useful just in LDE study.

First of all, principal components maintain additivity under independent action of the agents. Actually, the transformation of data vector x into principal component z is carried out by linear unitary transform U, and if Eq.8 *is valid for all observable values* x_i, it is true for the coefficients $a = [a_n]_{n=1}^{N}$ in expansion (2), as well:

$$\mathbf{a}(D_A, D_B) = \mathbf{U}^{+}(\mathbf{x}^{(0)} + \delta\mathbf{x}(D_A) + \delta\mathbf{x}(D_B)) = \mathbf{a}^{(0)} + \delta\mathbf{a}(D_A) + \delta\mathbf{a}(D_B) \quad (12)$$

Hence, in this case the influence of external agents is additive for every principal component. With this result, let the coefficient of synergy in relation to principal component z_n be defined as:

$$K_s^{(z_n)}(A; B) = \frac{E^{(z_n)}(A; B)}{E^{(z_n)}(A) + E^{(z_n)}(B)} = \frac{\delta a_n(A, B)}{\delta a_n(A) + \delta a_n(B)} \quad (13)$$

Contrary to the estimation by individual observable value (10), coefficient $K_s^{(z_n)}(A; B)$ is a certain generalized characteristic of combined action, because it takes into consideration (with certain factor loads) all observable parameters. Hence, this criterion provides a way for establishment and estimation of such effects in LDE study, as it reduces experimental statistical error (approximately by a factor of $\sqrt{N_{\text{eff}}}$, where N_{eff} is the number of effective factor loads), as a result of statistical averaging over the ensemble of independent observable data. But yet, the coefficients $K_s^{(z_n)}(A; B)$ determined for different components z_n describe combined effects relative to different groups of experimental parameters, because different components have different factor loads. In addition, different components can represent simultaneously both antagonistic and synergetic reaction, if such influence of external agents holds for different groups of object indices. Moreover, coefficient (13) (as for individual coefficient (11) too) may not be definitely positive, since the "effect" itself may be both positive and negative, $\delta a_n \gtrless 0$. Logical interpretation of negative-valued coefficient K_s is not complicated, but its system definition (the opposite tendency of agent action in individual and combined application) invites further investigations in every specific case.

Finally, global estimation of combined influence of external agents could be determined by data vector as follows. In sign space a total reaction of the object is determine by M-dimensional vector $\delta\mathbf{a}_M = \sum_{n=1}^{M} \delta a_n z_n$ that is, the projection of vector $\delta\mathbf{a}$ on the M-dimensional sign subspace, spanned over the most effective principal components. The overall amount of observed influence could be estimated as a norm of this vector, in other words, as a distance between the points of the object under consideration and the object of control group:

$$E^{(int)}(A) = \|\delta\mathbf{a}_M(D_A)\| \equiv \sqrt{\sum_{j=1}^{M} \delta a_j^2(D_A)}, \quad M \leq N. \tag{14}$$

If agents A and B acts independently, calculated additive effect is estimated similarly as the norm of vector sum $\delta\mathbf{a}_M(A) + \delta\mathbf{a}_M(B)$:

$$E_{add}^{(int)}(A,B) = \|\delta\mathbf{a}_M(A) + \delta\mathbf{a}_M(B)\| \equiv \sqrt{\sum_{j=1}^{M} [\delta a_j(A) + \delta a_j(B)]^2} \tag{15}$$

Consequently, global quantitative estimation of combined influence of external agents in LDE study may be done by integral synergy coefficient $K_s^{(int)}$:

$$K_s^{(int)} = \frac{E^{(int)}(A,B)}{E_{add}^{(int)}(A,B)} = \frac{\|\delta\mathbf{a}_M(A,B)\|}{\|\delta\mathbf{a}_M(A) + \delta\mathbf{a}_M(B)\|} \equiv \sqrt{\frac{\sum\limits_{j=1}^{M} \delta a_j^2(A,B)}{\sum\limits_{j=1}^{M} \delta a_j(A) + \delta a_j(B)^2}}. \tag{16}$$

Coefficient $K_s^{(int)}$ (Eq.16) describes only the absolute value of the non-additive effect, irrelative of increment signs. In addition, its numerical value increases monotonically with the dimensionality M. This property is useful in the study of asymptotic behavior of the system with respect to the number of experimental parameters under consideration. 5. Results and discussion.

5. Sign-space analysis of the influence of ionizing radiation and cadmium on organisms in PCA-representation

It should be remembered, that chronic influence of ionizing radiation and cadmium on organisms is characterized by a set of specific peculiarities, caused by data normalization procedure and principal component formation. The point is that object clasterization in sign space is due not only to the external influence of detrimental agents, but to time dynamics of principal components too. This fact may play an essential role in correct interpretation of the obtained results.

Usually data processing in PCA needs the pretreatment of primary experimental values, as they have essentially different magnitudes and are measured in different units. Any type of such normalization is ambiguous, as this procedure mathematically is equivalent to nonunitary transformation of correlation matrix.

Two special cases of data preprocessing are of main interest: 1) normalization in every time group to the magnitudes of indices for the control group of the same age; and 2) normalization of all data to the control of the very early time group. When used, normalization of the first type practically eliminates all qualitative age-related changes in laboratory animals. Hence, all regular deviations in normalized parameters will be caused by the influence of external agents, and not by time-depended changes in the status of organism. It is precisely this representation in sign space which has revealed the trend in the agent's effectiveness to forming classes, associated with LDE under individual and combined action of agents. The scattering graphs inside these classes indicate object status in relation to absorbed dose.

The normalization of the second type retains a twofold information: time-dependent and dose-dependent changes of the object. In that case the possibility exists of simultaneous tracing the status dynamics under natural control conditions and chronic influence of radiation and other external agents.

In addition, principal components in chronic experiment may be formed in different ways, with respect to the definition of sample correlation matrix and corresponding statistical model. In particular, one can construct the model, whereby every time group forms its own principal components. In this case principal components and factor loadings are different for different groups: they take into account sample correlation in one time group with fixed absorbed doses only. This model is especially useful in the analysis of time-dependent effects.

The studies of LDE under chronic influence of ionized radiation and cadmium on organism by application of PCA were conducted with due regard to these peculiarities. 49 experimental indexes (Table 1) of the rats' organs were measured in three time groups (42, 84 and 145 days), corresponding to accumulated doses 0.3, 0.6, and 1.0 Gy respectively.

These findings after appropriate renormalization were used as a training set for the determination of sample correlation matrix, which we omit here for reasons of inconvenience. It makes sense here merely to point out that some of measured parameters are reliably correlated (anticorrelated) with respect to modification of experimental conditions.

After diagonalization, only few eigenvalues of the correlating matrix of renormalized parameters were found to be significant (see Table 2). Since these values by their nature determine average contribution to overall correlation, only few principal components have to be taken into account in analysis: in our case the first three components exhaust over 80% of total dispersion. Therefore, the analysis below is carried out if only three principal components (sign space in coordinates a_1, a_2, a_3) are accounted for.

Table 2. Eigenvalues of correlation matrix

λ_1	λ_2	λ_3	λ_4	λ_5	λ_6	λ_7	λ_8	λ_9
0.58	0.29	0.17	0.13	0.09	0.06	0.04	0.03	0.02

The results of the analysis for every time group with its own principal components are shown in the Figure1.

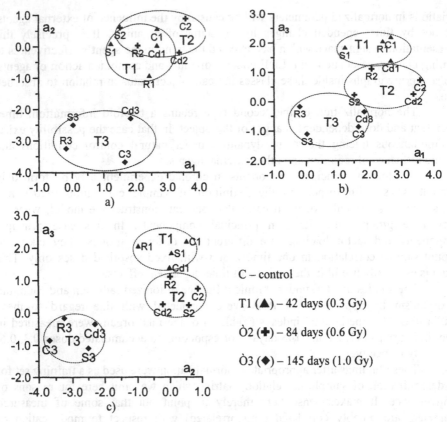

Fig. 1. The results of sign-space analysis for time groups with their own principal components. Classes in the space spanned over first three principal components. Scatter diagram shows time changes in object status.

In these diagrams coordinates a_n, $n=1\div3$, are the projections of corresponding data vector onto the principal component z_n for control group (C), groups of chronic irradiation (R) and influence of cadmium (Cd), and under their combined influence (S). As is evident from depicted patterns, well-defined clusterization (especially distinct in coordinates a_1–a_3, a_2–a_3) exists with regard to time-dependent groups, forming corresponding classes $\dot{O}_1$, $\dot{O}_2$, $\dot{O}_3$. The influence of external agents in this connection is only weakly expressed: the scatter diagram in every class T is relatively compact.

Hence, if we are to clearly recognize LDE under chronic combined influence, the elimination of time effects should be done with the help of proper data normalization in every time group to the magnitudes of indices for the control group of the same age. As indicated above, such normalization has to eliminate time dependence of principal components almost completely, and detect in detail general features in the action of external agents. One would expect that objects' separation in sign space in this case will be caused by absorbed doses and combined effects only. This assumption is actually supported by our experimental results (Figure 2).

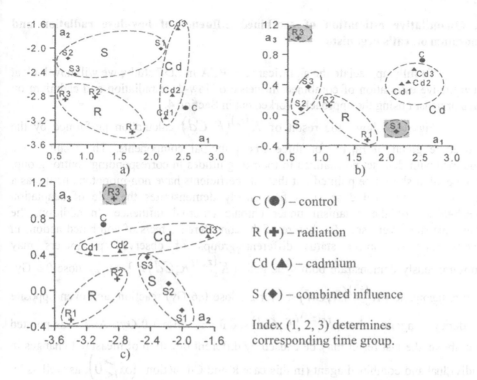

C (●) – control

R (✚) – radiation

Cd (▲) – cadmium

S (◆) – combined influence

Index (1, 2, 3) determines corresponding time group.

Fig. 2. The results of sign-space analysis for chronic influence. Normalization of all data is done to the magnitudes for control group of the same age. Scatter diagram shows clusterization of experimental data with different external influence.

The scatter diagram in this figure shows reasonable clusterisation of experimental data with different external influence (corresponding classes R, Cd, and S are qualitatively separated, the results for control group C are the same for all time intervals as a consequence of normalization). In particular, non-additive effect of combined influence of radiation and cadmium is clearly visible.

It must be emphasized that the careful analysis of factor loads makes it possible to choose essential measurable values, which mostly contribute to main principal components. In our experiments structure-functional state of intestine cell membranes and their proliferative activity were found to be the controlling factors in observed effects, the status of liver is of minor importance. As a practical matter, this result allows to reduce experimental complexity of investigation without loss in generality. Also it is significant to note that the magnitudes of factor loads actually have different signs for all (except one) principal components. This feature is of fundamental importance in the analysis of non-additive effects in LDE study as it permits to clarify the competition of different processes (and corresponding parameters) in principal component formation.

6. Quantitative estimation of combined influence of low-dose radiation and cadmium on rat's organism

To fully appreciate the significance of PCA in LDE study, we will now look at quantitative estimation of combined influence of low-dose radiation and cadmium on rat's organism using the approach, worked out in Section 4.

Figure 3 shows the result of $K_s^{(z_n)}(R;Cd)$ calculation performed by the procedure given in Eq.13 for three first principal components. The results were obtained with data set normalized to their magnitudes in corresponding control group. Above all, it should be pointed out that all coefficients have non-monotone nature as a function of absorbed dose. This fact clearly demonstrates the role of adaptation mechanisms of the organism under chronic external influence. In addition, the diagrams show very specific feature of the considered effects of combined action: in every class of object status different groups of observed parameters may simultaneously demonstrate both synergetic ($K_s^{(z_n)}(R;Cd) > 1$ for z_2, dose 0.6 Gy) and antagonistic ($K_s^{(z_n)}(R;Cd) < 1$ for z_3, dose 0.6 Gy) reaction, and even opposite tendency of agent action ($K_s^{(z_n)}(R;Cd) < 0$ for z_3, dose 1.0 Gy). As it was pointed out above, the last result may be caused by different direction of measured changes in individual and combined agent (in this case R and Cd) action, $\left(\delta x_i \gtrless 0\right)$, as well as by the competition of factor loadings with different signs $\left(u_{kj} \gtrless 0\right)$. It should be particularly emphasized that both of these situations indicate non-additive effects in combined interaction. If, for example, a combined influence significantly affects the parameters with positive factor loadings, and an individual influence mostly affects negatively loaded parameters, synergy coefficient $K_s^{(z_n)}(R;Cd)$ will be negative. A more sophisticated treatment of our experimental data and factor loadings indicates, that both of these reasons may be the case. Further accumulation of experimental data will make possible to create effective mathematical models of such mechanisms of non-additive interaction in LDE study.

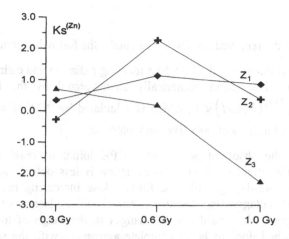

Fig. 3. Synergy coefficient $K_s(z_n)$ in relation to principal component z_n, n=1, 2, 3. All data are normalized to the values of corresponding control group.

For completeness the global synergy coefficient $K_s^{(int)}$ of combined influence of low-dose radiation and cadmium is shown in Figure 4. As it is evident from the graphs, chronic influence of these agents in low doses leads to distinctly antagonistic interaction in overall organism status. It may be a further important evidence of complicated nature of adaptation processes in radiation damage.

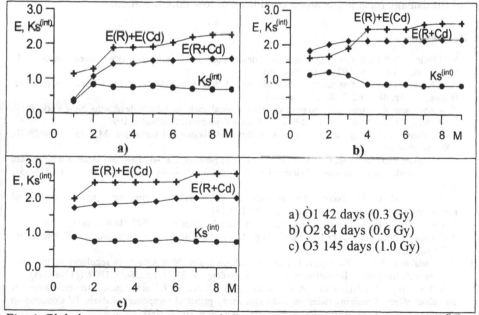

Fig. 4. Global quantitative estimation of effects (E(A)) and synergy coefficient ($Ks^{(int)}$) for combined low-dose radiation and cadmium under chronic influence.

7. Conclusion

On the grounds of the received results, we can make the following conclusions:

- The chronicle influence of the low-dose ionizing radiation and cadmium ions shows an antagonistic interaction numerically characterized by the integral synergy coefficient $K_s^{(int)}(R, Cd) < 1$, which is calculated by PCA of 49 morphological and biochemical indices of rat's liver and intestine.

- Increasing of the absorbed dose leads to the initial decreasing of the agents' antagonism (the effect of the combined action is less detectable for 84-day age group than for 42-day group), and further dose increasing raises it again (the coefficient of antagonism deviates from 1 to a lesser or greater extent correspondingly). We consider such changes in the range of low cadmium and radiation-absorbed dose to be in complete agreement with the scheme of stress-adaptation syndrome.

- The observed decreasing of antagonism with absorbed dose does not mean the equivalent direct decreasing of effect value itself, as the synergy coefficient is determined only by the effect's ratio. At the same time, effects themselves can decrease or increase with absorbed dose, as well. Exactly such situation is typical for the experimental data analyzed above (the graphs for effect values in Fig.4 are non-monotone with dose or time duration). Such a behavior, along with mentioned above, is in well agreement with the general concept of the adaptation mechanisms of the damage reparations under conditions of low-dose radiation.

8. References

1. V.G.Petin, V.P.Komarov. Quantitative description of radiosensitivity modification. M., Energoatomizdat, 1989, 192 pp. (In Russian).
2. A.M.Kuzin. Problem of synergy in radiobiology. Acadamy of science of the USSR, Izvestija, ser. Biology. ¹4, pp. 485-502, 1983. (In Russian).
3. T.W.Anderson. An introduction to multivariate statistical analysis. John Wiley&Sons, New York, 1958.
4. H.Martens, T.Naes. Multivariate calibration. John Wiley&Sons, Chichester, 1991.
5. O.V.Volkova, Iu.Eleckij Foundation of histology and histological technique. M., Nauka, pp.25-40, 1982. (In Russian).
6. W.E.Chijsen, M.D.De Jong, C.H.Van Os Dissociation between Ca^{2+}-ATPase and alcaline phosphatase activities in plasma membranes of rat duodenum. Biochem. et biophys. acta. Vol. 599, 2. pp.538-551, 1980.
7. V.Harms, E.M.Wright. Some characteristics of Na^+,K^+-ATPase from rat intestinal basolateral membranes. J.Membr.Biol., Vol. 53, ¹ 2. pp.119-128, 1980.
8. Modern methods in biochemistry / Ed. V.N.Orehovich. M., Medicina, 1977. (In Russian).
9. A.N.Klimov, N.G.Nikul'chev Lipid and lipoprotein metabolism and its derangement. SP: Piter Kom, 1999. (In Russian).
10. E.B.Burlakova, A.N.Goloschapov, G.P.Jijina, A.A.Konradov. New aspects in regularity of low-dose low-intensity irradiation. Radiation biology. Radioecology. V. 39, ¹1, pp. 26-33, 1999. (In Russian).
11. V.M.Voitsitsky, S.V.Hizhnyak, A.V.Kurashov, O.O.Kysil, N.E.Kucherenko. An estimation of combined effect of ionizing radiation and cadmium by principal component analysis. IV Congress on radiation investigations, Abstr, Vol. II, p.379.Moscow, Nov. 20-24. 2001.
12. L.E.Franks. Signal theory. Prentice-Hall, N.J., 1969.

CHARACTERIZATION AND TREATMENT OF ACTINIDE CONTAMINATED SOILS AND WELL WATERS

J. D. NAVRATIL

Environmental Engineering and Science, Clemson University, 342 Computer Court, Anderson, South Carolina 29625-6510 USA

1. Abstract

Soil and well water contaminated with actinides, such as uranium and plutonium, are an environmental concern at most U.S. Department of Energy sites, as well as other locations in the world. Remediation actions are on going at many sites, and plans for cleanup are underway at other locations. This paper will review work underway at Clemson University in the area of characterization and treatment of soil and water contaminated with actinide elements.

2. Introduction

Actinide contamination in soil and water has been identified at many sites worldwide including the United States. A variety of treatment methods are available to remove actinides from soil and water, and many methods are commercially available. This paper will review some work in progress at Clemson University on actinide contaminated soil remediation methods as well as characterization methods. The Clemson work on inexpensive methods to remove uranium from water will also be highlighted.

3. Soil treatment

In the United States, the Nevada Test Site (NTS) possesses widespread soil contamination caused by deposition of uranium, plutonium and other radionuclides from defence related nuclear test operations. Clean up efforts are ongoing using conventional remediation techniques. However, the U.S. Department of Energy (DOE) desires to obtain technologies that can further reduce risks, reduce clean up costs, and reduce the volume of remaining contaminated soil.

Clemson University and teaming partner Waste Policy Institute, through a cooperative agreement with the National Environmental Technologies Lab in Morgantown, West Virginia, are assisting the NTS in evaluating possible technologies that have the potential of reducing risks and clean-up cost. Physical, chemical and biological treatment processes are being evaluated. Our technology assessment includes the following sub-tasks:

- A description of soil contamination problems at the NTS including a description of contaminant distribution, soil characteristics and adhesion/absorption characteristics of contaminants on soil particles.
- An evaluation of physical, chemical and biological processes that have potential to remediate radioactive contaminated soils.
- Ranking of technologies based on technical merit, potential experience and ease of implementation.

F. Brechignac and G. Desmet (eds.), Equidosimetry, 403–407.

- An engineering evaluation of technologies to determine scale-up potential and cost effectiveness.
- Identification of secondary waste treatment needs for full-scale implementation.
- Identification of barriers and research needed to overcome technology limitations.
- Preparation of a comprehensive final report that will provide a road map for developing and demonstrating an optimal treatment approach for soils at the NTS.

The removal of actinides from NTS soils has been attempted using various combinations of attrition scrubbing, size classification, gravity based separation, flotation, air flotation, segmented gate, bioremediation, magnetic separation and vitrification [1-6]. Attrition scrubbing was used extensively as a pretreatment step to break up agglomerated materials, to remove surficial coatings from larger soil particles, and hopefully to make the contaminated soil more amenable to processing. Size separation, based on wet sieving of the contaminated soil, showed that the contamination was associated with the smaller particles. Attrition scrubbing and wet sieving of NTS soil was able to achieve a 70% volume reduction.

Gravity-based processes work on the principle of Stokes' law; heavy particles settle faster than light particles. However, the size of the particle also influences the rate of settling, thus, gravity based separation is not very effective for fine particles, typical of the NTS soils. Gravity and centrifugal separators utilize the terminal velocity of a particle for the basis of separation, which depends on the combination of density and size. Thus, the fine (size) and heavy (density) actinide oxide particles will be separated and report together to the same concentrate product stream with the large (size) and light (density) soil matrix particles.

Flotation separation is highly dependent on using the correct reagent in the slurry, which would permit air bubbles to attach to contaminated mineral grains. More work is needed to identify the best reagents to produce better separations of soil and contaminants. Carrier flotation has the advantage over air flotation of use of a carrier, which is especially important if the contaminant of interest is present at very low concentrations.

The segmented gate system separates contaminated soil from clean soil according to a preset radioactivity criterion. Poor results were obtained in a recent field test of the technology performed at the NTS: only 61% of the plutonium activity ended up in the "concentrate" with a volume reduction of 2:1 (weight of feed to weight of clean).

Bio-leaching of plutonium oxide occurs with sulphuric acid produced from elemental sulphur in the presence of sulphur oxidizing bacteria. This technology is also based on a precipitation of plutonium sulphate complex ($PuO_2(SO_4)_3^{4-}$) as plutonium oxide sulphur (PuO_2S) in the presence of sulphate reducing bacteria. Good results of field testing at the NTS were obtained and the technology has been evaluated as promising. The unit processes are based on sound scientific concepts that have been proven in the acid leaching of uranium oxides with sulphuric acid and oxygen, and in the precipitation of metal ions in wetland-treatment of acid-mine drainage.

Early studies with magnetic separation indicated that the magnetic susceptibility of fine soil is very low. Wet magnetic separation was being tested and indications were that wet magnetic separation might work. Although results to date have not been encouraging, there is the potential to further optimize treatment and reduce the amount of material that is held up.

Vitrification and fixation are expensive and do not achieve soil volume reduction. It does not meet DOE programmatic goals of volume reduction; instead, it provides only immobilization of the contaminant. However, it is possible that vitrification may be an acceptable form of treatment on certain locations, especially for uranium contaminated soils.

Leaching methods may be more attractive for uranium contaminated soils than plutonium soils. Carbonate leaching at pH above seven has shown to work extremely well to solubilize the uranyl ion as a carbonato complex while not significantly leaching any of the soil constituents [7].

In summary, size separation helps as an initial soil decontamination step and is especially useful for smaller sized particles. But there can be significant variability in contaminant distribution in the soils and any successful treatment process must address this drawback. Attrition scrubbing appears to help although there is limited data to compare results with and without scrubbing – most studies used either one or the other. Carbonate leaching may improve these systems as well as be totally effective as a primary method. None of these processes have been fully optimized, so significant improvement may be realized by more in-depth studies.

4. Water treatment

Naturally occurring uranium has contaminated numerous private domestic water wells in South Carolina at levels exceeding the U.S. Environmental Protection Agency's (EPA) drinking water standard of 30 µg/L for uranium concentration. This is based on nephro-toxicity, rather than on radiological hazards. Other areas of the U.S., such as California and New Mexico, also have natural uranium in well water that exceeds the EPA standards.

In Simpsonville-Greenville, South Carolina, high amounts of uranium (30 to 9900 µg/L) were found in 31 drinking water wells. The contamination of uranium in the well water was most likely the result of veins of pegmatite, running east of Greenville to southwest of Simpsonville. In addition to the elevated uranium concentration, elevated radon levels have also been discovered. All the wells that have been tested (currently 111) are above 300 pCi/L in radon. The South Carolina Department of Health and Environmental Control (SC DEHC) requested homeowners to discontinue the use of the well water since chronic ingestion of this water may result in an increased risk of cancer. In addition, SC DEHC is beginning a program to test the levels of radon in air in this community.

The purpose of our work at Clemson University was to determine the form of uranium present in the well water and to test the effectiveness of common household treatment devices to remove uranium and radium. Batch tests with activated carbon, iron powder, magnetite, anion exchange resin and cation exchange resin were used to characterize the form of uranium in the drinking water. In the tests, water and the separation materials were first equilibrated, filtered and then analyzed by alpha spectrometry. The results of the batch tests showed that it is possible to remove greater than 90% of the uranium and radium in the drinking water by using any of the sorbents listed above. Simple filtration with a 0.1 µm filter had little to no impact on uranium removal. Testing of two household treatment devices was performed and found not to be totally effective.

A magnetic field-enhanced filtration/sorption device is also being evaluated for removal of actinides from water [8-10]. The device consists of a column of supported magnetite surrounded by a movable permanent magnet. The mineral magnetite, or synthetically prepared iron ferrite ($FeO\cdot Fe_2O_3$), is typically supported on various materials to permit adequate water passage through the column. In the presence of the external magnetic field, enhanced capacity was observed in using supported magnetite for removal of actinides and heavy metals from wastewater. The enhanced capacity is primarily due to magnetic filtration of colloidal and submicron particles along with some complex and ion exchange sorption mechanisms. The loaded magnetite can easily be regenerated by the removal of the magnetic field and use of a regenerating solution.

One inexpensive device for containing supported magnetite is simply some type of static-bed column with screens at the entrance and exit for holding the support in place while permitting the flow of smaller magnetite particles in and out of the column. The first step in operating the column would be to pass a slurry of activated magnetite down the column containing a support material such as glass beads. Once the column is loaded with magnetite, the magnets would be placed around it to hold the magnetite in place. Then water to be treated would be passed upflow through the column until contaminant breakthrough is reached. Then the magnetic field would be removed and an additional amount of fresh magnetite slurry added to the top of the column to displace a portion of loaded magnetie and the captured contaminate particles. Then the magnets would be replaced around the column and more water passed upflow. The spent magnetite could be regenerated or discared depending on the contaminants and operational situation. Discarding the material would be useful for contaminants that sorb strongly onto the adsorbent material (thus perhaps serving as a favourable waste form for final disposal).

In some cases it may be advantageous to have the magnetite bonded to the support and to use a normal fixed-bed column operation with regenerating solution. There are several methods available to adhere or bond the magnetite particles onto supports. For example, crushed glass and glass wool may be mixed with the support in the presence of a solution of a alkali or alkaline hydroxide (this process would also activate the magnetite); heating may also assist the process, as well as raising the temperature to near the melting point of the glass without the solution present. Grinding the glass wool (or waste fiberglass insulation) with the magnetite may also bond the materials, as would the use of silica gel, resins and other bonding adhesives.

Many processes have utilized iron oxides for the treatment of liquid wastes containing radioactive and hazardous metals. These processes have included adsorption, precipitation and other chemical and physical techniques [11-12]. For example, a radioactive wastewater precipitation process includes addition of a ferric hydroxide floc to scavenge radioactive contaminants, such uranium [13-15]. Some adsorption processes for wastewater treatment have utilized ferrites and a variety of iron containing minerals, such as akaganeite, feroxyhyte, ferrihydrite, goethite, hematite, lepidocrocite, maghemite and magnetite. Ferrite is a generic term for a class of magnetic iron oxide compounds. Ferrites posses the property of spontaneous magnetization and are crystalline materials soluble only in strong acid. Besides ferrite treatment, wastewater could also be pretreated with standard flocculation/precipitation and filtration steps to remove gross amounts of actinides from solution. The magnetic filtration/sorption device would then be used as a polishing step for the water to remove colloids, small particles and complex ions not removed in the precipitation/filtration steps.

5. References

1. Cost/Risk/Benefit Analysis of Alternative Cleanup Requirements for Plutonium-Contaminated Soils On and Near the Nevada Test Site, DOE/NV-399, US DOE Nevada Operations Office, May 1995.
2. Murarik, T.M., "Characterization, Liberation and Separation of Plutonium at the Nevada Test Site: Safety Shot Residues", Proceedings of the Soil Decon '93: Technology Targeting Radionuclides and Heavy Metals, June 16 and 17, 1993, Gatlinburg, TN, ORNL-6769, Oak Ridge National Laboratory, Oak Ridge, TN, September 1993.
3. Cho, Eung Ha, et. Al., Soil Volume Reduction Technologies, Evaluation of Current Technologies, DOE Cooperative Agreement DE-FC26-98FT40396, Deployment Leading to Implementation, Final Report, submitted by West Virginia University to US DOE National Energy Technology Laboratory, July 2000.
4. Heavy Metals Contaminated Soil Project Technical Summary, DOE/EM-0129P, National Technology Transfer Centre, Wheeling, WV, 1994.
5. Misra, M., et. al., "Characterization and Physical Separation of Radionuclides from Contaminated Soils," Proceedings of the Soil Decon '93: Technology Targeting Radionuclides and Heavy Metals, June 16 and 17, 1993, Gatlinburg, TN, ORNL-6769, Oak Ridge National Laboratory, Oak Ridge, TN, September 1993.
6. Papelis, et. al., Evaluation of Technologies for Volume Reduction of Plutonium-Contaminated Soils from the Nevada Test Site, DOE/NV/10845-57, US DOE Nevada Operations Office, June, 1996.
7. Bonnema,B.E., Navratil, J.D. and Bloom, R.R., "SOIL*EX Process Design Basis for Mixed Waste Treatment, *Proceedings of WM'95*, Tucson, AZ (1995).
8. Kochen R.L. and Navratil J. D., U.S. Patent 5,595,666, January 21, 1997, and U.S. Patent 5,652,190, July 29, 1997.
9. Navratil J. D., Kochen R. L. and Ritter J. A., *Proceedings of WM'95*, Tucson, AZ (1995).
10. Ebner A. D., Ritter J. A. , Ploehn H. J., Kochen, R. L. and Navratil J. D, Sep. Sci. Tech., 34, 1277, 1999.
11. King C.J. and Navratil J. D., Chemical Separations, Litarvan, Denver, 1986.
12. Macasek F. and Navratil J. D., Separation Chemistry, Horwood, New York, 1992.
13. Boyd T. E., Kochen R. L., Navratil J. D., and Price M. Y., Radioactive Waste Manag. Nucl. Fuel Cycle, 42, 195, 1983.
14. Boyd T. E., Cusick M. J. and Navratil J. D., in Li N. N. and Navratil J. D. (Eds.), Recent Developments in Separation Science, Vol. VIII, CRC Press, Boca Raton, FL, 1986.
15. Kochen R. L. and Navratil J. D., Lanthanide/Actinide Research, 2, 9, 1987.

polishing step for filter water to remove colloidal, small particles and complex ions not removed in the precipitation/filtration steps.

References

THE ENDOGENIC AND EXOGENIC FACTORS OF THE REALIZATION OF PHENOTYPIC ADAPTATION

A. N. MIKHYEYEV, M. I. GUSCHA, Y. V. SHILINA
Institute of Cell Biology and Genetic Engineering, National Academy of Sciences of Ukraine,
148 Zabolotny st., Kyiv, 03143, UKRAINE

One of actual branches of modern biology is the research of biological objects resistance to extreme environmental factors (low and high temperatures, various types of electromagnetic radiation, drought, hypoxia etc.). In this article the attempt is made to consider some aspects of adaptation problem and, in particular, mechanisms of phenotypic adaptation (PA).

Hardening of plants, change enzyme spectrum of bacteria passing on different growth medium, virulence change of pathogenic bacteria and fungi, which parasite on the hosts with different levels of resistance and at cultivating of pathogens on growth medium *in vitro* is represented as the most known examples of PA. These examples of PA can be continued. So, the lag phase on a growth curve of bacterial culture is frequently defined as the adaptation phase to new growth medium conditions [1]. Under influence of host changes the physiological characteristics of parasite can be shown as virulence increase as a result of passages [2]. The increasing of phytopathogenic bacteria virulence by the passage through the appropriate plants is also a well-known fact and is frequently used for restoration of pathogenicity properties in museum cultures [3]. Plant hardening phenomenon is investigated well enough [4 - 7]. All above-mentioned facts have created empirical an base for transition to a stage of fundamental researches concentrated on disclosing of resistance mechanisms and on a theoretical substantiation of updating of the methodology [8 - 12].

A few definitions of adaptation phenomena are known. At first, the adaptation is necessarily considered as the actual process of the adaptation (process of adaptiogenesis, adaptive response) to changed conditions. For example, temperature hardening (low or high-temperature) is realized not immediately, but gradually, passing through some phases (stages). Second, the adaptation can mean an already generated adaptation state of an object (result of adaptatiogenesis, adaptability, resistance) to the influence of extreme factors in the environment. In other words adaptation in this case is considered as the final result of the process. An example of this aspect of adaptation is the existence of certain current levels of the resistance of biological objects to stressor influence (coldhardiness, radioresistance and so on). The concept « adaptive response» more unequivocally specifies the dynamic character of adaptation.

Thus, the adaptation of biological objects can be considered as the process and the result of their resistance increase to influence of external and, probably, internal stressors.

Stressors are physical, chemical or biological factors, which act on biological system with dose levels that exceed limits of its sensitivity. An example of plants and microorganisms adaptation can be the increase of their resistance to salting, drought, low or high temperatures, xenobiotics, ultra-violet or ionising radiation. Plant organisms also acquire a system of resistance as a reply to the I nfluence of specific or unspecific pathogens There may also be the adaptation of

F. Brechignac and G. Desmet (eds.), Equidosimetry, 409–417.

plants and microorganisms populations to their coexistence with individuals of other plants or animals species [1, 4, 5, 8, 12-15].

The changes caused by a stressor are appearing in the transition process of a biological object into a temporarily or a constantly new state, which is characterized by the changed resistance level to the next influences. This can be revealed in conditions of repeated (testing, challenge) action of stressor in an acute or prolonged mode. The final or the current result of interaction of the object with the factor (stressor) can be a state of adaptation or desadaptation. In the first case after action of the certain stressor dose an increase of the resistance of a biological object to the subsequent action of that or an other factor is observed. The first factor is referred to as adaptative, and its action transits the object to adaptation state (state of eu-stress by H. Selye) [16]. In the second case the system decreases its resistance level for the next influences, passing into a condition of desadaptation (distress by H. Selye).

The essence of the adaptation phenomenon as a process is the transition of a biosystem under influence of the stressor into a new state: a stress state. In its turn, the transient process is temporary or constant state of the biological object in a condition of the re-regulation of its parameters [4, 17-20]. This condition is represented by phases of hypo- and hyperrecovery (hypo- and - or hypercompensation). For example, hyperrecovery of growth rate of the main root is observed at gamma- irradiation of pea seedling in doses about 2 - 5 Gy [21] or at a high temperature treatment [22]. In these phases the biological object can be characterized by increased resistance to repeated influence of the same or an other factor (cross or multiple adaptation) [15, 23-31]. Just the presence of re-regulation state (hyperrecovery in particular) testifies that biological objects are in a adaptation state [13, 26]. Further, if the adaptive response stops, the biological object more or less full restores to the initial values of the parameters. In case of incomplete recovery we deal with an irreversible component of stressor action on the biological object. The state of the increased resistance (adaptation) can be testified as a persistence of the biological system in the phase of hyporecovery (inhibition) of parameters modified by stressor, staying thence in a condition of hyperrecovery (stimulating action, hormesis) [32, 33]. In the first case the increase of resistance is caused by passive mechanisms. For example, the increase of abscisic acid level in root meristem cells after radiation as a stressor promotes a more complete recovery of cells due to lengthening of time necessary for realization of reparation processes [34]. In the second case the biological system, being in hyporecovery state (hypobiosis) [35], has an opportunity to use except of constitutive recovery systems, also induced components. The character of the transition process at PÀ is universal and can have a variable duration [36]. We are obtained results that illustrated this concept (Figure. 1).

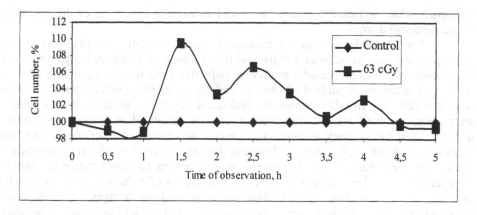

Fig. 1. Effect of gamma-irradiation on cell number dynamic Escherichia coli.

The results of the study of the adaptation phenomenon specify that inertia of biological regulation systems lays at its basis. Biological systems during their development and function are characterized by increased or lowered resistance, which arise spontaneously, under influence of internal factors, and are caused by inertia of biological regulation systems. This circumstance allows us to speak of the existence of constitutive processes and results of adaptation, which are brought about under the influence of endogenous factors, and also about the existence of inducible processes of adaptation, which are caused by the action of exogenous factors. It is possible to consider that the induced adaptation is increased constitutive adaptation. It also concerns processes of decrease (reduction) of initial resistance, i.e. desadaptation processes.

One can distinguish the ontogenetic (phenotype modification, epigenetic) adaptation and phylogenetic (genetic) adaptation. An attribute of phenotype adaptation is the increase of resistance in time limited to the ontogenesis period. During ontogenetic adaptation the occurrence of new properties is not related to a change of the primary genome structure, i.e. this type of adaptation is not inherited, and its mechanism is based on epigenetic reorganizations, which can be inherited by non-genetic mechanisms, for example, due to mechanisms of DNA methylation [37]. A first step of adaptation by procaryotae to external stressors is the initiation of DNA transcription by participation of regulator proteins and change of the S-factors [38]. The phylogenetic adaptation is an adaptation, which is characterized by the genetic inheritance of the increased resistance. The base of its mechanism is the inherited change of the primary genome structure owing to direct or indirect action of various physical, chemical and biological stressors. For example genetic adaptation of microorganisms is based on dot and block reorganizations in certain locuses of its genome [39].

The range of the stressor characteristics, in which limits the adaptive response of biological object is possible, refers to as an adaptive zone. It is necessary to distinguish an adaptive zone on a stressor dose and adaptive zone on its capacity. For example, the adaptation to high temperatures is impossible, if their value exceed a threshold of protein coagulation [23].

The stressor can acutely act during a time interval, which should essentially be the less ontogenetic period of object, repeatedly or chronically (in specific case - is constantly). The acute influence can be defined as this type of influence, which exposition comes to an end before the beginning of the biological object response display.

For the understanding of biological object adaptation mechanisms it is a principal necessity to account for its multilevel structural-functional organization. It predetermines the existence of a multilevel inductive system, which provides both for multilevel reactions of biological objects, and for multilevel mechanisms of adaptive reactions [26, 40]. In other words, the biological object has as many mechanisms of adaptation, as there are reactive mechanisms on action of stressors available and accordingly it has as many restoration systems of structural and functional parameters' values as modified by the stressor. Ignoring this circumstance results in a reduction of the variety of mechanisms of adaptive reactions to only mechanisms, which function on intracellular level, for example, to only the mechanism of induction of stress proteins synthesis [41, 42]. Not always thus presence and advantages of hierarchical organization of a biological object are taken into account, and that each structurally functional level, apparently, is capable, under the appropriate conditions to ensure conditions for the adaptive response of certain subsystem of the object.

It is shown, that the adaptive changes caused by influence of any factor, irrespective of biological organization level, have a phase character [36]. For example, the ecosystem reaction as a consecutive number of communities represents the adaptive response on ecosystem level [43]. At an essential disturbance of biocenosis, its structure appears unstable, changing its mutual processes of the communities -, which in an ideal case leads to the restoration of the initial ecosystem type. Such ecological succession is one of the most expressive displays of adaptations at the biocenosis level [43]. From the point of view of multilevel adaptation mechanisms we should also take into account the existence of adaptive zones which are appropriate to each level of integration.

As a consequence of the interaction of certain levels of exogenous factors with endogenous factors, which arise on the basis of the realization of genetic and epigenetic programs, on the moment of the tested– influence, the biological system is in a concrete state, which is characterized by a certain resistance. The reaction of the biological system to this influence is determined by the state of this system at the moment of influence [36]. In particular, the system of DNA protection at the a level of reparation of its damages, is one of the endogenous factors, which determine the reaction to the influence of various exogenous factors (ionizing radiation, UV, chemical compounds). The increase of the efficiency of DNA repair processes can be considered as one of the ways of realization of the adaptive response of the organism [31, 44, 45]. It is also possible to consider the effective functioning of antioxidant systems as one of the endogenous factors of adaptation [46, 47, 48, 49, 50, 51]. Recently there was an enough actively studied group of «stress» proteins and proteins of «the early period of adaptation» which are synthesized in reply to action of extreme levels of the external factors (high or low temperatures, influence of chemical compounds) [52]. It suggested, that under certain conditions proteins can be protected not only against lethal action of heating, but also against consequences of ionizing irradiation [53].

It is obvious, that on the organism level there is an opportunity to use induction mechanisms to increase proliferation and growth (regeneration) processes in the adaptive response. It is possible to assume with a high probability, that there are gamma-radiation doses, which raise the radioresistance of critical tissue of plants not only by the stimulation of reparation activity in meristem cells, but also by the stimulation of proliferation activity. Thus, it was established, that chronic radiation exposure of pea plants caused by radionuclide incorporation in their tissue (Cs-137), lead to increased radioresistance. This effect could be realised at the expense of increased work of reparation systems, i.e. by increase of the proliferation activity of apex meristem [54, 55].

The formation of the system acquired resistance in plants owing to action of the exogenous biological factors such as by pathogenic microorganisms, is a example of the inductive adaptive response in plants. Development of system resistance in plants is induced by the action of a number of fungi, bacterial and virus pathogens, and also by some chemical compounds (salicylic acid, 2,6-dichlorisonicotinic acid etc.). There are then changes in expression of the appropriate genes and induction of protein synthesis with antimicrobial properties [56].

As was already marked above, besides the exogenous factors, which only at the certain values of the characteristics are capable to ensure the adaptive response, also endogenous determination of reaction to influence of stressors in biological system is inborn. The endogenous features of the adaptive response are formed due to structurally functional memory of the biosystem on the influences of a previous stressor. This memory sufficiently determines the subsequent biosystem reactions to action of similar or others stressors. Precisely this memory of interaction with exogenous factors allows biological system to transit to a re-regulation state and to ensure at the certain stage increased resistance to the next influences. As to endogenous mechanisms this opportunity arises not only due repair mechanisms, but also owing to mechanisms of restoration at higher levels of integration, which also participate in «storing» of stressor influence. Thus, at first, stability is shown, (endogenous component), and, secondly, is formed under influence of external factors (exogenous component).

With the purpose of classification of types of adaptation mechanisms it is possible to schematically present the existence of the biological object as one of elements of the system «object - environment», that we further shall designate accordingly as SO (system - object) and SF (system - factor). According to the basic law of one of variants of the general theory of systems, which was developed by Y.À. Urmantsev [57], any SO owing to interaction with SF is capable to be reconstructed A) or in itself, B) or in other SO - by one of seven various types of changes (reorganizations):

1) quantities of elements, which make SO,
2) their qualities,
3) relations between elements SO,
4) quantities and quality of elements,
5) quantities and relations between elements,
6) qualities and relations of elements and
7) quantities, quality and relations all or part of elements SO.

Among this listed types of reorganizations there are four types of primary reorganizations, which cover first four, including identical reorganization. Taking into account that each cooperating systems (SO and SF) is capable only by 8 ways to change the initial state, the total quantity of joint transformations of two systems makes 64 (8 x 8). Simultaneously it makes that quantity of mechanisms, by means SO is capable to adapt to prolongation action of SF.

414

We put as a basis of classification of primary types adaptive reorganizations (adaptive mechanisms) of plants and in general of biological objects (SO) in case of acute action SF, the classification of types of SO reorganizations, proposed by Y.À. Urmantsev (Table 1). Thus, the change of competitive relation between elements of SO, for example, between seedling organs in autotrophic phase, also modifies the resistance not only of the separate organs, but also of the whole seedling [58].

Table 1. Classification of possible types of biological objects adaptive reorganizations in conditions of acute stressor action.

Adaptive transformation (mechanisms) of biological systems		
Proliferative (caused by modification of biological object initial elements numbers)	**Functional** (related to quality change of biological object elements functioning)	**Structural** (caused by change of the relations between primary elements (subsystems of biological object)

All the above allows us to make an essential specification of the concept of the adaptation potential or resistance potential of a biological object. The adaptation potential can be considered, on the one hand, as the greatest possible level of resistance (for example, cold-resistance, radioresistance, salting resistance etc.), and on the other hand, as the not yet realized opportunity of the adaptation, which quantity value can be expressed as the difference between the greatest possible resistance value and its current value. Since the resistance is determined by a system of multilevel mechanisms, it is necessary to take into account the presence of several levels of resistance potential, i.e. to distinguish the adaptation potential on molecular, cellular, tissue, organism, population levels. We consider that is possible to postulate that the more the above level of organization of the adaptation factor (or its dose), which acts on biological object is realised, the more fully its adaptation potential is realized. There are two modes of adaptation to an influence: - preventive and therapeutic. In the latter case the «adaptation» action occurs after testing (Figure. 2).

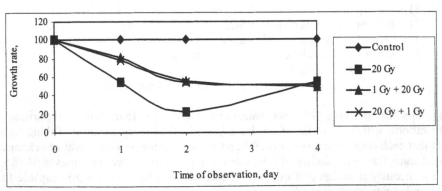

Fig. 2. Effect of gamma-irradiation on Oenothera biennis L. seedlings.

Thus, the given experimental and theoretical results allow to formulate a number of theses, which reflect the basic laws and mechanisms of phenotype adaptation formation:

1) the phenotype adaptation, as opposite to the genetic one is the result of multilevel epigenetic reorganizations, which are realized under influence of the external and internal factors upon biological object features;

2) in the same time with existence of inductive adaptation processes, which are caused by exogenous factors action, there exist also constitutive processes and results of adaptation, which are brought about under influence of the endogenous factors;

3) phenotype adaptation is possible at all levels of biological integration;

4) the phenotype adaptation is always relative in the spatial - temporary relation and characterized by a certain unspecificity in relation to the adaptation factor, which may be biotic or abiotic nature;

5) the value of phenotype adaptation depends not only on the characteristics of an initial state of biological object, but also on the qualitative and quantitative (dose) of the stressor characteristics, and also from its dose rate;

6) irrespective of a structural-functional level of organization, on which the phenotype adaptation is fixed, on the basis of its primary mechanism is laying a change of quantitative and - or qualitative structure of elements of the biological system, and also a change of the relations between elements;

7) the adaptation potential is formed and is realized by multilevel endogenous recovery system of biological objects and by the action of various exogenous factors.

References

1. Gromov B. V., Pavlenko G.V. Ecology of bacteria. Leningrad, Ed. of Leningrad University, 1989. 248 p. (in Russian).
2. Beylin I.G. Parasitism and epyphytology. M., Nauka, 1986. 352 p. (in Russian).
3. Voronkevich I.V. Surviving of phytophathogenic bacteria in nature. M.: Nauka, 1974. 270 p. (in Russian).
4. Veselova T.V., Veselovsky V.A., Chernavsky D.S. Stress at plants (Biophysical approach). M., Ed. of Moscow University, 1993. 144 p. (in Russian).
5. Genkel P.A. Physiology of plant high temterature and drought resistance. M., Nauka, 1982. - 280 p. (in Russian).
6. Kaplan E. Y., Tsirenjapova O.D., Shatalova L.N. Optimization of adaptive processes of organism. M., Nauka,, 1990. 94 p. (in Russian).
7. Petrushenko V.V. Adaptive reactions of plants – Kiev, High school, 1981. 184 p. (in Russian).
8. Drosdov S.N., Kurets V.K., Titov A.F. Thermoresistance of active growth plants. L., Nauka, , 1984. - 168 p. (in Russian).
9. Karasevich Y. N. Experimental adaptation of microorganisms. - M., Nauka, 1975. - 180 p. (in Russian).
10. Sergeeva S.A., Semov A.B., Abramov V.I., Shevchenko V.A. Study of radiation damages reparation in plants, that growing in conditions of chronic beta-irradiation // Radiobiology. - 1985. 25, 6. P. 774 - 777. (in Russian).
11. Chernikova S.B., Gotlib V.Y., Pelevina I.I. Influence of small dozes of ionizing radiation on sensitivity to the next irradiation // Radiat. biology. Radioecology. - 1993. - 33, 1 (4). - P. 537 - 541. (in Russian).
12. Cortes F., Domingues J., Mateos S., Pinero J., Mateos J.C. Evidence for adaptive response to radiation damage in plant cells conditioned with X-rays or incorporated tritium // Int. J. Radiat. Biol. - 1990. - 57, 3. - P. 537- 548.
13. Altergot V.F. Action of increased temperature on plants in experiment and environment. - M., Nauka, 1981. - 56 p. (in Russian).
14. Voynikov V.K., Ivanov G.G., Rudikovsky A.V. Plant proteins of thermal shock. Physiol. Plant. 1986. 31, 5. P. 970 - 980. (in Russian).

416

15. Solovyan V.T. The adaptation of cells to the adverse factors. The characteristic of adaptive response. // Biopolymers and cell. 1990. 6, 49. P. 32-42. (in Russian).
16. Selye H. Stress without distress. M., Progress, 1982. 128 p. (in Russian).
17. Antonomov Y. G. Modelling of biological systems. - Ê., Nauk. dumka, 1977. 260 p. (in Russian).
18. Grodinz F. Theory of regulation and biological systems. - M., Mir, 1966. 256 p. (in Russian).
19. Savin V.N. Action of ionizing radiation on whole plant organism. M., Energoatomizdat, 1981. - 120 p. (in Russian).
20. Chlebovich V.V., Berger V.Y. Some aspects of study of phenotype adaptations // J. General Biology. 1975. 36, 1. P. 11-25. (in Russian).
21. Edited by D.M. Grodsinsky. The forms of postirradiation restoration of plants. Kiev, Nauk. Dumka, 1980. 188 p. (in Russian).
22. Îusienko N.N., Daskaluk T.M., Kaplya A.V. Growth reaction of wheat seedling on high temperatures action // Physiol. Plant. 1986. 33, 1. P. 134 -141. (in Russian).
23. Àlexandrov V.Y. Reactivity of cell and proteins. L., Nauka, 1985. 318 p. (in Russian).
24. Genkel P.A. About connected and convergent resistance of plants // Physiol. Plant. 1979. 26, 5. P. 921-931. (in Russian).
25. Gikoshvili T.J., Vagabova M.E., Vilenchik M.M., Kuzin A.M. // Radiobiology. 1985. 25, 6. P. 806 - 809. (in Russian).
26. Êîstyuk A.N., Mikhyeyev A.N. The problem of phenotype stress and adaptation at plants // Physiology and biochemistry of cultural plants. 1997. 2. P. 81-92. (in Russian).
27. Nevsgladova O.V., Shwartsman P.Y., Soydla T.R. The pleutropic resistance at yeast and its connection with multiple resistance at high eukaryotic // Mol. Biol. 1992. 26, 3. P. 501 - 511. (in Russian).
28. Urmantsev Y.A., Gudskov N.M. A problem of specificity and unspecificity of plant response to damaging influence // J. General Biology. 1986.- 47, 3. P. 337 - 349. (in Russian).
29. Efferth T., Volm M. Multidrag-resistenz von tumoren // Biol. unserer Zeit. - 1990. 20, 3. - P. 149 - 153.
30. Rostchupkin B.V., Kuznetsov V.V. Role of heat shock proteins and protein kinase system in complex resistance of plant cells // Physiol. Plant. 1990. 79, 2. P. 46.
31. Salachetdinova O.L., Pshenichnov M.R., Nesterova L.Y. et al. The topologic DNA changes as reflection of Escherichia coli adaptation to oxidative stress in condition of glucose starvation and passage to growth // Microbiology. 2000. 69, 1. P. 70-74. (in Russian).
32. Kuzin À.Ì. A problem of small dozes and ideas of hormesis in radiobiology // Radiobiology. 1991. 31, 1. P. 16-21. (in Russian).
33. Ìàrkovskaya E.F., Sisoeva M.N., Char'kinaT.G., Sherudilo E.G. Thermomorphogenetic reaction of cucumber plants // IV Congresses of Russian Plant Physiology Society: Abstr. Moscow, 1999. - P. 629 - 630. (in Russian).
34. Gudkov I.N. Dynamics and radioresistance of meristem // The forms of postirradiation restoration of plants. Kiev, Nauk. Dumka, 1980. P. 82-115. (in Russian).
35. Goldovsky À.Ì. The bases of doctrine about organism's conditions. L., Nauka, 1977. - 116 p. (in Russian).
36. Êàlendo G.S. Early reactions of cell on ionizing radiation and its role in protection and sensibilisation. M., Energoizdat, 1982. - 96 p. (in Russian).
37. Holliday R., Ho T., Gene silencing and endogenous DNA methylation in mammalian cells // Mut. Res., 1998. V. 400 (1-2). P. 361 - 368.
38. Shumilov V.Y. The helical phasing between promoters and operator in the Escherichia coli genome // Plant under environmental stress. Moscow, 2001. P. 270 - 272.
39. Belyakov V.D., Yafaev R.H. Epidemiology. M.: Medicine, 1989. 416 p. (in Russian).
40. Hochachka P., Somero D. Biochemical adaptation. M., Mir, 1988. 568 p. (in Russian).
41. Kulikov N.V., Alshits A.K. Increase of radiostability of plant cell genetic structures as a result of preliminary gamma-irradiation // Ecology. 1989, N 1. P. 3-8. (in Russian).
42. Serebryany À.Ì., Zoz N.N. Increasing of adaptive response by antioxidants at radiating mutagenesis at wheat // Radiobiology. 1993. 33, 1. P.81-87. (in Russian).
43. Shilov I.A. Ecology. - M., High School, 1998.- 512 p. (in Russian).
44. Filippovich I.V. The phenomenon of adaptive response of cells in radiobiology // Radiobiology. 1991. 31, 6. P.803-813. (in Russian).
45. Verbenko V.N., Krupyan E.P., Shevelev I.V., Kalinin V.L. The mutant allele gam18, in RecF-dependent repair in Escherichia coli // Genetica. 1999. 35, 3. P. 309-313. (in Russian).
46. Kolosova N.G., Shorin Y.P., Seletitska B.G., Êulikov V.Y.The mechanism increased of organism reliability by antioxidants // Reliability and homeostasis of biological systems. Kiev, Nauk. Dumka, 1987. P. 162 - 167. (in Russian).

47. Sokolosky V.Y., Gesser N.N., Sokolov A.V., Belozerskaya T.A. Stress-response mechanisms in mycelial fungi // Plant under environmental stress. Moscow, 2001. P. 275-276.

48. Sriprang R, Vattanaviboon P, Mongkolsuk S. Exposure of phytopathogenic Xanthomonas spp. to lethal concentrations of multiple oxidants affects bacterial survival in a complex manner. // Appl Environ Microbiol 2000 Sep;66(9):4017-21.

49. Buchmeier NA, Libby SJ, Xu Y, Loewen PC, Switala J, Guiney DG, Fang FC. DNA repair is more important than catalase for Salmonella virulence in mice.// J Clin Invest 1995 Mar;95(3):1047-53. Comment in: J Clin Invest. 1995 Mar;95(3):924-5.

50. 50. Buchmeier NA, Lipps CJ, So MY, Heffron F. Recombination-deficient mutants of Salmonella typhimurium are avirulent and sensitive to the oxidative burst of macrophages. // Mol Microbiol 1993 Mar;7(6):933-6

51. Klotz MG, Hutcheson SW. Multiple periplasmic catalases in phytopathogenic strains of Pseudomonas syringae. // Appl Environ Microbiol 1992 Aug;58(8):2468-73

52. Éravets V.S. Development of idea about plants adaptation to low temperatures // Physiology and biochemistry of cultural plants. 1996. 28, 3. P. 167 - 182. (in Russian).

53. Grodzinsky D.M. Reliability of plant systems. Kiev, Nauk. dumka, 1983. 386 p. (in Russian).

54. Antropogenic radionuclide anomaly in plants. Edited by D. Grodzinsky. Kiev. Lybid, 1991, 160 p. Russian).

55. İikhyeyev A.N., Guscha M.I., Malinovsky Y.Y. The role of proliferation activity in the radioadaptive response of plants // Dokl. of Nat. Acad of Science of Ukraine. - 1998. ¹ 10. P. 177 - 174. (in Russian).

56. Kessmann H., Staub., Hofman C. et al. Induction of systemic acquired disease resistance in plants by chemical // Annual review of phytopathology. 1994. 32. P. 439 - 458.

57. Urmantsev Y.À. The general theory of systems: state, application and prospect // System. Symmetry. Harmony./ Edited by V.S.Òyuchtin, Y.À. Urmantsev . M., Idea, 1988. P. 38 - 130. (in Russian).

58. İikhyeyev A.N. The role of cells, tissue and organs interaction in formation of plants radiobiological reactions. Kiev, 1983. 24 p. (in Russian).

BIOLOGICAL APPROACH TO EVALUATING THE ECOLOGICAL SAFETY OF RADIOACTIVE WASTE DISPOSAL SYSTEM: STUDY OF SMALL RODENTS

V. SYPIN, A. OSIPOV, A. ELAKOV, O. POLSKY, S. DMITRIEV, V. EGOROV, P. PUCHKOV, A. MYAZIN
Moscow Scientific and Industrial Association "Radon", 7th Rostovsky lane 2/14, Moscow, 119121, RUSSIA;
E-mail: aosipov@radon.ru

G. KOLOMIJTSEVA
A.N. Belozersky Institute of Physique-Chemical Biology M.V.Lomonosov Moscow State University, Moscow, 119899, RUSSIA

V. AFONIN
Institute of Genetics & Cytology, NASB, Akademicheskaya,. 27, Minsk, 220072, BELARUS

B. SYNZYNYS, E. PRUDNIKOVA
Institute of Nuclear Power Engineering, Studgorodok-1,Obninsk, 249020, Kaluga region, RUSSIA.

1. Introduction

A major issue in evaluating the ecological acceptability of a disposal system for long-lived solid radioactive waste (SRW) is that doses or risks may arise from exposures in the distant future [1]. There is uncertainty surrounding any estimate of these doses or risks due to lack of knowledge about future conditions [1]. To solve this problem, complex radioecological investigations must be performed.

Environmental surveillance at the Sergievo-Posadsky radioactive waste disposal site of the Scientific and Industrial Association "Radon" in addition to a standard complex radiological tests includes also sampling of different biological objects. It is believed that small rodents are adequate model subjects for ecological estimates [2]. In present report the results obtained on small rodents (mice and voles) from the strict mode and sanitary-protective zones of this disposal system are displayed and discussed.

2. Materials and methods

The studies were carried out in 1996-2001 years. At the moment of the beginning of the studies on the territory of the Strict Mode Zone (SMZ), the Sergiev-Posadsky radioactive waste disposal site, was inhabited by approximately the 30-th generation of rodents. Pubertal mice (*Apodemus agrarius)* and bank voles (*Clethrionomys glareolus*) in the age of 2-3 months and higher were sampled from the territory of the SRW disposal site, with external radiation levels measuring around 40-500 μR/hr. Doses of the collected animals were from 0,5 to 3,0 cGy (from external and internal irradiation). Control rodents were sampled from the

F. Brechignac and G. Desmet (eds.), Equidosimetry, 419–425.

territory of the sanitary - protective zone of the SRW disposal site. External radiation levels on control areas were between 8-12 µR/h.

The DNA-protein cross-links and single-strand DNA breaks levels in spleen lymphocytes were determined using the K/SDS assay [3] and the DNA alkali unwinding assay [4], accordingly. The myeloperoxidase activity in blood neutrophiles was estimated by the cytochemical method [5]. The metallothionein content in rodent kidneys was determined by the cadmium-haemoglobin affinity assay [6].

The frequency of leukocytes with apoptotic markers (condensation and fragmentation of chromatin, budding) and the frequency of leukocytes with a micronucleus were counted on slides from bone marrow and thymus of rodents.

Measurements of the total α-activity of animals' ashes were made using a UMF-1500 M (Russia) and a HT 1000 W Alpha/Beta System (Canberra Nuclear, USA) devices. The metal contents in ash of animals were determined on a Model C-115 M1 (Russia) atomic adsorption spectrometer. A non-parametric test (Mann-Whitney) served to compare groups of animals from different areas using ANOVA testing.

3. Basic results

3.1. ACCUMULATION OF β-RADIONUCLIDES AND HEAVY METALS IN RODENTS.

The results of measurements of the total beta-activity of animal bodies have shown that animals inhabiting on territory of the strict mode zone of the SRW disposal site have contents of β-radionuclides which are 2-4 times higher compared to animals from control zones (Table 1). It is important to note that the voles accumulate more radionuclides than mice.

Table 1. The total β-activity of soil, plants and rodents sampled from the areas of the strict mode zone (SMZ) and the sanitary-protective zone (SPZ) of the SRW disposal site (The data are presented as the Mean±SD)

Areas	Total β-activity, Bq/kg			
	soil	plant	rodents	
			mice	bank voles
SMZ	1928±408*	575±43*	176±25*	381±114*
SPZ	417±113	262±57	95±10	98±19

* - statistically significant with p<0.05

To estimate the possible contribution of heavy metals on their biological effects in rodents we have studied the contents of heavy metals in animals. The results are presented in Table 2. This table show that no significant differences between heavy metals contents in rodents sampled from the strict mode zone and the sanitary-protective zone were registered.

Table 2. The heavy metals contents in rodents sampled from the areas of the strict mode zone (SMZ) and the sanitary-protective zone (SPZ) of the SRW disposal site (The data are presented as the Mean±SE)

Areas	Metals contents, mg/kg					
	Pb		Cd		Zn	
	mice	voles	Mice	voles	mice	voles
SMZ	1.06±0.15	1.42±0.21	0.21±0.09	0.30±0.03	28.4±3.8	42.95±4.17
SPZ	0.90±0.14	1.69±0.26	0.44±0.10	0.34±0.03	29.5±1.5	49.47±11.54

3.2. PATHOLOGY-ANATOMIC STUDIES.

Pathological-anatomical studies of the rodents caught during 6 years did not reveal any morphological anomalies or teratisms. The clinical status of animals sampled from different zones was normal . In 1998 and 2001 weighing of spleens and livers of rodents was performed. The organ mass to body mass ratio was calculated. The results showed that rodents inhabiting the territory of the strict mode zone have a tendency to decrease their spleen mass to body mass ratio and increase their liver mass to body mass ratio, without being significant though (Table 3).

Table 3. Organ mass to body mass ratio for rodents sampled from the areas of the strict mode zone (SMZ) and the sanitary-protective zone (SPZ) of the SRW disposal site (The data are presented as the Mean±SE)

Areas	Rodents	liver mass/body mass $\times 10^3$	spleen mass/body mass $\times 10^3$
SMZ	Voles	53.7±2.9	3.0±0.31
	Mice	58.3±2.6	3.8±0.6
SPZ	Voles	46.5±2.9	3.2±1.0
	Mice	53.8±4.7	4.0±0.5

3.3. HEMATOLOGICAL STUDIES

Tables 4 and 5 presents the data on the total numbers of red and white blood cells of rodents as well as the cellular composition of blood (leukoformula). The tendency to increase the total number of blood erythrocytes of rodents living in the the SPZ has been noted. However, statistical significant changes in the studied haematological parameters were also not observed.

Table 4. The total numbers of red and white blood cells and the cellular composition of blood (leukoformula) of bank voles (Clethrionomys glareolus) sampled from different areas.

Statistical Parameter	Mass g	Erythro-cytes, $n10^{-12}/l$	Leuko-cytes, $n10^{-9}/l$	Leukoformula						
				Neutrophile			E	I	B	L
				Y	L	S				
Strict mode zone (n=87)										
Mean	16.71	10.6	3.92	0.4	5.9	23.5	1.4	3.5	2.9	62.2
±SD	4.16	0.9	1.75	0.84	5.06	14.5	2.1	2.8	3.6	20.3
Sanitary-protective zone (n=74)										
Mean	15.97	9.97	4.12	0.33	2.67	23.3	2.3	3.0	1.7	66.6
±SD	3.93	1.67	1.64	0.58	3.05	16.6	1.5	1.0	1.5	20.0

Table 5. The total numbers of red and white blood cells and the cellular composition of blood (leukoformula) of mice (Apodemus agrarius) sampled from different areas.

Statistical Parameter	Mass g	Erythro-cytes, $n10^{-12}/l$	Leuko-cytes, $n10^{-9}/l$	Leukoformula						
				Neutrophile			E	ì	B	L
				Y	L	S				
Strict mode zone (n=55)										
Mean	16.0	9.8	5.4	0.5	2.7	15.5	2.0	0	0.2	79.0
±SD	3.2	0.8	1.9	1.0	3.4	12.3	0.0	0	0.5	15.3
Sanitary-protective zone (n=46)										
Mean	19.2	9.5	4.6	0.6	5.4	18.4	2.0	3.0	1.4	69.2
±SD	5.8	1.2	2.5	0.9	6.1	9.2	1.5	1.0	2.2	9.1

Interesting data have been obtained in studies on the myeloperoxidase activity of rodents' blood neutrophiles (Table 6). It was shown that bank voles inhabiting the more strict zone show an increase in the myeloperoxidase activity of blood neutrophiles. This fact indicates the development of adaptive processes and the increase in the natural resistance of these animals.

Table 6. The myeloperoxidase activity of rodents' blood neutrophiles (The data presented as the Mean±SD).

Areas	Mice	bank voles
SMZ	1.29±0.41	1.84±0.16*
SPZ	1.48±0.29	1.35±0.45

* - statistically significant with p<0.05

3.4. DNA BREAKS AND DNA-PROTEIN CROSS-LINKS IN RODENTS

The results of quantitative studies of the DNA-protein cross-links in white blood cells of bank voles for 6 years are presented in Fig. 1. These studies demonstrated that the DPC level in the leukocytes of rodents inhabiting territories of the strict mode zone the SRW disposal site are 1.3-1.8 times higher in comparison to animals from the SPZ. There is the positive correlation between the DPC level and the beta-emitted nuclide contents in body animals from the SMZ. Values of Pearson's correlation coefficient are within 0.6 ± 0.14 to 0.86 ± 0.13. A statistical significant difference between the single-strand DNA breaks levels in blood leukocytes of animals sampled from studied areas was not found.

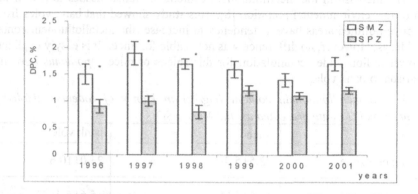

*Fig. 1. The levels of DNA-protein cross-links in white blood cells of voles sampled from the areas of the strict mode zone (SMZ) and the sanitary-protective zone (SPZ) of the SRW disposal site for several years (The data presented as the Mean± SE).   * - statistically significant with p<0.05.*

The metabolic processes in cells depend on the state of the antioxidant defence systems of cells. For the estimation of the antioxidant systems in spleen lymphocytes of bank voles we use the treatment of cell suspension by hydrogen peroxide [7].

The results showed that spleen lymphocytes of bank voles sampled from the SMZ were more resisted to hydrogen peroxide compared to lymphocytes of animals from the SPZ (Fig 2). It is possible, that living of these animals in radionuclide contaminated areas results in a formation of an animals population with higher activity of the antioxidant defence systems.

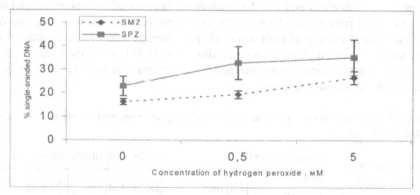

Fig. 2. The hydrogen peroxide - induction of the single strand DNA breaks in spleen lymphocytes of bank voles inhabited at the strict mode zone (SMZ) and the sanitary-protective zone (SPZ) of the SRW disposal site. (The data presented as the Mean± SE).

3.5. METALLOTHIONEIN CONTENT IN KIDNEY OF RODENTS

The increase in the metallothionein content in animal tissues is a well known indicator for environmental pollution [8]. This study showed that bank voles from the radioactive polluted areas have a tendency to increase the metallothionein content in their kidneys. However, no difference was noticeable for mice. It is easyly explained by the lower radionuclide accumulation for this spices of mice (*Apodemus agrarius*) in comparison to bank voles.

Table 7. The metallothionein content (mg/kg) in kidney of rodents sampled from different areas (The data presented as the Mean± SD).

Areas	mice	bank voles
Strict mode zone	12.5±5.4	23.0±10.6
Sanitary-protective zone	13.5±12.0	15.6±4.4

3.6. STUDY OF APOPTOSIS AND CYTOGENETIC DISTURBATIONS IN RODENTS TISSUES

In collaboration with Institute of Genetics & Cytology (Belarus) we have studied the frequencies of the apoptotic cells and the cells with a micronucleus in rodents tissues. Bank voles were sampled from Belarus state biosphere natural reservation (BSBNR – control areas), from Polessky state radiation ecological reservation (PSRER) as well as from the strict mode zone of the Sergievo-Posadsky SRW disposal site

The analysis of the frequency of the cells with micronucleus and apoptosis markers in bone marrow showed the increase in the number of leukocytes with cytogenetic distributions and the decrease in the number of the apoptotic leukocytes (Table 8). A different ratio between these criteria in bone marrow was shown. In estimating the frequency of cells with different damage markers in thymus an other pattern have been noted. A positive correlation (r=0,76 P<0,01) between the frequency of cells with micronucleus and apoptotic cells in thymus was marked for rodents from the radioactive contaminated areas.

Table 8. The frequencies of apoptotic cells and cells with micronucleus in bone morrow and thymus of bank voles.

Areas of rodents sampling	Apoptotic cells, %		Cells with micronucleus, %	
	Bone marrow	Thymus	Bone marrow	thymus
Strict mode zone	0.07±0.05	1.04±0.20	0.40±0.04	0.44±0.13
PSRER	0.26±0.05	1.41±0.09	0.24±0.05	0.10±0.02
BSBNR (control)	0.51±0.10	0.79±0.06	0.09±0.03	0.08±0.02
	P1-3<0.01 P2-3<0.02	P2-3<0.01 P2-3>0.05	P1-3<0.05 P2-3<0.05	P1-3<0.05 P2-3>0.05

4. Conclusions

In summary, constant inhabiting of rodents at the SMZ of the Sergievo-Posadsky radioactive waste disposal site induces some genetic effects. However, these changes are not typical only for ionizing factors. Pathological-morphological changes in internal organs, muscle and osteal tissues were not noticed. Statistical significant changes in the haematological parameters and clinical status of animals were not detected. The data on the myeloperoxidase activity in blood neutrophiles and the sensitivity of spleen lymphocytes to hydrogen peroxide prove a development of some adaptive processes in bank voles from the SMZ of the Sergievo-Posadsky radioactive waste disposal site.

5. References

1. Radiation protection recommendations as applied to the disposal of long-lived solid radioactive waste. A report of The International Commission on Radiological Protection. *Ann ICRP*, 1998, V. 28(4), P. 1-25

2. V. Maslov About carrying out of the complex radioecologic studies in biogeocenoses with a high radioactivity. In "Radioecological studies in natural biogeocenoses". Ed. I. N. Verchovskay. M., Nauka, 1972, P. 9-21 (in Russian).

3. A. Zhitkovich, M. Costa A simple, sensitive assay to detect DNA-proten cross-links in intact cells and in vivo. *Carcinogenesis,* 1992, V. 13, P.1485-1489.

4. P. Kanter, H. Shwartz A fluorescence enhancement assay for cellular DNA damage. *Mol. Pharmacol.,* 1982, V. 22, P.145-151.

5. R. Nartsissov Estimates of the mieloperoxidase activity in blood neutrophiles by Cytochemical assay for *Laboratornoe delo*, 1964, No 3, P. 150-153 (in Russian).

6. D. Eaton, M. Cherian Determination of metallothionein in tissues by cadmium-haemoglobin affinity assay.
Methods. Enzymol. 1991, V. 205. P. 83-90.

7. A. Osipov, A. Elakov, P. Puchkov et al. Estimation of the molecular and cytogenetic effects in mice chronically exposed to low dose-rate gamma-radiation. *Russian Journal of Genetics ,* 2002 (in press).

8. G. Rott, B. Romantseva, B. Synzynys The metallothionein content in freshwater molluscs inhabited in reservoirs of the Middle Russia *Russian Journal of Ecology*, 1999, No. 4. P. 306-308.

USE OF PHOTOSYNTHESIS AND RESPIRATION OF HYDROPHYTES FOR WATER TOXICITY DEFINITION

E. PASICHNA
Institute of Hydrobiology, NASU, Geroyiv Stalingrada St., 12, Kiev, 04210, UKRAINE

1. Introduction

The problem of pollution of natural waters by heavy metals and other chemical elements is very acute today. It is therefore necessary to investigate the effects of such substances on different species of water organisms. Besides, it is very important to study the possibility of environmental toxicity estimation using physiological and biochemical processes of these organisms.

Hydrophytes are the production basis and the energy providing link in trophic chains of water ecosystems. Besides, the processes of photosynthesis and respiration of aquatic plants reflect a very important aspect of metabolism and an ability for generation of energy and synthesis of ATP [1]. A large sensitivity of gas exchange functions of hydrophytes to increased concentrations of toxic substances in water medium has been observed previously [2]. It is known that heavy metals are the ones of the prime pollutants of natural waters [3]. So, our intention is to investigate the effect of increased concentrations of copper and manganese on photosynthesis and respiration of water plants. Also, we tried to define a possibility of using their gas exchange functions for water toxicity biotesting.

2. Materials and methods

The hydrophytes used were submerged the higher aquatic plants *Najas guadelupensis* L. and *Ceratophyllum demersum* L. which have a significant surface of contact with water.

Investigations of gas exchange in the processes of photosynthesis and respiration were carried out by the method of electrometric oxygen determination. For this purpose, we use the system allowing the study of the physiological state of aquatic plants under a wide range of environmental conditions [4].
Copper (Cu^{2+}) was added to the water medium, in which aquatic plants were grown, in concentrations 0.0005, 0.005, 0.02 mg/l, manganese (Mn^{2+}) – 0.005, 0.02, 0.1, 0.2 mg/l.

Photosynthesis and respiration intensities were determined after 14-days experiments, when organisms had passed all phases of toxicities and adaptation (sometimes indeed, in short-term experiments we can observe a stimulation of physiological functions under the influence of toxic substances as a protective reaction of aquatic organisms to the destructive effect of the toxins).

3. Results and discussion

The results of the investigations show some regularity in changes of photosynthesis and respiration intensity under influence of increased concentrations of Cu^{2+} or Mn^{2+}.

F. Brechignac and G. Desmet (eds.), Equidosimetry, 427–431.

As we can see (Table 1) addition to water 0.0005 mg/l Cu^{2+} or 0.005, 0.02 mg/l Mn^{2+} has no toxic effect on gas exchanges processes and even increases their intensity. At the same time, presence of 0.02 mg/l Cu^{2+} or 0.1, 0.2 mg/l Mn^{2+} in water medium leads to inhibition both photosynthesis and respiration.

Table 1. Influence of Cu^{2+} and Mn^{2+} on photosynthesis and respiration intensity of aquatic plants (n=4–5, P<0.05)

Medium samples	Test organism	Oxygen (mg/l·hour)		$\dfrac{P}{R}$	Toxicity (T)
		Photosynthesis (P)	Respiration (R)		
Control	*Najas*	4.78±0.05	1.41±0.05	3.39	0
	Ceratophyllum	3.64±0.17	1.32±0.03	2.76	0
+0,0005 mg/l Cu^{2+}	*Najas*	5.92±0,04	1.53±0,04	3.87	-0.14
	Ceratophyllum	5.14±0,32	1.62±0,07	3.17	-0.15
+0,005 mg/l Cu^{2+}	*Najas*	3.34±0.07	1.89±0.07	1.77	0.48
	Ceratophyllum	2.97±0.30	1.73±0.09	1.72	0.38
+0,02 mg/l Cu^{2+}	*Najas*	1.48±0.05	1.10±0.05	1.35	0.60
	Ceratophyllum	1.59±0.27	0.98±0.11	1.62	0.41
+0,005 mg/l Mn^{2+}	*Najas*	5.52±0.30	1.62±0.05	3.41	-0.01
	Ceratophyllum	4.53±0.27	1.57±0.08	2.89	-0.05
+0,02 mg/l Mn^{2+}	*Najas*	6.52±0.33	1.83±0.08	3.56	-0.05
	Ceratophyllum	4.95±0.34	1.90±0.15	2.60	0.06
+0,1 mg/l Mn^{2+}	*Najas*	3.97±0.30	1.31±0.04	3.03	0.11
	Ceratophyllum	2.93±0.08	1.15±0.10	2.55	0.08
+0,2 mg/l Mn^{2+}	*Najas*	2.78±0.35	1.14±0.08	2.44	0.28
	Ceratophyllum	2.41±0.17	1.05±0.09	2.29	0.17

It is observed that copper ions in concentration of 0.005 mg/l depress photosynthesis simultaneously while enhancing respiration. Apparently, it is a protective reaction of the organisms against stress as far as respiration is a source of energy requested for the reparation of damage appearing under the action of the metal [5].

Based on the results of the investigations, we can draw the conclusion that ions of copper are much more toxic for water plants than ions of manganese. Also, it is obvious that *N. guadelupensis* is more acceptable as a test object because this species is more sensitive to the action of toxic substances than *C. demersum*.

So, we can see that the reaction of plant organisms to the influence of increased concentrations of the metals show some regularities: small quantities of Cu^{2+} and Mn^{2+} stimulate the processes of vital activity and great quantities of the ions depress them. Such two-phase impact on gas exchange functions is conditioned by the fact that these metals are necessary for the structure and activity of many physiological systems including those which take part in the processes of photosynthesis and respiration. In the same time, it is proven that high concentrations of heavy metals inhibit the enzymes because of connection with their active centres [6].

Thus, the changes in the oxygen content in the water (because of gas exchange functions) gives us a possibility to assess the concentration of the metal in the solution. However, absolute values of photosynthesis and respiration intensities depend on some factors, including the physiological state of the plant, the conditions of the experiment, etc. So, we can make a more informative and precise definition of water quality by using the correlation between photosynthesis and respiration intensities.

In this connection the following formula for estimating water toxicity is proposed (T) [4]:

$$T = 1 - \frac{Pex / \mathrm{Re}\,x}{Pc / Rc}$$

where P_{ex} and R_{ex} are the registered values of photosynthesis and respiration in the experiment under influence of the investigated concentrations of the substances; P_c and R_c are the photosynthesis and respiration values for control water without any toxins.

This method gives us a possibility to estimate toxicity not only of the solution with well-known substances (Table 1), but also to define the quality of any type of water. It is very important, as often we do not know the chemical composition of the water solution. Also, we must take into account that substances which are toxic to organisms when they exist individually often produce unexpected effect when they are present in combination. So, this method allows us to estimate toxicity of solution of any chemical composition to water plants.It can be proven by the results of quality determination of different kinds of water with use of N. guadelupensis and C. demersum. As we can see (Table 2), the toxicities of the water tested, measured by means of photosynthesis and respiration registration, varied considerably. Aquarium water as a control medium was not toxic (T=0). Water from waterconduct compared with the aquarium water affected test organisms, as it apparently contains residual amounts of chlorine. River water sometimes greatly influences the oxygen response of aquatic plants. We can observe that if contamination of water is within the limits of maximum concentrations of toxic substances permissible in fish industry (or water is rich in biogehic elements), the toxicity (T) is negative (water from the river of Dnieper). On the other hand, when the water contains great quantities of toxic components T shows a steady deviation on the positive side (water from the river of Shostka). Industrial wastes may kill the water plant, consequently, such water must be diluted.

Table 2. Toxicity of different waters, measured by means of the proposed method with use of N. guadelupensis and C. demersum (n=5, P<0.05)

Medium samples	Test organism	Oxygen (mg/l·hour)		$\frac{P}{R}$	Toxicity (T)
		Photosynthesis (P)	Respiration (R)		
Aquarium water	*Najas*	4.93±0.08	1.45±0.06	3.40	0
	Ceratophyllum	3.79±0.12	1.37±0.05	2.77	0
Water-pipe	*Najas*	4.35±0.10	1.57±0.08	2.78	0.18
	Ceratophyllum	3.41±0.22	1.44±0.11	2.36	0.14
Dniepr river water (medium part)	*Najas*	6.20±0.14	1.67±0.09	3.72	-0.10
	Ceratophyllum	4.48±0.26	1.52±0.07	2.95	-0.07
Shostka river water (nearby the city of Shostka)	*Najas*	1.01±0.35	1.06±0.14	0.95	0.72
	Ceratophyllum	1.46±0.24	1.11±0.11	1.32	0.52

Thus, natural and industrial waters can be classified in the following way:

$T>0$ water is toxic

$T<0$ water is eutrophic

$T\approx0$ water is not toxic.

4. Conclusion

The results of the water toxicity measurements by using the proposed method demonstrate that it can be helpful for both the study of photosynthesis and respiration of aquatic plants in general, and for water toxicity measurements as well.

Further researches must be directed for revealing the most sensitive species of plants to a chemical stress (to use them as test organisms), and for studying effects of individual and simultaneous actions of different chemical substances on physiological functions of hydrophytes.

5. References

1. R.Polystchuk. To the question of choose of tests for estimation of the toxic substances action on algae. Byologya morya, Kiev, 1975, V. 35, pp. 58–65 (in Russian).
2. L.Braginsky, I.Velitchko and E.Stcherban. Freswater plankton in toxic medium. Kiev, Naukova dumka, 1987, 178 p. (in Russian).
3. A.Bozhkov, S.Mogilyanskaya. Relationship between Cu^{2+} ion concentration in medium, bioaccumulation and toxicity for the microalgae Dunaliella viridis Teod. Algologia, 1994, V.4, N 3, pp. 22–29.
4. A.Mereshko, E.Pasichnaya, A.Pasichnyi. Biotesting the toxicity of aquatic environment by functional characteristics of macrophytes. Hydrobiological j., 1998, V. 34, N 2–3, pp. 103–110.
5. R.Polystchuk. Reaction of overgrowing macrophytes on influence of heavy metals ions. In: The biological bases of fight against overgrowing. Êiev, Naukova dumka, 1973, pp. 155–193 (in Russian).

6. Physiology of plant organisms and the role of metals. Edited by N.Chernavina. Moskow University Publishing House, 1989, 155 p. (in Russian).

"Ecological standardization and equidosimetry for
radioecology and environmental ecology"

Kiev, Ukraine, 14-20 April 2002

Key Speakers from NATO Countries:

Henri Maubert	CEA- Cadarache. Radioprotection Service (France) Henri Maubert, SPR Bt 355, 13108 Saint Paul lez Durance, France
Randall C. Morris	TREC, Inc., and the Environmental Science and Research Foundation, USA
James D. Navratil	Environmental Engineering and Science Clemson University, USA
George Shaw	Department of Environmental Science & Technology Faculty of Life Sciences Imperial College at Silwood Park Ascot, Berkshire, SL5 7PY, United Kingdom
Guillaume Echevarria	Laboratoire Sols et Environnement UMR 1120 ENSAIA-INPL/INRA, BP 172 2, avenue de la Forкt de Haye 54505 Vandoeuvre-les-Nancy, France
Gerassimos Arapis	Agricultural University of Athens, Laboratory of Ecology and Environmental Sciences, Iera Odos,75-botanokos, Gr-11855, Athens, Greece
Michael Karandinos	Agricultural University of Athens, Laboratory of Ecology and Environmental Sciences, Iera Odos,75-botanokos, Gr-11855, Athens, Greece

F. Brechignac and G. Desmet (eds.), Equidosimetry, 433–436.
© 2005 *Springer. Printed in the Netherlands.*

434

Other Participants from NATO Countries:

Mustafa Alparslan   Canakkale Onsekiz Mart University,Faculty of Fisheries Terzioglu Campus, 1700 Canakkale/Turkey

Jiy.K. Kim   Korea Atomic Energy Research Institute, 150 Dukjin-dong, Yusong-gu, Taejon, 305-353 Korea

Marie Svadlenkova   University of South Bohemia, Dept.of Physics, Jeronymova10. 371 14 Ceske Budeiovoce, Czech Republik

Key Speakers from Partner Countries:

Alexander Dvornik   Forest Institute of NAS Belarus,BY 246654, Gomel, Belarus

Vassili Davydchuk .   Institute of Geography NASU, UA 252034, Kiev, Ukraine

Yuri Kutlakhmedov   Taras Shevchenko National Kiev Univeristy 03022, Kiev, Ukraine

Valeriy Kashparov   Institute of Agricultural Radiology , UA 255205,Kiev, Ukraine

Alexandr Orlov   Polesskya Agro-forest-ameliorative scientific research station,UA 262004, Zhitomir, Ukraine

Georgiy Perepelyatnikov   Ukraine Radiation Training Centre, RU 252022, Kiev, Ukraine

Staniclav Geraskin   Russian Institute of agricultural radiology and Agroecology, Ru-249020, Obninsk, Russia

Vladimir Georgievsy   Institute of Atomic Energy named Kurchatov, Moscow, Russia

Dmitry Grodzinsky   Institute of Cell Biology and Genetic Engineering NASU, UA 252022, Kiev, Ukraine

Vladimir Kovalevsky   Ministry of Defense of Ukraine, Kiev, Ukraine

Mikhail Kuzmenko   Institute of Hydrobiology NASU, Kiev, Ukraine

Vladislav Petin   Russian Institute of medical radiology, Ru-249020, Obninsk, Russia

Galina Talalaeva	Institute of Plant and Anomal Ecology Urals Division Academy Sciences of Russia, Biophysical Station , Zarechny, Russia
Victoria Tsytsugina	Institute of Biology of the Southern Seas NASU, Sebastopol, Ukraine
Vladimir Voitsitsky	Kiev Taras Shevchenko National University, Ukraine
Eugeniy Yakovlev	State Geological Commetee Of Ukraine, Kiev, Ukraine

Other Participants from Partner Countries:

Alexandr Dmitriev	Institute of Cell Biology and Genetic Engineering NASU, UA 252022 , Kiev, Ukraine
Vyacheslav Sypin	Scientific and Industrial Association "Radon", 2/14, 7th Rostovsky lane, Moscow, 119121, Russia
Andreyan Osipov	Scientific and Industrial Association "Radon", 2/14, 7th Rostovsky lane, Moscow, 119121, Russia
Elena Rovdan	Institute for Problems of Natural Resources Use & Ecology National Academy of Sciences of Belarus 10, Staroborisovsky trakt , 220114 Minsk, Belarus
Sergey Kirpich	National Institute for Higher Education Belarusian State University, Nikiphorov Street, 7-160, Minsk 220141, Belarus
Nataliya Terestchenko	IBSS NANU, prosp. Nakchimova, 2, Sevastopol 11, 99011, Ukraine
Alla Dvorzhak	Institute of Atomic Energy named Kurchatov, Moscow, Russia
Asamat Tynybekov,	Manas 22a, Bishkek, 720017, Kyrghyzstan
Dustov Abdusamad	734061, 2 majakovski str., Dushanbe,Tajikistan
Ludmila Perepelyatnokova	Institute of Agricultural Radiology , UA 255205,Kiev, Ukraine
Tatyana Ivanova	Institute of Agricultural Radiology , UA 255205,Kiev, Ukraine

Petr Yushkov — 202, 8 Marta Str., Institute of Plants & Animals Ecology, 620144 Ekaterinburg, Russia

Natalia Grytsyuk — Institute of Agricultural Radiology , UA 255205,Kiev, Ukraine

Vladimir Krasnov — Polesskya Agro-forest-ameliorative scientific research station,UA 262004, Zhitomir, Ukraine

Georgiy Geletukha — Research Institute of Mechanics of Quick-Proceeding Processes, 1, Cosmonavt Komarov Avenue, 03058, Kyiv-58, Ukraine

Alexandr Mikheev — Institute of Cell Biology and Genetic Engineering NASU, UA 252022 , Kiev, Ukraine

Larissa Ptizkya — Institute of radiation Safety and Ecology, Kazakhstan

Anatoliy Kachinsky — Ministry of Defenses of Ukraine, Kiev, Ukraine

Elena Turos — Kiev, Ukraine

Vladimir Starodubtsev — National Agricultural University of Ukraine, Soil science and conservation Dept. 15 Geroiv Oborony st., Kiev - 03041, Ukraine

Ludmila Sorokina — Institute of Geography NASU, UA 252034, Kiev, Ukraine

Fedir Demydyuk — Development Foundation, Address: 57 Illinska Str., Apt. 5, Kharkiv, 61093, Ukraine

Anatoliy Boiko — National Taras Shevchenko University of Kyiv, Biological Faculty, Department of Virology, Kiev, Ukraine

Valeriy Polishuk — National Taras Shevchenko University of Kyiv, Biological Faculty, Department of Virology, Kiev, Ukraine

Ludmila Kalinenko — Institute of Agricultural Radiology , UA 255205,Kiev, Ukraine